CONSERVATION

In ecological economics, economic throughput *is all of the energy and material that enter human society as "resources" (e.g., petroleum, timber, foodstuffs) and leave the system as "waste" (e.g., air pollution, garbage, sewage).*

The "Limit of Nature's Services" *represents the amount of resources that nature can supply and the amount of pollutants that it can absorb without the natural systems breaking down. For example, when the fish in a river die because of an overload of sewage input, the limit of nature's services (in this example purification) have been exceeded. Another example is desertification resulting from overgrazing.*

On the time scale *of the figure, the boom and bust cycles would have an interval of 30 to 50 years.*

Both Management to Maximize Growth *and* Management for Quality of Life *begin with an exponential growth in economic activity. As* Management for Quality of Life *approaches the limit of nature's services, feedback signals such as fish die-off and soil erosion are heeded, and remedial action is taken. Remedial actions may result in a tapering off of the rate of growth with a consequent increase in sustainability. In contrast, in economics where management is designed to maximize growth, resources (including clean air and water) are considered either infinite or replaceable, and the economy is seen as unhindered by any limits to growth.* Management to Maximize Growth *ignores negative environmental feedback signals and seeks only to expand. Recessions or crashes are considered to result from a maladjustment of the money circuit.*

Ecological economics recognizes the importance of the money circuit, but views it as a subset of a larger system. That system is the movement of resources from the earth, through the economic system, and back to the earth. The resource system drives the economic system. The economic cycle functions only as long as the resource cycle functions (see Daly 1993).

Ecological economics also recognizes that technology can increase the efficiency of resource use, and through this means increase the limit of nature's services (see Daly 1991a).

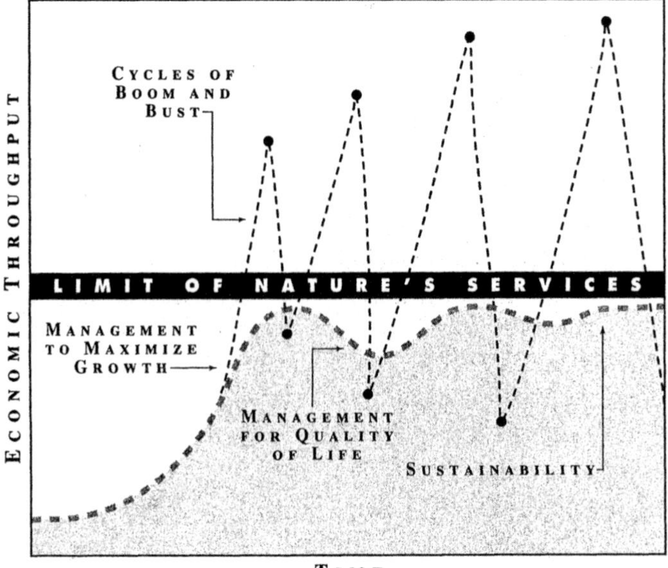

CONSERVATION

Replacing Quantity with Quality as a Goal for Global Management

Carl F. Jordan

School of Ecology
University of Georgia

John Wiley & Sons, Inc.

ACQUISITIONS EDITOR Sally Cheney
MARKETING MANAGER Rebecca Herschler
PRODUCTION EDITOR Deborah Herbert
TEXT DESIGN Carolyn Joseph
COVER DESIGN Carol C. Grobe
DESIGN DIRECTION Karin Gerdes Kincheloe
MANUFACTURING MANAGER Susan Stetzer
ILLUSTRATION Rosa Bryant

Library of Congress Cataloging in Publication Data:
Jordan, Carl F.
Conservation / Carl F. Jordan
p. cm
Includes bibliographical references (p. 301) and index.
ISBN 0-471-59515-2 (cloth)
1. Conservation of natural resources. 2. Natural resources--Management. 3. Human ecology. I. Title.
S938.J67 1995 94-31732
393.7—dc20 CIP

Printed in the United States of America

10 9 8 7 6 5 4 3 2

To Carmen

with love

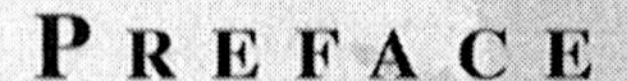

PREFACE

In the past, conservation was concerned with the management of natural resources and the prevention of species extinctions. Attention was focused on forests, prairies, mountains, oceans, and agricultural land, as well as on the species that inhabited them. Although the practice of conservation was justified in terms of benefiting humankind, human beings and their social institutions were not a central concern.

Today, economic, political, and cultural man* is at the center stage of conservation. This new perspective results from a growing realization that lack of management techniques no longer generally limits the sustainable use of natural resources and the preservation of species. Rather, it is social institutions that usually are the stumbling blocks. One problem is an economic system that fails to adequately value a healthy and productive environment. A second is governmental policy that relies on command and control regulations instead of incentives for preserving environmental quality. A third is a lack of appreciation for the role of culture in development.

Most important in this change from resource-centered conservation to man-centered conservation is the replacement of *quantity of resource production* with *quality of life* as a goal of global management. The new conservation is no longer preoccupied with techniques of resource production. Instead, it is concerned with how the use of resources affects the well-being of humankind.

**Note that "man" and "mankind" are occasionally used generically to include all people and both genders in this preface and throughout the text of this book.*

This critical shift is not emphasized in most textbooks on conservation. The changing paradigm, if stated at all, is usually mentioned as an afterthought. In contrast, this book emphasizes an understanding of ecological economics, environmental policy, and culture as being paramount to achieving the goals of conservation.

The humanities, too, play an increasingly important role in modern conservation. The chapter on environmental ethics explores the issue of whether human beings have a moral imperative to protect the environment.

A question that almost always inspires furious debate between some economists and most conservationists is, Are things really getting worse? Conservationists often assume that they are. However, the chapters on conservation history and environmental trends show that there is very little data on a global scale to verify that assumption. But, although we cannot prove that *things are rapidly getting worse,* the data also leave little doubt that *things could be a lot better.*

Despite this text's emphasis on human beings' role in the environment, traditional conservation is not neglected. Resource management is brought up to date with concepts of how energy subsidies in resource systems are often at the root of ecological problems and how for remediation, the services of nature can be substituted for the services of fossil fuels. The chapter on biodiversity reviews the current controversy over the species approach versus the habitat approach in conservation biology. It also emphasizes that management to *preserve existing biodiversity* on a global scale requires an approach that is different from *management to increase biodiversity* on a local scale.

To give regional detail to the practice of conservation and to provide concrete examples of some of the problems and proposed solutions facing conservationists today, this book features four photo-essays: Deforestation in the Amazon Basin; Social Forestry in Thailand; Trivializing Indigenous Resistance to Deforestation Sarawak, Malaysia; and The Farmer First Approach to Agricultural Development in the Uplands of Mindanao, The Philippines. Each essay is placed appropriately within the chapter related to the essay's presentation.

Many books currently available focus on such specialized fields of conservation as conserving biodiversity, resource management, ecological economics, environmental policy, environmental ethics, and cultural anthropology. Few of these books are broad enough however, for use as a general introduction to conservation, or for use when examining the range of social sciences, natural sciences, and humanities that contribute to solving environmental problems.

This book differs from specialized texts in its comprehensive approach, which is suitable for a general introduction to a curriculum in conservation, or for an overview course, such as "Man and the Environment". It is appropriate for an environmental literacy requirement for liberal arts majors and preprofessional curriculums, and also a reference for the layperson. The artist wishing to portray human beings' impact on the environment, or the business person concerned about social responsibility will find here an overview of conservation problems. In addition, at the conclusion of each chapter a list of suggested readings gives direction to readers who wish to pursue in greater depth more aspects of the discussion that they find most interesting.

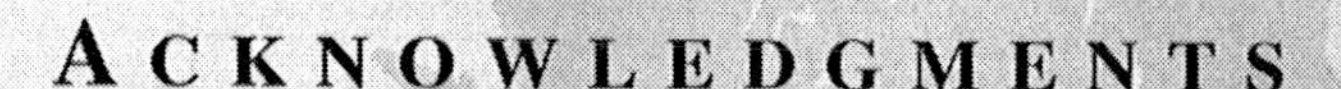

ACKNOWLEDGMENTS

In the late 1980s, Garo Batmanian encouraged me to write a text that would cover the breadth of modern conservation, and in 1992, Dr. Paul Colinvaux opened a door that made that undertaking possible. Once the book was started, many of my colleagues were generous in their contributions. Drs. Vernon Meentemeyer, Frederick Ferre, Peter Morton, and Eugene Odum made suggestions on individual chapters. Anabel Jordan helped me with style. Christian Castellanet, Rodney Vargas, and Barrett Walker took an early interest in the entire project, and their help was particularly significant throughout its development. I deeply appreciate the contributions of all these individuals.

The photographs for the photo-essay "Deforestation in the Amazon Basin" were taken while I was supported by the National Science Foundation (U.S.) and Companhia Vale do Rio Doce (Brazil). "The Social Forestry In Thailand" photo-essay resulted from research sponsored by the Program in Science and Technology Cooperation of the U.S. Agency for International Development, and was coordinated through Chulalongkorn University, Bangkok. Photographs for "The Farmer First Approach to Agricultural Development in the Philippines" were taken as part of the U.S. Agency for International Development's Program entitled "Sustainable Agriculture and Natural Resource Management Collaborative Research Support Program." The University of Georgia, Athens, Georgia provided institutional support throughout the two decades during which this work was carried out.

The following individuals reviewed the book: Kamaljit S. Bawa, *University of Massachusetts–Boston;* Lyn C. Branch, *University of Florida–Gainesville;* Peter F. Brussard, *University of Nevada–Reno;* William R. Chaney, *Purdue University;* Robert Hamilton, *Louisiana State University;* Richard D. Laven, *Colorado State University;* David W. Orr, *Oberlin College;* Rex D. Pieper, *New Mexico State University;* David S. Woodruff, *University of California–San Diego.* I thank them for their efforts and valuable suggestions.

May 1994

CARL F. JORDAN
School of Ecology, University of Georgia
Athens, Georgia, U.S.A.

To The Student

"To be an effective Conservationist, you must first build a soapbox. Then you can stand up on it, and people will listen to you." That has been the counsel of eminent ecologist, Eugene Odum, whenever a student asked his advice on a career in conservation.

Conservation is like politics. Anyone with an opinion can claim to be a conservationist or a politician. However, to be a politician that is elected, or to be a conservationist that is influential, one usually has to be established first in another field. Often, politicians start as business people or lawyers. In earlier days, conservationists began in basic sciences, such as zoology and ecology, or applied sciences, such as forestry or agronomy.

Although basic and applied life sciences still are good options, today there are many more opportunities for aspiring conservationists. Geographers look at conservation from a landscape perspective. Engineers can specialize in pollution control. Scientists in fields such as climatology and marine science monitor and model environmental trends. For those inclined toward the social sciences, there are opportunities in resource economics, environmental law and policy, and anthropology. The humanities contribute to conservation through environmental ethics and environmental history.

How can a budding conservationist know where to begin?

The overarching goal of this book is to provide an introduction to the fields relevant to conservation. It is intended to help the student with a broad interest in conservation select a specialty. Regardless of the specialty selected, it is important that he or she be aware of the broad array of factors that contribute to solving environmental problems. Having a vision of the many facets of conservation will help the conservationist build a better soapbox.

C.F.J.

CONTENTS

Conservation

INTRODUCTION

Conservation has two roots, one in resource management and the other in natural history. In the past, most textbooks with the word *conservation* in the title were about resources, and there were two basic paradigms: *nonrenewable* resources were limited and *renewable* resources had to be managed with care to prevent them from becoming extinct. Managing resources in a way that was not wasteful and that ensured a sufficient supply for future generations was the major theme of conservation.

Conservation still is about resources and their management, but in recent years its emphasis has changed. Although there are always periodic shortages of certain nonrenewable resources, scarcities often lead to discoveries of new sources, more efficient extraction, technological substitution, and recycling. Today there is little evidence that scarcity of resources on a global basis is a threat to the economy and to human subsistence. The economic recession of the early 1990s was triggered more by global geopolitical events and world economic imbalances than by an actual scarcity of resources. The localized shortages of food that have occurred recently in Africa appear to have resulted not because of insufficient food in the global market but because of problems in getting existing food to the people or the people's inability to buy food.

Resources for fending off starvation on a global scale and for producing manufactured goods in developed countries do not at present seem limited. What does appear to be in danger of disappearing, however, are the resources for ensuring the *well-being* of the human race. Such resources include clean air, clean water, and refuges from the stress caused by a continuously close proximity to other people. Emerging "resource" problems include degradation of the global commons, namely, the oceans, atmosphere, and Antarctica; degradation of the local commons, including open land and parks close to urban centers; loss of species diversity, particularly in tropical forests; and loss of ecosystems with functions of particular importance to survival of life, such as wetlands. The limited ability of the environment to absorb waste and the stress caused by crowding are more critical factors than resource availability in setting boundaries for economic growth.

The other root of conservation is found in the writings of naturalists and philosophers who loved the natural landscape and the species that inhabited it. Their joy consisted in the spiritual solace offered by the wilderness and the satisfaction in knowing that a multitude of wild species shared the landscape with them. For these conservationists the degradation and loss of rural and natural landscapes has been a greater concern than resource scarcity. For them, population growth and economic expansion are twin plagues that threaten all the things they hold most dear in life.

Many recent books on conservation have been written by the spiritual descendants of the early naturalists. For these conservationists new urgencies are replacing former priorities. Although the preservation of landscapes remains important, the loss of biodiversity (species extinction) has become critical. Landscapes lost can never be rehabilitated, if the species that comprise the landscapes are lost.

Resource conservationists and naturalists have often been at odds over strategies to manage the forests, lakes, and other ecosystems of the earth, but they are beginning to agree on one thing: improving quality of life should be given importance equal to or greater than increasing quantities of resources. Polluted air, poisonous water, urban congestion, and other manifestations of unbridled economic expansion and population explosions are unhealthy for humans as well as for wild species. The strategies of the two conservation groups are beginning to merge, now that they recognize that the well-being of humankind and of nature are inextricably intertwined.

Not only are conservation's priorities changing but also its emphasis is shifting from *techniques* of resource management and landscape preservation to *policies* that will promote conservation. Today, in general, it is not lack of technical expertise that is limiting efforts to use resources less wastefully, to mitigate pollution, and to save species from extinction. In many cases, it is already known what should be done, but action is stymied because of political, economic, and cultural barriers. This book reflects the changing paradigm.

TEACHING CONSERVATION

What does this book do? How can it be used to teach students "conservation"?

It is easiest to start by telling the reader what this book is *not,* and what it does *not* do.

- It is not a compendium of environmental laws and regulations, although selected laws and regulations are discussed to illustrate certain points.
- It does not list conservation organizations and their activities, although organizations are mentioned when their activities illustrate a theme.
- It is not gloom-and-doom sensationalism; when pessimistic projections are cited, optimistic opinions are also aired.
- It is not a history of conservation and of leading figures in the field, although a review of both is given in order to set the stage.
- It is not a book on environmental ethics, but it does present an overview philosophy on human beings and their place in nature as a prelude to a call for action.
- It is not a manual on management of natural resources; books on specialized techniques of wildlife management, silviculture, fish culture, and the like are readily available elsewhere.

- It is not a text on ecology, although it explains principles of ecology when they are relevant to conservation.
- It is not an atlas of resource geography, although geography is included when it is essential for discussion.
- It is not a treatise on population biology and the genetics of species preservation, although theories from conservation biology are used in discussions of biodiversity.
- It does *not* pretend to give a complete review of economics, policy, and culture as they pertain to conservation, although it does contain major chapters on these subjects. Readings are suggested for those who are interested in further pursuit.

What then, does this book do?

It teaches a set of principles.

Conservation is not a basic science like physics, chemistry, and biology. In basic science courses, the student is expected to learn certain fundamentals. For example, in physics, the student will learn the laws of thermodynamics; in chemistry, the periodic table of elements; and in biology, the photosynthetic cycle. The student will come away from such courses with these fundamentals ingrained, and this knowledge will give the student a certain confidence that he or she actually knows something. When a parent asks a physics student, "What did you learn?", the student can answer, "Every action has an equal and opposite reaction."

Conservation is sometimes thought to be an applied science. When a parent asks a student in an applied field such as engineering what he or she learned, the student can answer, "I learned to build a bridge," or "I learned to design a computer." A student in agronomy might answer, "I learned to increase the production of corn," and a forestry student might answer, "I learned how to grow taller trees."

But, what, exactly, should a student of conservation learn? "To conserve" is a circular answer that says nothing. The problem may be that, although conservation incorporates aspects of applied science, it is not in itself an applied science. Upon receiving a degree or completing a course, a conservationist does not go out and start "conserving." Conservation is not a technical skill in the same sense as engineering, medicine, and agronomy. *Conservation is a philosophy of managing the environment in a way that does not despoil, exhaust, or extinguish.*

A conservationist may get a job as a lawyer, engineer, forester, teacher, professor, park guard, or bureaucrat for the Department of Interior. The conservationist employs certain technical skills relevant to his or her job and, at the same time, when decisions are necessary, seeks to make them in a way that will help achieve the ends of conservation.

It is impossible to anticipate all the technical problems a conservationist will face during his or her career, and what facts will be necessary to solve the problems. Instead of prescribing the "right" answers for particular technical situations, this book focuses on *principles* that will help guide decisions, regardless of the conservationist's particular profession or occupation.

One objective of this book is to help the student both to understand and to absorb these principles. When the principles are so deeply ingrained that they become part of the student's subconscious, then they will serve as moral guideposts for a professional career, regardless of whether the career is that of environmental lawyer, pollution engineer, forest manager, high school teacher, college professor, or any other pursuit that touches on conservation.

The principles make up the theme of each chapter of this book: they are as follows:

1. *Conservation History.* Pragmatism and idealism form the two historic roots of conservation. Utilitarian conservationists have been concerned with scarcity of natural resources, whereas idealistic conservationists have worried primarily about the preservation of nature. We now know that both problems are interrelated and that neither can be solved without attention to economics, politics, and culture.
2. *Environmental Trends.* The *ability of the earth to maintain ecological integrity* has replaced the *supply of resources* as the central problem of conservation.
3. *Environmental Ethics.* One of our most important responsibilities is to pass on to future generations an environment whose health, beauty, and economic potential are not threatened.
4. *Ecological Economics.* The market system of a frontier economy puts little or no value on the services of nature as a supplier of natural resources and as a sink for pollution. This leads to inefficient use of resources and a degradation in nature's capability to provide services. As an economy matures and sustainability becomes a social goal, the total utility value of resources and services of nature must be incorporated into the economic system.
5. *Policies for Conservation.* Neither the state nor the market alone can meet all human needs.
6. *Management of Natural Resources.* Resource production systems (farms, forests, etc.), which resemble the natural ecosystems of a region, require fewer subsidies than do systems that are quite different. As a result, they are more stable and more resilient, both ecologically and economically.
7. *Conservation of Biodiversity.* Protection of species depends on protection of habitat. Because it is impossible to protect *all* habitats, we must choose those that will best contribute toward maximizing global diversity.
8. *Culture and Development.* Development can be sustainable only in the context of culture.

Thus, although conservation is an applied science, it does not teach the student a particular technical skill, in the same sense that a medical school teaches prospective doctors how to set broken bones. Rather, it teaches students a philosophical approach to solving environmental problems.

To encourage students to develop a conservation perspective, the following experiment is suggested: Upon entering the class the first day, the students should be given a sheet of paper listing these eight principles. The students should be told that for the final examination they will be asked to write an essay on the meaning of one of the principles and that the principles will be randomly assigned to the students at the beginning of the examination period. This approach, instead of forcing students to memorize facts, will encourage them to understand principles, an achievement much more beneficial for their career.

CHAPTER 1

CONSERVATION HISTORY

CHAPTER CONTENTS

CHAPTER OVERVIEW

Throughout history there have been two groups, each of which considered themselves to be the true Conservationists. One group has been resource managers such as agronomists, foresters, and managers of wildlife, range, and fisheries. These conservationists have been concerned with preventing future scarcities and have sought to ensure that exploitation did not destroy the resource base. The other group has been the natural historians. These conservationists have sought to preserve nature for its inherent beauty, or for its own sake, regardless of any utility to humans. As world population has grown and the intensity of resource use has increased, conflicts between these groups have increased over the

PRINCIPLE

Pragmatism and idealism form the two historic roots of conservation. Utilitarian conservationists have been concerned with scarcity of natural resources, whereas idealistic conservationists have worried primarily about the preservation of nature. We now know that both problems are interrelated and that neither can be solved without attention to economics, politics, and culture.

proper use of the world's remaining forests, lakes, rangelands, and other ecosystems.

Both types of scientists have tended to assume that the solution to environmental problems is technical. For example, whereas the resource managers have felt that the solution lays in finding out how to produce faster growing trees or drought-resistant crops, the natural historians have felt that the solution lays in determining the minimum area required for a viable population of endangered species. Recently, scientists of all types have begun to realize that environmental problems are neither technical problems of resource supply nor matters of nature preservation. Rather, they are part of a tangled web of interrelated social and economic problems, and solutions must be sought through political means.

ANTIQUITY

Genesis

The first chapter in the Old Testament of the Christian Bible contains the seeds of the fundamental split that has divided conservationists throughout the ages into two often-conflicting camps.

> **Genesis 1:26** *And God said, Let us make man in our image, after our likeness: and let them have dominion over the fish of the sea, and over the fowl of the air, and over the cattle, and over all the earth, and over every creeping thing that creepeth upon the earth.*
>
> **Genesis 1:27** *So God created man in his own image, in the image of God created he him; male and female created he them.*
>
> **Genesis 1:28** *And God blessed them, and God said unto them, Be fruitful and multiply, and replenish the earth, and subdue it: and have dominion over the fish of the sea, and over the fowl of the air, and over every living thing that moveth upon the earth.*

The split among conservationists occurred over the interpretation of the word *dominion.* In one camp are those who interpret *dominion* to mean ownership, in the sense that the fish of the sea and the other creatures of nature have been placed on earth merely for the convenience of human beings for use in their efforts to multiply. Thus, the creatures are seen as having no other purpose or value than to serve humankind.

The other camp interprets *dominion* to mean stewardship. They interpret the Bible as saying that human beings have responsibility for preserving the well-being of all living creatures, ensuring that they not perish from the face of the earth. In other words, the creatures have an intrinsic value that exists apart from any use they might have for people.

The command that humans should "replenish the earth, and subdue it" has also sparked dispute. Those who put humans above nature emphasize the command to subdue, whereas those who see humans as part of nature argue for stewardship and replenishment. Throughout most of history, the subduers have been dominant over the restorers. Most people, states, or nations found it more convenient to search for new frontiers than to replenish areas that had already been conquered and degraded.

Of course, the conquest of new frontiers was possible as long as there were new frontiers to conquer. However, for the most part the world has now run out of frontiers, or at least of frontiers that hold potential for development. Although a few areas such as northern Alaska and part of Amazonas in Brazil remain, there is little further need to subdue nature. Instead, the present need is for replenishment. Nonetheless, the frontier ethic retains considerable momentum.

The ethic to subdue has been both glorified and vilified—glorified as humankind's manifest destiny to conquer the wild and dangerous frontier and vilified as an excuse to exploit and plunder the beauty of the world in order to satisfy an immense greed. While conquest benefited the conquerors, it degraded the environment. Indeed, environmental history is a history of degradation of resources.

Mesopotamia

Approximately 3000 years before the Christian era, the city-states of Sumer, between the Tigris and Euphrates rivers, were militaristic, hierarchical societies. Based on the detailed administrative records that they kept in their temples, we now know that environmental problems contributed to the ultimate collapse of those societies.

In the early dynastic period, the major city-states had a food surplus that enabled them to build their bureaucracies and armies and to extend their influence. They maintained this surplus, despite the region's hot, dry climate, because of water storage and irrigation projects. Records of the declining amount of wheat cultivation and its replacement by the more salt-tolerant barley indicate that, over the centuries, irrigation resulted in salinization of the region's soil. Because of the hot, dry climate, evaporation from the soil surface was high. The upward movement of the water from depth between periods of irrigation carried dissolved salts to the surface where they accumulated and damaged the crops. Between 3500 and 2500 B.C.E., wheat fell from 50 percent of the crop to 15 percent.

In about 2200 B.C.E., a marked increase in aridity and wind circulation, subsequent to a volcanic eruption, induced severe degradation of land-use conditions in Subir, immediately north of Sumer (Weiss et al. 1993). As land in Subir was abandoned, populations immigrated southward at the same time that southern irrigation agriculture was suffering from the reduced flow of the Euphrates. The growing population of Sumer necessitated cultivation of new areas. But the amount of new land that could be cultivated was limited, even with the more extensive and complex irrigation works that were becoming common at the time. Consequently, the size of the bureaucracy and the army that could be fed and maintained fell rapidly, making the state vulnerable to external conquest. The decline and fall of Sumer closely followed the decline of its agricultural base (Ponting 1990).

The Mediterranean

In the millennia before Christ, the natural vegetation of the Mediterranean basin was a mixed evergreen and deciduous forest of oak, beech, pine, and cedar. Bit by bit, the forest was cleared to provide land for agriculture and wood for cooking, heating, and construction. Sheep, cattle, and goats as well as fire suppressed regeneration, and gradually the region was transformed into scrubland.

One of the first areas to suffer deforestation was the hills of Lebanon and Syria. The natural climax forests there were particularly rich in cedars; the cedars of Lebanon were famous throughout the ancient Near East for their height and erectness.

In Greece, the first signs of widespread environmental problems appeared in about 650 B.C.E., as the population grew and cities expanded. Although the Greeks were well aware of techniques such as manuring and terracing to preserve soil, the pressure from a continually rising population proved too great. The hills of Attica were stripped bare of trees within a couple of generations. The only tree that would grow on the badly eroded land was olive because it had roots strong enough to penetrate the underlying limestone rock.

The Greek Philosophers

The Man-Nature dualism was a problem of great importance to the philosophers of ancient Greece (Hughes 1975). The Greek philosophers had two attitudes toward nature. One school regarded nature as the theater of the gods. They believed that the gods had their abodes in nature; they appeared from nature, were often clothed in natural forms, and withdrew into nature when they left human contact. For these philosophers, the natural environment was the scene of divine activities that could be witnessed by mortals. When men obstructed justice in the courts, Zeus responded with a disastrous flood, but when a king ruled his people with wise care for their welfare, the gods could respond by making the fields fruitful, the herds fertile, and the seas bountiful. According to this perception of the natural environment, nature was sacred. Thus many human activities that affected nature such as agriculture were surrounded by religious precautions.

The second school of thought espoused a teleological interpretation of nature. It established an anthropocentric and utilitarian attitude toward animals and plants. Aristotle for example, taught that all things are created for a specific purpose or end and that when an object fulfills its end, it is both useful and beautiful. Therefore, no animal lacks beauty because all animals are formed for their proper end. And their proper end is the service of human beings.

Over the ages Aristotle and his followers exerted the predominant influence on Western thought, so that even today animals are considered to be of a lower order, subservient to human needs. They were thought to be put on earth to satisfy these needs both through domestication and through uses as food. It was considered right and proper that humans could use or kill a plant or animal just as they could use water from a stream or a stone from a quarry. All animals, and indeed all other things, were created for the human good and so the entire natural world was to be used for human purposes. This viewpoint led Eugene Hargrove (1989), a modern environmental ethicist, to state that "Greek philosophy is the primary source of the philosophical perspectives that have historically inhibited the development of appropriate environmental and preservationist attitudes" (p. 33).

Rome

The overgrazing and overintensive cultivation that devastated Greece would be repeated a few centuries later in Rome. In 300 B.C.E., the land comprising modern Italy and Sicily was still well forested, but the increasing demand for land and timber resulted in rapid clearing. The growth of the Roman empire increased the pressure on the environment in other areas of the Mediterranean. Many of the empire's provinces were turned into granaries to feed the population of Rome. Even North Africa contains many Roman remains such as the great city of Leptis Magna in what once was a highly productive agricultural province.

The North African provinces declined through a gradual process of increasing overexploitation of resources and consequent environmental deterioration. Soils eroded and the desert slowly encroached. The process intensified after the fall of Rome, when Berbers and other tribes moved into the cultivated areas with large flocks of grazing animals (Ponting 1990).

THE MIDDLE AGES

Forest ownership and laws governing forest use in Europe can be traced back to at least the beginning of the Middle Ages. Some woodlands were royal and feudal manorial forests and woodlands, while others were owned in common. Traditions of forest management were probably passed on to the medieval period from the Greeks and Romans. Written traditions included local forest ordinances, royal forest ordinances, and forestry literature.

Feudalism encouraged the conversion of forestlands to agriculture, for purposes of increasing production, accumulating capital, and supporting the feudal lords and royal courts through tributes. Agriculture provided a livelihood for much of the rural populace. As populations grew and the privatization of land through enclosure increased, the loss of forest lands, timber, fuelwood, fodder, and game caused some concern. The kings and gentry worried that their exclusive hunting grounds were in danger of disappearing. Therefore, game reserves were set aside for the kings of the Middle Ages to ensure a population of deer and boars adequate for sustainable hunting. One of these reserves, the Bialowieza Park on the Polish-Russian border, is now one of UNESCO's Man and the Biosphere reserves and the site where the European forest bison was reintroduced to the wild from zoos (Falinski 1986).

During the Middle Ages Europe's ruling classes recognized the potential economic opportunities of controlled forest management. However, the idea that forests should be managed for the state by agents of the state was not developed until the early eighteenth century (in France and Germany). Foresters and gamekeepers had been employed on the lands of royalty, the gentry, and the clergy for centuries, but it was not until 1787 that the first university training program in forestry was established, at the University of Freiburg. Other German universities followed suit in the early nineteenth century, and a national school of forestry was founded in Nancy, France, in 1824. Foresters from all over Europe and the United States attended these schools to learn their science. When they returned home or traveled to the European colonies in Africa, Asia, and Latin America, they carried with them the philosophy and methods of state-controlled or centralized forest management (Peluso, 1992).

THE INDUSTRIAL AGE

The Industrial Age was born as a result of the confluence of laissez-faire capitalism and science and technology. It was in this period that humans finally mastered the forces of nature that for so long had dominated them. Although the Industrial Age presented an opportunity for an improved material existence, not everyone was enamored with the marvels of modern technology.

The Romanticists

Romanticism was an artistic and philosophical movement that began in the late 1700s and flowered in the 1800s. In Germany, Friedrich von Schelling formulated a *Naturphilosophie* that romanticized nature and inspired man to find a new sense of personality in nature (Bronowski 1973). In England, the movement grew among poets and artists, especially Coleridge, Shelley, and Byron (Youngs 1984). Romanticism was a reaction against the Industrial Age; specifically, it deplored the scientific and technological forces, which it regarded as dehumanizing man and degrading nature. In particular, the Romantics hated industrialization for making the beautiful ugly. The Romantics also rejected the vulgarity of those who made money in trade. In reaction to a new class of wealth based on industrialization, the Romantics sought their own self-definition by eschewing the commodity society. They separated themselves from the vulgar bourgeoisie and the working-class proletariat not only by using physical means, but also by promoting the idea that their own labor could not be reduced to commodity values. Their labor, rather, was intellectual and artistic.

Their feelings of respect for nature were in contrast to the scientific framework, in which beauty had no objective existence and nature had no intrinsic value. In Baconian science, nature was merely a compilation of millions of organisms, and what they were and how they functioned could be fully understood by reductionist science. The Romantics rejected this view. For them, nature was more than the sum of its parts and had a higher level integrity. They saw purpose and meaning in nature. In today's parlance, the Romantics were idealistic conservationists.

The Naturalist Writers

Although the Romanticists opposed reductionism and were appalled by a science that lacked spiritual value, many scientists saw a holistic value in nature. Reports by scientists on colonial expeditions during the seventeenth and eighteenth centuries contained elements of a nature philosophy, although such views were sometimes rooted more in economic than in aesthetic concerns. Whereas reductionistic science was concerned only with the efficient exploitation of natural resources, these scientists saw that efficient exploitation alone would be economically disastrous if provision were not made for conservation of the resource.

The European powers colonized and exploited tropical lands throughout the seventeenth and eighteenth centuries. After the military forces and the colonists established a foothold, invariably the scientists followed. Many of them were medical surgeons or custodians of the early colonial botanical gardens, and they quickly became aware of the environmental devastation caused by colonial exploitation. For example, in 1827, a Dr. Wallich was sent to report on the natural resources of Tenasserim Province, which at that time was part of Burma (now Myanmar) under British rule. He was so alarmed by the overcutting that he recommended strict government control. A few years later, a medicobotanist was sent to examine the forest and wrote a report in which he discussed girdling, disease, fire, and the lack of natural regeneration (Blanford 1958).

Artists who traveled to the colonies played an important role in the origins of Western environmentalism by communicating to Europeans the extent of the environmental degradation caused by colonialism. For example, artistic renditions of Mauritius in 1677 forcefully depicted the stark reality of felled ebony forests. As a result, an awareness of the ecological impact of capitalism and colonial rule began to emerge (Grove 1992).

Alexander von Humboldt

Alexander von Humboldt (1769–1859), in the accounts of his travels in South America between 1799 and 1804, emphasized the need to conserve the continent's natural resources (E. L. Jordan 1981). To the fishermen of Araya concerned about the destruction of the once lucrative pearl fisheries, he pointed out that it was not the sound of the oars that had driven away the oysters, as they imagined: it was their ruthless overfishing of the oyster beds. When the worried landholders around Lake Victoria in Venezuela wondered why the lake was shrinking year after year, he told them that the phenomenon was of their own doing. By denuding the nearby mountain sides and cutting the trees that surrounded the lake, they themselves had caused the erosion that was diminishing the water supply of the springs that fed the lake.

Similarly, von Humboldt told the padres of Caripe that their "mine of fat," which involved the annual killing of thousands of guacharo birds, would lead to the near extinction of the species. He warned the Franciscan missionaries who supervised the turtle egg harvest on the islands of the Orinoco about exhausting a source of food without giving a thought to the future. If they did not leave part of the island beaches untouched, so that a sufficient number of turtles were hatched every year, before long there would be no egg harvest.

As an admirer of the beautiful groves of quinine-producing cinchona trees on the Cordilleras of Peru, von Humboldt was dismayed that thousands of trees were felled for their bark without any attempt at reforestation. As a result, the Spanish colony lost an important resource.

Charles Darwin

Charles Darwin's (1809–89) *Natural History and Geology of the Countries Visited During the Voyage of H.M.S. Beagle Round the World* (1845) laid the groundwork for his ideas on evolution which would revolutionize biology. In the conclusion to his work he wrote,

> *When I say that the scenery of parts of Europe is probably superior to anything which we beheld, I except, as a class by itself, that of the intertropical zones. The two classes cannot be compared; but I have already often enlarged on the grandeur of those regions. As the force of impressions generally depends on preconceived ideas, I may add, that mine were*

taken from the vivid descriptions in the Personal Narrative of Humboldt, which far exceed in merit anything else which I have read. Yet with these high-wrought ideas, my feelings were far from partaking a tinge of disappointment on my first and final landing on the shores of Brazil.

Among the scenes which are deeply impressed on my mind, none exceed in sublimity the primeval forests undefaced by the hand of man; whether those of Brazil, where the powers of Life are predominant, or those of Tierra del Fuego, where Death and Decay prevail. Both are temples filled with the varied productions of the God of Nature: — no one can stand in these solitudes unmoved, and not feel that there is more in man than the mere breath of his body. (p. 496)

North American Naturalists

Throughout the colonization of North America, the frontier ethic held sway. "Buffalo Bill" was deified for his wanton slaughter of North American bison, and thousands of other hunters followed his example. Even today he is idolized in the name of an American Football League team.

Nevertheless, at an early date there were others who rejected the conquer and extinguish approach to the North American wilderness. Henry David Thoreau (1817–62) was one. Although he is best known for his eloquent description of nature at Walden, his views on conservation are well expressed in his ironic essay "Life Without Principle" (1863, reprinted in Glick, 1973):

If a man walk in the woods for love of them half of each day, he is in danger of being regarded as a loafer; but if he spends his whole day as a speculator, shearing off those woods and making earth bald before her time, he is esteemed an industrious and enterprising citizen. As if a town had no interest in its forest but to cut them down! (p. 157)

In 1864 George Perkins Marsh (1801–82), a scientist and congressman from Vermont, published a book titled *Man and Nature.* It presumed an innate order in which all things moved according to a natural law, maintaining the most delicate and perfect balance. When man entered with all his ignorance and presumption, there followed a succession of "disturbed harmonies" in the natural order, with implications of destruction extending ultimately to man himself (McConnell 1960).

Marsh accepted the man-as-steward interpretation of Genesis.

Man has too long forgotten that the earth was given to him for usufruct alone, not for consumption, still less for profligate waste. Nature has provided against the absolute destruction of any of her elementary matter, the raw material of her work; the thunderbolt and the tornado, the most convulsive throes of even the volcano and the earthquake being only phenomena of decomposition and recomposition. But she has left it within the power of man irreparably to derange the combinations of inorganic matter and of organic life, which through the night of aeons she had been proportioning and balancing, to prepare the earth for his habitation, when in the fulness of time, his Creator should call him forth to enter into its possession. (cited on p. 190 in McConnell, 1960)

Marsh's writings have had a tremendous influence on succeeding generations of conservationists and ecologists.

John Muir (1839–1914), a pioneer of conservation and founder of the Sierra Club, called public attention to the environmental destruction caused by development in California. In the February 5, 1875, edition of the *Sacramento, California Record-Union,* he wrote an article titled "God's First Temples: How Shall We Preserve Our Forests" (Wolfe 1945). In this piece, Muir appealed to "practical men" by stressing the economic results of forest destruction in floods, droughts, and river channels choked with silt that covered lowland fields with detritus. Waste and destruction due to sawmills, fires set by stockmen and sheep hordes annually invading the Sierra, were proceeding at an accelerating rate, he said, so that within just a few years the forests would be gone. He questioned whether the government was really able or willing to do anything about it.

THE CONSERVATION MOVEMENT

Theodore Roosevelt and Gifford Pinchot

The voices of Thoreau, Marsh, Muir, and others like them were basically solitary voices. In their day, there was little institutional support for the idea that some of the American frontier should be preserved or at least managed in a way that did not destroy it. As the nineteenth century drew to a close, however, there arose within the North American establishment a group that shared the idea that the continent was not an endless frontier with unlimited resources. If forests continued to be destroyed, they warned, soon they would no longer exist.

Their focus on the practical means to use nature without destroying it resulted in a social and political phenomenon called the conservation movement (Graham 1971; Pinkett 1970). When Theodore Roosevelt (1858–1919) became president in 1901, he spelled out the conservation philosophy in his first State of the Union Message. In the process, he established the basis of a policy for conservation of natural resources in the United States, a philosophy that has lasted almost a century.

Specifically, Roosevelt addressed issues of forest management. He stated:

> *The fundamental idea of forestry is the perpetuation of forests by use. Forest protection is not an end in itself: it is a means to increase and sustain the resources of our country and the industries which depend upon them. The preservation of our forests is an imperative business necessity. We have come to see clearly that whatever destroys the forest, except to make way for agriculture, threatens our well-being. (Graham 1971, p. 105)*

That philosophy was distilled into the words *wise use* and has been taught as the guiding principle of conservation in schools of forestry and natural resource management throughout most of the twentieth century. The cornerstone of this movement was not preservation in the sense of locking up the resource, but rather, as Roosevelt described it, by *using* them, but using them without destroying them. Gifford Pinchot (1865–1946), the first chief of the Forest Service, played a leading role in implementing the philosophy that the government had a responsibility to preserve forests for their economic value both for present and future generations. Wise use of the forest was interpreted to mean multiple use. Pinchot's views were summarized in his definition of conservation as the use of natural resources for the "greatest good of the greatest number in the long run" (Pinkett 1970, p. 59).

The view that resources should be sustained through "wise use" was applied to management of all natural resources, and not only the forests. Such an approach was theoretically satisfactory as long as the resources were sufficient to satisfy all interests—that is, providing enough game for the hunters; enough fish for the fishermen; enough trees for the loggers; enough wilderness for the nature lovers; enough trails for the ski-mobilers; enough water for the farmer. Even in Pinchot's time, however, conflicts arose between wise-use advocates and those frontiersmen who felt that every American (Native American Indians excepted) had a "right" to exploit at will the national forests (then called forest reserves). Regulations that interfered with this supposed "right" became known as "locking up the land" or "Pinchotism," a deprecation akin to socialism or even communism.

Pinchot's dictum of making "wise use" of resources no longer applies today, for what is "wise" depends on the perspective of the interest group that defines wisdom. For example, many environmentalists believe that the wisest use of the last remaining old-growth forests in the Pacific Northwest would be as an inviolate preserve for rare and endangered species. A different perspective comes from politicians who proclaim that the best use of these forests is to continue logging them, to keep alive the economy of towns dependent on the logging industry.

By 1992, use of the term *wise use* had been completely perverted. A coalition of groups that pushed for repeal of environmental legislation that they saw as interfering with their "right" to exploit national parks and forests for private gain has outlandishly dubbed itself the "wise use movement" (O'Callaghan 1992).

Franklin D. Roosevelt

In the early 1930s the United States was mired in the Great Depression, and its natural resources were suffering from years of overexploitation. The election of Franklin D. Roosevelt (1882–1945) to the presidency of the United States in 1932 marked an important epoch of change in the nation, in both its social and conservation policy. The Roosevelt administration initiated a number of projects that addressed economic, social, and conservation problems. The Tennessee Valley Authority (TVA) and the Civilian Conservation Corps (CCC) are outstanding examples.

The TVA

The idea of building a dam on the Tennessee River where it dips into northern Alabama at a place called Muscle Shoals actually originated in the early 1900s (Graham 1971). However, continual bickering between private and public interests had kept the dam from ever being completed. Roosevelt saw much more potential in the Southern Appalachian region than just a mere dam. It was a region that had been hit hard by the depression, and excessive logging had resulted in a badly degraded landscape. In an address to Congress in 1933 Roosevelt stated:

> *It is clear that the Muscle Shoals development is but a small part of the potential usefulness of the entire Tennessee River. Such use, if envisioned in its entirety, transcends mere power development: it enters wide fields of flood control, soil erosion, afforestation, elimination from agricultural use of marginal lands, and distribution and diversification of industry. In short, this power development of war days leads logically to national planning for a complete river watershed involving many states and the future lives and welfare of millions. It touches and gives life to all forms of human concerns.*
>
> *I therefore, suggest to the Congress legislation to create a Tennessee Valley Authority—a corporation clothed with the power of government but possessed of the flexibility and initia-*

tive of a private enterprise. It should be charged with broadest duty of planning for the proper use, conservation and development of the natural resources of the Tennessee River drainage basin and its adjoining territory for the general social and economic welfare of the nation. (cited in Graham 1971, pp. 250–251)

TVA was a crowning achievement of the conservation movement. The nitrate plants at Muscle Shoals became the center of the world's research on fertilizers. Forest fires and soil erosion on the slopes of the river valleys were checked, and flood damage was mitigated. In addition, the production of farms and forests increased, and navigation and recreation flourished on the transformed river.

The CCC

The Great Depression of the 1930s threw millions of men out of work. To provide relief, Roosevelt established the Civilian Conservation Corps (CCC), which gave young men an opportunity to live away from the cities, earn a little money, and, at the same time, do something that was important. They planted trees, developed parks and recreation areas, restored silted waterways, provided flood control, controlled soil erosion, and protected wildlife. The lodges, trails, and cabins that they built gave the public an opportunity to visit the national parks and to experience nature at first hand.

With the beginning of World War II, many young men went into the armed services, while others worked in the factories that built the supplies for war. The CCC faded, but it had served its purpose. Its "work for welfare" had provided an economic bridge for part of a generation, and the work they did created a lasting mood that was most favorable to conservation (Nash 1990).

The mood they created was perhaps more important than their actual work. Modern ecologists sometimes chuckle over the stories of how the CCC removed old dead logs from forests to tidy them up. Modern ecology has shown that old dead logs perform a critical function in the forest by providing an energy source for the microbes that recycle nutrients. But the CCC boys did not know that.

Aldo Leopold

By the mid-twentieth century, the image of the U.S. Forest Service was changing. The agency that was supposed to regulate logging and grazing on national forests was instead serving the loggers and ranchers it was supposed to regulate. Those who sought to protect national forests no longer considered the Forest Service an ally but an enemy because it had forsaken them.

A North American forester/game manager by the name of Aldo Leopold became so dismayed by the destructive effect of Forest Service policies on the national forests that in 1949 he wrote an essay that many consider to be the first attempt in modern Western literature to develop an ethical theory dealing with the relationship of humans to nature. In that work titled "The Land Ethic," Leopold considered three phases in the development of human ethics: (1) the relation between individuals; (2) the relation between the individual and society; and (3) the individual's relation to land and to the animals and plants that grow on the land. He said:

The extension of ethics to this third element in human environment is, if I read the evidence correctly, an evolutionary possibility and an ecological necessity. It is the third step in a se-

quence. The first two have already been taken. Individual thinkers since the days of Ezekiel and Isaiah have asserted that the despoliation of land is not only inexpedient but wrong. Society, however, has not yet affirmed their belief. I regard the present conservation movement as the embryo of such an affirmation. (Leopold, 1949, p. 203)

Because Leopold was a professional forester, his assertion that people have an ethical responsibility to protect nature was particularly influential.

Echo Park Dam

With the economic expansion and increased population that followed World War II, pressures on federal lands grew severe. Private conservation organizations that had previously supported the concept of multiple use now followed Leopold's lead. They became skeptical of the government's increasing emphasis on timber harvesting and the increasing abuse of grazing privileges on federal lands by western ranchers. Prior to the 1950s, each of these conservation organizations had pursued its own particular interest—the Audubon Society in studying and protecting birds, the Sierra Club in preserving the giant redwoods and sequoias, and the Wilderness Society in protecting forests, rivers, deserts, and shorelands. However, the Echo Park plan to dam the Colorado River galvanized the private organizations into an organized and effective political force that fought against the ruin of public property for the gain of a privileged few (Graham 1971).

The federal government had proposed a billion-dollar dam-building program in eleven western states which would be carried out by the Bureau of Reclamation and called the Colorado River Storage Project. A dam 525 feet high was to be built in Echo Park on the Green River in Colorado and Utah. If the dam were completed, it would flood the lovely unique canyons in Dinosaur National Monument. David Brower of the Sierra Club described what would happen:

There would be construction roads in the canyons and above it, tunnels, the whole power installation and transmission lines, the rapid build-up of silt at the upper end of the reservoir to enable it to fulfill its function—a fluctuation that would play hob with fish and wildlife. The piñon pines, the Douglas firs, the maples and cottonwoods, the grasses and other flora that live on the banks, the green living things that shine in the sun against the rich colors of the cliffs—these would also go. The river, its surge and its sound, the living sculptor of this place, would be silent forever, and all the fascination of its movement and the fun of riding it, quietly gliding through these cathedral corridors of stone—all done in for good. (cited in Graham 1971, p. 296)

The conservation groups were successful, and on April 11, 1955, President Eisenhower signed a bill that prohibited the construction of a dam in any national park or monument. Although the conservationists won the Echo Park battle, the frontier ethic remained an important part of the American psyche so that in the following years, exploitation rather than preservation won most of the battles.

Agriculture

Agriculture was an important base of the American economy in the nineteenth century. As the United States moved into the Industrial Age, it became clear that just as technology was helping to improve industrial production, so technology could also help to improve agricultural production. Consequently, in 1862 an act to establish agricultural and mechanical arts colleges was passed—the land grant colleges (Smith 1971). The colleges almost entirely emphasized increasing the production of crops, or increasing the efficiency of crop production.

Agriculture as practiced by the early American settlers had evolved in the brown forest soils of Europe. Because of ample rainfall, these soils have a leached topsoil and a darker subsoil in which the materials moved down from the topsoil are deposited. The soil surface is characterized by a layer of litter and humus. The technique of deep plowing tends to mix together the litter and humus, the light topsoil, and the deep nutrient-rich subsoil. With care and with the standard European practices of crop rotation, the productivity of these soils could be sustainable. This type of farming was taught at the agricultural colleges of both the East and Midwest.

When the wave of settlement reached the Great Plains, the pioneers encountered a different type of soil. There they found deep soils with excellent structure and high natural fertility. However, the key to maintaining these soils was the native grassland vegetation. Plowing broke the thick layer of roots and exposed the soil. In years with abundant rainfall, crops covered the soil and prevented excessive erosion. However, the region was subject to periodic droughts. Severe ones occurred in 1890 and 1910, and the losses of soil due to wind erosion caused some farmers to move on. Many remained, however, and when the rains returned, they again cultivated the soil. World War I increased the demand for wheat, and the intensity of cultivation increased.

Then in 1931, an exceptionally severe drought began. In 1933 high winds carried away tons of unprotected soil. The dense clouds that obscured the sky gave rise to the name "the Dust Bowl" for western Kansas and Oklahoma, and parts of Texas, New Mexico, and Colorado. In the process, millions of acres of farms were damaged, and drifting dunes buried roads, fences, and farmhouses. Ruined farmers and their families left for California or for the breadlines in the city. Finally, the government was spurred to action, and in 1935 the Soil Conservation Service was created to provide aid and technical assistance to the farmers. Grass was planted where the plow should never have been used, and on soils better suited to farming, the use of conservation techniques such as contour plowing and trees for wind breaks was initiated (Dasmann 1968).

After World War II, the lessons of the Dust Bowl were seemingly forgotten, and emphasis again shifted toward increasing production. The government now put scientists to work to increase the productive capacity of resource ecosystems: farms, forests, rangelands, and fisheries. Often the scientists were spectacularly successful, especially in agriculture. Fertilizers increased production, in spite of soil erosion. Insecticides kept down the pests, in spite of the expansion of even-aged monocultures that were highly susceptible to exponential growth of the pests. And when a commercially important strain of crops lost its resistance to pathogens, another could be found or bred to replace it. Advances were made in other fields too—in plantation forestry, fisheries management, and range management. All increased the output from the land.

THE POSTINDUSTRIAL AGE

Silent Spring

By the 1960s, it was becoming apparent that the continued increase in productivity had created undesirable environmental side effects. For many years these effects were hardly noticeable—a polluted stream here, a dead eagle there—but every year the destruction accumulated, until it was difficult to find a stream that was not polluted or ever see an eagle in the wild. Rachel Carson's 1962 landmark book *Silent Spring* dramatically brought to the public's attention the fact that as population grew, the natural resource base was no longer able to supply all the goods and services demanded, without a gradual diminishing of the base of their productive capacity. Thus people became acutely aware of the loss of soil due to erosion; the poisoning of terrestrial and aquatic ecosystems from pollution; and the loss of species that form the basis of future food supplies and pharmaceuticals as well as give aesthetic pleasures. Carson's book was highly successful, stimulating legislation to prohibit the worst of the pesticides and resulting in increased scientific efforts to find less harmful methods of agriculture.

Carson's book was a conservation milestone in that it brought to public attention the fact that conservation problems were not limited to national parks and forests. Philosophical questions about whether nature should be managed primarily for human beings or for wildlife paled in the face of the world's mounting environmental problems resulting from population increase and economic growth. Problems such as air pollution, water pollution, congested highways and cities, and disposal of waste were beginning to affect the health and lives of everyone.

The Broadening View

In 1963 the U.S. secretary of the interior, Stewart Udall, wrote *The Quiet Crisis,* a book that described the worsening environmental conditions. In this work Udall made a new, compelling plea for a conservation ethic:

> *We must develop a land conscience that will inspire those daily acts of stewardship which will make America a more pleasant and more productive land. If enough people care enough about their continent to join in the fight for a balanced conservation program, this generation can proudly put its signature on the land. But this signature will not be meaningful unless we develop a land ethic. Only an ever-widening concept and higher ideal of conservation will enlist our finest impulses and move us to make the earth a better home both for ourselves and for those as yet unborn. (pp. 190–191)*

A few years later, the microbiologist René Dubos (1968) wrote in his book *So Human an Animal:*

> *This book should have been written in anger. I should be expressing in the strongest possible terms my anguish at seeing so many human and natural values spoiled or destroyed in affluent societies, as well as my indignation at the failure of the scientific community to organize a systematic effort against the desecration of life and nature. Environmental ugliness and the rape of nature can be forgiven when they result from poverty, but not when they occur in the midst of plenty and indeed are produced by wealth. The neglect of human problems by the scientific establishment might be justified if it were due to lack of resources or of methods of approach, but cannot be forgiven in a society which can always find enough money to deal with the issues that concern selfish interests. (pp. 1–2)*

He concludes "Balance involves man's relating to his total environment. Conservation therefore implies a creative interplay between man and animals, plants, and other aspects of Nature, as well as between man and his fellows. The total environment, including the remains of the past, acquires human significance only when harmoniously incorporated into the elements of man's life" (p. 199).

Although these works set the stage for modern conservation, the book that really put the environment in the public eye was Paul Ehrlich's *The Population Bomb* (1969). The book begins with a fantasy 900 years hence when the world's population of 60 million billion people are housed in a continuous 2000-story building covering the entire planet. The upper 1000 stories contain only the apparatus for running the gigantic "warren."

Then Ehrlich points out that, long before such a warren could exist, other manifestations of environmental crises would stop growth—crises such as industrial pollution, the extinction of species, the loss of wilderness, and the inexorable spread of urban blight which creates a breeding ground for youthful disaffection, crime, and violence. Ehrlich's alternative scenarios end in nuclear war or starvation for a fifth of the world's population in Africa and Asia. The affluent nations would then group under an enlightened U.S. leadership to control population growth, introduce controlled agricultural and industrial development, and eventually stabilize the world population. Ehrlich concludes that many observers believe that population control by government, not just voluntary planning, is our only hope.

Until recently, Great Britain's conservationists had a different emphasis than their American counterparts (Young 1990). They were concerned less about the psychological or genetic values of wilderness, and more about the amenities of the cultivated British countryside, which seemed threatened by the demands of industry and the vulgarities of excessive tourists. Therefore, in the early 1970s, a group of ecologists, supported by 37 eminent chemists, zoologists, medical scientists, microbiologists, and botanists, together with an archaeologist and an economist, got together and compiled the *Blueprint for Survival* (Goldsmith 1972).

The publication begins with a stern warning about "the extreme gravity of the global situation today" and forecasts the impending breakdown of society and the "irreversible disruption of the life-support systems on this planet possibly by the end of this century" if current trends are allowed to persist. It sought to found a "movement for survival" based on a new philosophy of life in which man would learn to live as part of nature rather than as its antagonist.

Like Ehrlich, Goldsmith pointed to overpopulation as the root of the problem. Indefinite growth confronted with finite resources, he emphasized, must inevitably lead to disaster. If resources did prove more extensive than expected, that would produce graver problems of waste disposal. Another concern was the increasing inclusion of traditional societies in the Western economy, resulting in heightening the material aspirations in these societies. For the health of the planet, the Blueprint recommended that these aspirations must remain unsatisfied.

The solution the Blueprint outlined lay in a proposal for a utopian society. Utopia was to consist of small village-type settlements of about 500 people, forming part of larger communities of around 50,000 in regions of about half a million. Decentralization would facilitate a reduction in capital costs through community self-sufficiency in such things as sewage disposal and water supplies. No new roads would be built. The money saved would be used to subsidize public transport.

In addition to the 37 scientific supporters of the Blueprint, 180 other scientists embraced the ideas of the Blueprint in general but felt unable to subscribe to it completely "because it contained scientifically questionable statements of fact and highly debatable short term and long term policy statements" (*The Times,* cited in Young 1990, p. 9).

This resistance to utopian solutions was among the first indications that the environmental crisis was not susceptible to a solution by the application of a single formula or even by a combination of specific remedies such as those recommended by the *Blueprint for Survival.* The problem was more than one of applying technical solutions to pollution, resource depletion, famine, and landscape degradation.

The Limitations of Science

A popular question of the 1960s was, "If we can send a man to the moon, why can't we solve the problems of (crime), (poverty), (threat of war), or pertinent to the discussion here, (the environment)?" Until that time, it had been assumed that the environmental crisis was basically a technical one. All that scientists had to do then was to determine how to control pollution, how to manage forests sustainably, and so on, and as soon as that information became available, governments would proceed to do it. It proved difficult to recognize that this was not the case and that environmental problems were as much social as they were technical.

Barry Commoner's *The Closing Circle* (1971) stressed that the spectacular advance of science and the increase in knowledge during and after World War II were a result of the scientific method which isolated problems and employed specialized skills. There was a basic difference between the technical problems of getting a man on the moon and the social problems of crime and poverty. Social problems were not nearly as easy to solve, in part because they involved compromises between groups, each of which disagreed or saw itself as threatened by any proposed action.

In one area after another, Commoner pointed out, American society was unable to deal with social problems because it believed that the scientific method, which was so effective in dealing with isolated technical problems, could also solve the overlapping problems of society in general. Especially important was the notion that conservation of natural resources and preservation of the environment had more in common with social problems than with technical problems such as increasing agricultural productivity.

Commoner concluded his book by saying that

> *In our progress-minded society, anyone who presumes to explain a serious problem is expected to offer to solve it as well. But none of us, singly or sitting in Committee, can possibly blueprint a specific "plan" for resolving the environmental crises. To pretend otherwise is only to evade the real meaning of the environmental crises: that the world is being carried to the brink of ecological disaster not by a singular fault, which some clever scheme can correct, but by the phalanx of powerful economic, political and social forces that constitute the march of history. Anyone who proposes to cure the environmental crisis undertakes thereby to change the course of history.*
>
> *But this is a competence reserved to history itself, for sweeping social change can be designed only in the workshop of rational, informed, collective social action. That we must act is now clear. The question which we face is how. (p. 300)*

Many scientists, seeing Commoner's book as an attack on science itself, issued a strong response. Some countered that the problems that science created needed not less but more science for their solution, and they labeled Commoner an "anti-intellectual." John Maddox, a theoretical physicist, responded in 1972 with *The Doomsday Syndrome,* a book in which he argued that the most obvious solution to the problem of famine was to make fuller use of "unproductive" land. But this argument did not repudiate Commoner's point. It merely vindicated it, inasmuch as making fuller use of unproductive land is more an economic than a technical problem.

A New Understanding

Scientists and humanists, conservatives and liberals, human-centered and nature-centered conservationists—all are beginning to realize that environmental problems are not isolated technical problems, but rather are interrelated social and economic problems. And while science-based technologies can provide short-term, stopgap solutions, long-term solutions must be sought through political means.

This understanding is still far from universal. During the 1980s, the premier scientific journals featured articles in which technical fixes were still being heralded as the solution to problems of development in remote regions such as the Amazon (Sanchez et al. 1982). They gave no consideration to the economic costs or to the cultural appropriateness of high tech in the Amazon. Even as late as 1992, geneticists were working on increasing cold hardiness in crops so that they could be cultivated in the tundra, in the belief that such a technical accomplishment would decrease world hunger (Moffat 1992).

Nevertheless, increasing numbers of scientists from agronomists to ornithologists are beginning to understand that conservation of resources and biodiversity is a political, social, economic, and cultural dilemma as well as a technical problem. Technical solutions will work only when they are economically justifiable, politically feasible, culturally acceptable, and environmentally sustainable. This book is organized based on this perspective.

SUGGESTED READINGS

For an environmental history of the world:

Ponting, C. 1991. A Green History of the World. Sinclair-Stevenson Ltd., London.

For a history of conservation in the United States as seen through the lives of its most important figures:

Strong, D. H. 1971. Dreamers and Defenders. University of Nebraska Press, Lincoln.

Udall, S. 1988. The Quiet Crisis and the Next Generation. Peregrine Smith Books, Salt Lake City.

For an American environmental chronology, through excerpts from or about the leading players:

Nash, R. F. 1990. American Environmentalism: Readings in Conservation History. McGraw-Hill, New York.

For a comprehensive coverage of the resource management side of conservation, including history in the United States:

Burton, I., and R. W. Kates, eds. 1965. Readings in Resource Management and Conservation. University of Chicago Press, Chicago.

For a history of conservation in Africa:

Anderson, D., and R. Grove. 1987. Conservation in Africa: People, Policies and Practice. Cambridge University Press, New York.

For a perspective on forest conservation in Southeast Asia with historical views of Java, the Philippines, and Thailand:

Poffenberger, M. 1990. Keepers of the Forest. Kumarian Press, West Hartford, Conn.

CHAPTER

ENVIRONMENTAL TRENDS

CHAPTER CONTENTS

PRINCIPLE

The *ability of the* earth to maintain ecological *integrity* has replaced the supply of resources as the central problem of conservation.

CHAPTER OVERVIEW

Warnings that the world is on the verge of running out of resources are as old as recorded history but have increased since Malthus and the beginning of the Industrial Age. Nevertheless, lack of resources *has never been an insurmountable problem for the world's economy, owing in part to technology and substitution. Even today there is no evidence that it is a* lack of resources *that gives rise to poverty, local incidents of starvation, and social conflict. There is evidence, however, that environmental deterioration is resulting in a lower quality existence for most of the world's inhabitants. As a result, quality of life is replacing supply of resources as a priority for conservation.*

POPULATION AND RESOURCES

The Early Doomsayers

In 1798 the English political economist Thomas Robert Malthus (1766–1834) wrote his famous dictum: "Population, when unchecked, increases in a geometrical (exponential) ratio. Subsistence increases only in an arithmetical ratio." Malthus's assessment of the world's future was bleak. He predicted that because the amount of land available for agriculture was limited, population would quickly outstrip the ability of land to supply that population with food. The inevitable consequence was starvation. The conviction that population will exceed the supply of resources necessary to feed that population has been a recurrent theme since the time of Malthus. His name has become synonymous with the belief that soaring population and decreasing resources will plunge the world into disaster.

One nineteenth-century Malthusian was William Stanley Jevons, a British economist who in 1865 published an article titled "The Coal Question: An Inquiry Concerning the Progress of the Nation, and the Probable Exhaustion of Our Coal Mines." "The conclusion is inevitable", he wrote, "that our present happy progressive condition is a thing of limited duration" (Tierney 1990, p. 76). The factor that Malthusians did not foresee was the ability of technology to overcome the limitations of preindustrial agriculture. The predictions of disaster on a global scale were not realized.

The rapid increase in population growth after World War II prompted new attention to Malthusian theory. "POPULATION OUTGROWS FOOD, SCIENTISTS WARN THE WORLD" was a front-page headline in the *New York Times*, on September 15, 1948. The article warned that "overpopulation and the dwindling of natural resources" boded a "dark outlook for the human race." "NO ROOM IN THE LIFEBOATS" summarized a story in the *New York Times Magazine,* on April 16, 1978, warning that "the cost of natural resources is going up" as increasing population ushers in the "Age of Scarcity."

Recent Cassandras

In 1969 Paul Ehrlich in *The Population Bomb,* wrote: "The causal chain of the deterioration is easily followed to its source. Too many cars, too many factories, too much detergent, too much pesticide, multiplying contrails, inadequate sewage treatment plants, too little water, too much carbon dioxide—all can be traced easily to *too many people.*" (p. 157).

The concern has not diminished in the succeeding decades. In 1991 *Time* magazine wrote that the human population is so large and growing so rapidly that there is now a possibility of a "demographic winter" (Daily and Ehrlich 1992), an analogy to "nuclear winter".

In the early 1970s, a group of international businessmen, statesmen, and scientists organized the Club of Rome and commissioned a study to investigate the long-term causes and consequences of growth in population, industrial capital, food production, resource consumption, and pollution (Meadows et al. 1992). The book that resulted from the effort, *The Limits to Growth,* asked a number of questions such as

What will happen if growth in the world's population continues unchecked?

What will be the environmental consequences if economic growth continues at its current pace?

What can be done to ensure a human economy that provides sufficiently for all and that also fits within the physical limits of the earth?

The summary conclusions, written in 1972, were

1. If the present growth trends in world population, industrialization, pollution, food production, and resource depletion continued unchanged, the limits to growth on this planet would be reached within the next 100 years. The most probable result would be a sudden and uncontrollable decline in both population and industrial capacity.
2. Second, it is possible to alter these growth trends and to establish a condition of ecological and economic stability that is sustainable far into the future. The state of global equilibrium could be designed so that the basic material needs of each person on earth would be satisfied and each person would have an equal opportunity to realize his or her individual potential.
3. Finally, if the world's people decided to strive for this second outcome rather than the first, the sooner they began working to attain it, the greater would be their chances of success.

The first conclusion got most of the attention. Newspaper headlines (see Meadows et al., 1992, p. viii) announced:

A COMPUTER LOOKS AHEAD AND SHUDDERS

STUDY SEES DISASTER BY YEAR 2100

SCIENTISTS WARN OF GLOBAL CATASTROPHE

The warnings were not taken lightly. In 1977 the president of the United States, Jimmy Carter, directed the U.S. Council on Environmental Quality and the Department of State, along with other federal agencies, to make a one-year study of the probable changes in the world's population, natural resources, and environment through the end of the century. The product of that study, "The Global 2000 Report to the President" (Barney 1982), presented the following findings.

The world's population would grow from 4 billion in 1975 to 6.35 billion in 2000.

The gross national product in the most populous nations would remain below $200 per capita per year. The gap between the rich and the poor would widen.

Arable land would increase only 4 percent by 2000, so that most of the increased output of food would have to come from higher yields dependent on fertilizer, pesticides, fuels, and other inputs heavily dependent on oil and gas.

During the 1990s, world oil production would approach geological estimates of maximum production, and prices would increase, making oil-based fuel unavailable to the poor.

Nonfuel mineral resources would probably be sufficient to meet demands through 2000, but further discoveries and investments would be needed.

Regional water shortages would become more severe.

Significant losses of the world's forests would continue.

Serious deterioration of agricultural soils would occur.

Atmospheric concentrations of pollutants would increase.

Extinctions of plant and animal species would increase dramatically.

The study concluded that at the present projected growth rates, the world's population would reach 10 billion by 2030 and would approach 30 billion by the end of the twenty-first century.

Already the populations in sub-Saharan Africa and in the Himalayan hills of Asia have exceeded the carrying capacity of the immediate area, triggering an erosion of the land's capacity to support life. The resulting poverty and ill health have further complicated efforts to reduce fertility. Unless this circle of interlinked problems is broken soon, population growth in such areas will unfortunately be slowed for reasons other than declining birth rates. Hunger and disease will claim more babies and young children, and more of those surviving will be mentally and physically handicapped by childhood malnutrition. (p. 3)

The news media sensationalized the report.

"It reads like something out of 'The Empire Strikes Back.'" said *Newsweek* (cited in Barney 1988). "The time: the year 2000. The place: Earth, a desolate planet slowly dying of its own accumulating follies. Half of the forests are gone; sand dunes spread where fertile farm lands once lay. Nearly 2 million species of plants, birds, insects and animals have vanished. Yet man is propagating so fast that his cities have grown as large as his nations of a century before. " (first page of preface)

In the 1980s, a group of international leaders headed by the prime minister of Norway, Gro Harlem Brundtland, convened the World Commission on Environment and Development. Their report, Our Common Future, warned that

Poverty, injustice, environmental degradation, and conflict interact in complex and potent ways. One manifestation of growing concern to the international community is the phenomenon of "environmental refugees". The immediate cause of any mass movement of refugees may appear to be political upheaval and military violence. But the underlying causes often include the deterioration of the natural resource base and its capacity to support the population.

Events in the Horn of Africa are a case in point. In the early 1970s, drought and famine struck the nation of Ethiopia. Yet it has been found that the hunger and human misery were caused more by years of overuse of soils in the Ethiopian highlands and the resulting severe erosion than by drought. A report commissioned by the Ethiopian Relief and Rehabilitation Commission found: "The primary cause of the famine was not drought of unprecedented severity, but a combination of long-continued bad land use and steadily increased human and stock populations over decades." (World Commission on Environment and Development 1987), (p. 291)

Dire predictions continue in the 1990s. "The Worst Is Yet to Come" is the theme of *Preparing for the Twenty-first Century* (Kennedy 1993). The study alerts readers to the new consequences of continued economic and social growth. Kennedy particularly points to the threat posed by the ability of multinational corporations to place their high-tech production facilities in low-wage areas of the world. This, he says, diminishes the capability of the nation-state to deal with problems that threaten its sovereignty. Kennedy believes that nation-states are too large and slow-moving to deal with the specific strategies of the supranational corporation and too small to deal with their collective impact. Controlling environmental piracy will be beyond the capability of any one national government.

Does the trend in world population really justify the alarms sounded by all these Cassandras? When plotted on a linear scale over the last 40 years, population growth does not seem alarming (Fig. 2.1), even though it continues to grow at an increasing rate (Fig. 2.2). But the exponential significance of the increase is more apparent when population is plotted on a longer time scale (Fig. 2.3).

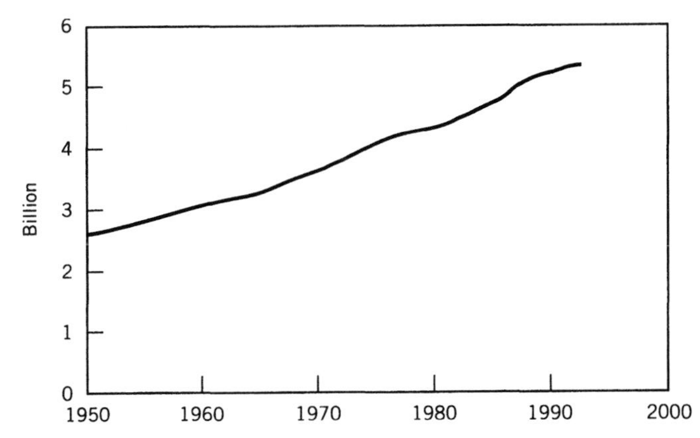

FIGURE 2.1 World population, 1950–92. *Source:* Brown, Kane, and Ayres (1993), p. 95.

The nonintuitive nature of exponential growth can be illustrated by considering a newly formed lake in which algae grow at a rate sufficient to support a population of a million minnows. The lake is initially empty of fish, but one day during a flood, a pair of minnows swim over from a nearby lake. They begin to breed at a rate whereby the population doubles every month. At this rate, it will take a little over 19 months to reach the limits of growth. At the end of 18 months, each fish still has more than twice the algae that it needs. There seems to be no cause for alarm. The minnow, with its limited mental capacity, does not understand that while it took 18 months to use up half the resources, the other half will be consumed in just one more month.

The people of the world are in a situation analogous to the fish in the pond. By the mid-1980s, almost half (40 percent) of all the terrestrial net primary production was directly consumed, co-opted, or eliminated by human activity (Vitousek et al. 1986). That means we have only one generation left before we experience disaster. The momentum is too great to ease population to an asymptote of steady state. Population will shoot through the limits imposed by the carrying capacity of the earth, and then widespread disease and starvation will cause it to crash dramatically.

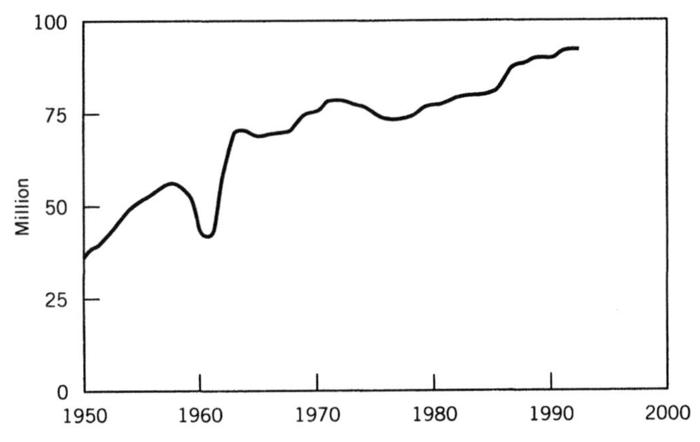

FIGURE 2.2 Annual addition to world population, 1950–92. *Source:* Brown, Kane, and Ayres (1993), p. 95.

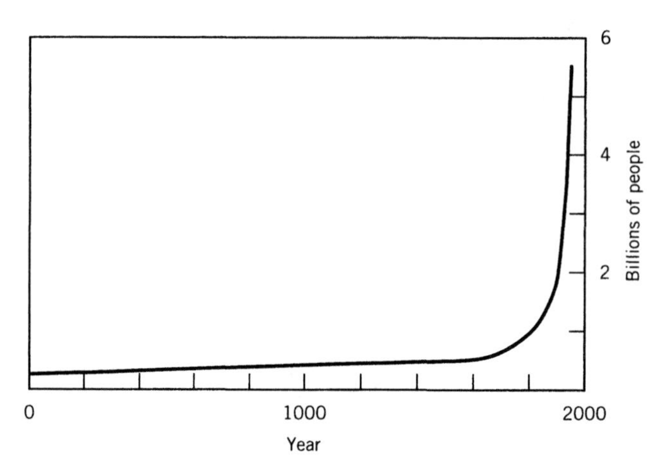

FIGURE 2.3 The world's population growth over the past two millennia.

Dr. Pangloss

Julian L. Simon is a professor of economics at the University of Maryland. His books carry jacket blurbs from Nobel laureate economists, and his views helped shape policy in Washington in the 1980s. Professor Simon has another view of the world. In an essay titled "There Is No Environmental, Population, or Resource Crisis" (Simon 1990), he observed that 10,000 years ago the earth sustained only a million or so people, while today there are more than 5 billion. In this connection, he stated:

> *One would expect lovers of humanity—people who hate war and worry about famine in Africa—to jump with joy at this extraordinary triumph of the human mind and human organization over the raw forces of nature. Instead, they lament that there are so many human beings, and wring their hands about the problems that more people inevitably bring. (p. 24)*

In *The Resourceful Earth* (pp. 2–3) (Simon and Kahn 1984), a panel of scientists presented the following conclusions:

- Many people are still hungry, but the food supply has been improving since at least World War II, as measured by grain prices, production per consumer, and the famine death rate.
- Land availability won't increasingly constrain world agriculture in the coming decades.
- In the United States, the trend is toward higher quality cropland and less erosion than in the past.
- The widely published report of increasingly rapid urbanization of U.S. farmland was based on faulty data.
- Trends in world forests are not worrisome, though in some places deforestation is troubling.
- There is no statistical evidence for rapid loss of plant and wildlife species in the next two decades. An increased rate of extinction cannot be ruled out if tropical deforestation is severe, but no evidence about linkage has yet been demonstrated.

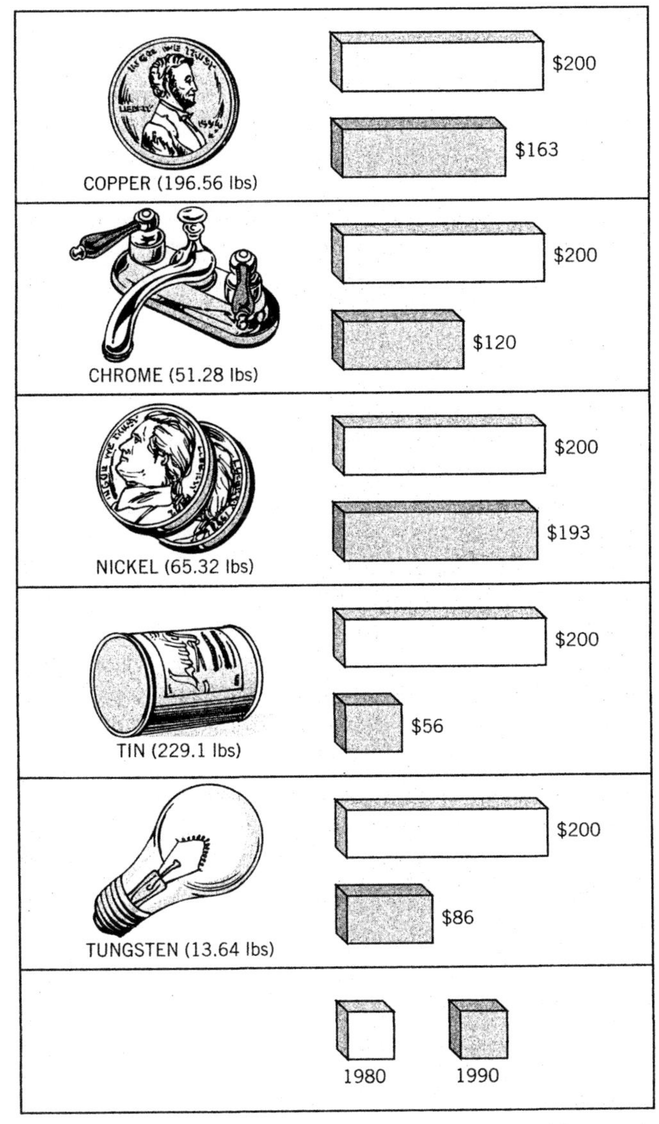

FIGURE 2.4 Five easy pieces. The amount in pounds of five metals that could be purchased for $200 (U.S.) in 1980, and the price for that same amount in 1990. *Source:* Tierney (1990), p. 81. Copyright © 1990 by The New York Times Company. Reprinted by permission.

- Water does not pose a problem of physical scarcity or disappearance, although the world and U.S. situations do call for better institutional management through more rational systems of property rights.
- There is no persuasive reason to believe that the world oil price will rise in coming decades. The price may fall well below what it has been.
- Compared to coal, nuclear power is no more expensive and is probably much cheaper, under most circumstances. It is also much cheaper than oil.
- Nuclear power gives every evidence of costing fewer lives per unit of energy produced than does coal or oil.

- Solar energy resources (including wind and wave power) are too dilute to compete economically for much of humankind's energy needs, although for specialized uses and certain climates they can make a valuable contribution.
- Threats of air and water pollution have been vastly overblown. The air and water in the United States have been getting cleaner rather than dirtier.

Simon (1980) has argued that population growth, rather than constituting a crisis, will in the long run be a boon that will ultimately mean a cleaner environment, a healthier humanity, and more abundant supplies of food and raw materials for everyone. This progress can continue indefinitely, he said, because the planet's resources are actually not finite.

In 1980 Simon issued a challenge to all Malthusians (Tierney 1990). He offered to let anyone pick any natural resource, whether it be grain, oil, coal, timber, or metals, and at any future date. If the resource really were to become scarcer as the world's population grew, then its price should rise. Simon predicted that the price would instead decline by the appointed date. Ehrlich derisively announced that he would "accept Simon's astonishing offer before other greedy people jump in." He then formed a consortium with two colleagues and bet $1000 on five metals: chrome, copper, nickel, tin, and tungsten. The amount of each metal that could be bought for $200 was calculated. In 1990 the bet was settled. The results are shown in Fig. 2.4. Ehrlich lost on *all five* metals.

What happened? Was Simon really right? Is this the best of all worlds and getting better as the population increases? Simon believes that it is. He maintains that as the population increases, human creativity for solving resource shortages also increases. The lower prices of the metals is proof of the point.

Or is it?

In the next section, we look at recent trends in resource availability, population, and pollution in order to determine whether we can tell who is right—the Malthusians who in their pessimism are convinced that the world's population is on the brink of outstripping the earth's resources; or the technological optimists who believe that we are nowhere near the limit of our resources.

THE DATA

Food Supply

Although world grain production is tapering off (Figs. 2.5, 2.6) and the world fish catch has declined slightly (Figs. 2.7, 2.8), soybean production and meat production continue to rise (Figs. 2.9–2.12). Between 1960 and 1980, meat and soybean production increased more rapidly than world population growth.

Energy

Energy indicators also present mixed signals. Oil production remains flat (Figs. 2.13, 2.14), but estimated reserves have increased from slightly over 600 trillion barrels in 1980 to almost 1000 trillion barrels in 1992 (Institute of Petroleum 1993). Nuclear power is at a standstill (Figs. 2.15, 2.16), and production of natural gas is increasing (Fig. 2.17). Alternative sources of energy such as wind power (Figs. 2.18, 2.19) and solar energy (Figs. 2.20, 2.21) are becoming more important, and energy conservation measures have been increasingly effective, as suggested by energy efficiency trends (Figs. 2.22, 2.23).

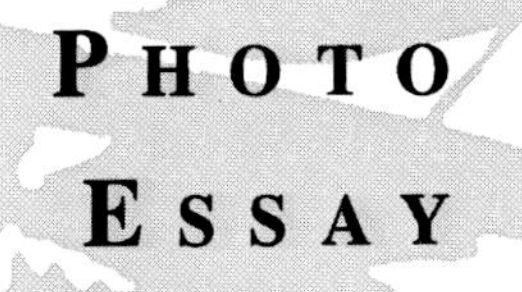

PHOTO ESSAY

DEFORESTATION IN THE AMAZON BASIN

The Amazon basin contains the largest remaining tract of tropical forest in the world, but even the Amazon forest is not safe from the devastation associated with large-scale development. This photoessay depicts the causes of forest loss and describes some attempts to conserve the forest through management which is less destructive.

Fig. P.1 There are many forest types within the Amazon. This moist forest, near Marabá, Pará, Brazil, is dominated by Brazil nut trees (*Bertholletia excelsa*), the emergent on the horizon.

Fig. P.2 Another type is the igapó forest along blackwater streams, which floods during the rainy season. Trees are adapted to survive with roots and some leaves underwater.

Fig. P.3 This orchid hangs from a branch in the igapó forest and is reflected in the water.

Fig. P.4 The Amazon forest supports a tremendous diversity of fauna including the magnificent harpy eagle, which is threatened in some regions.

Fig. P.5 This young ocelot was taken by hunters after they shot the mother.

Fig. P.6 The forest also supports an indigenous culture that has been sustainable for hundreds or even thousands of years. Here a Waica tribesman prepares for a ceremony. (*Source:* Gonzalez Niño, undated.)

Fig. P.7 Shifting cultivators such as these near Marabá, Pará are often blamed for loss of tropical forests. They will cultivate the land for two or three years and then move on as the soil loses its fertility.

Fig. P.8 However, huge projects such as this pulp mill and plantation at Jari contribute much more to the loss of tropical forests.

Fig. P.9 To prepare land for the fast-growing exotic trees at Jari, native forest was cut and burned.

Fig. P.10 Logging usually develops after roads are built for large-scale development projects.

Fig. P.11 The Tucuruí dam on the Tocantins River near Marabá flooded hundreds of square miles of forest. Erosion following deforestation is silting up the reservoir.

Fig. P.12 The open pit mine at Carajás, Brazil is one of the largest in the world. Iron ore is passed through the crushers shown here, and is loaded into railroad hoppers below.

Fig. P.13 Smelters are being built along the railroad that stretches from the mine to the sea. Kilns to supply charcoal to the smelters will use up thousands of square kilometers of forest.

Fig. P.14 Mercury flushed from gold-mining barges has severely contaminated the fish in many rivers.

Fig. P.15 Government propaganda encouraged settlement in western Amazonia by small farmers from the south of Brazil who had lost their land to large-scale soybean agribusiness.

Fig. P.16 The reality is quite different from the propaganda. Here a migrant farmer is explaining to an extension agent and a World Wildlife Fund representative that the weed competition here is just too severe to obtain a decent crop.

Fig. P.17 During the 1980s, clearing for ranches was an important cause of deforestation. The government gave big tax concessions to ranchers who established pastures in the Amazon.

Fig. P.18 After four or five years, pastures become degraded and scarcely productive.

Fig. P.19 To maintain production, pastures sometimes are burned and bulldozed. However, the soil eventually becomes so compacted and infertile that only scrubby brush can grow.

Fig. P.20 Viable alternatives for human use of the Amazon forest include strip logging and growing perennial crops such as peach palm shown here.

Fig. P.21 Another approach is to set aside extractive reserves for rubber and for Brazil nuts shown here. A collector cuts open the outer husk to expose the nuts inside.

Fig. P.22 If sustainable uses are not developed, the Amazon will end up like this pasture in Acre, where smoke dims the sun in September, 1989.

SUGGESTED READINGS

For perspectives on development in the Amazon basin:

Bunker, S. G. 1985. Underdeveloping the Amazon. University of Illinois Press, Urbana.

Hecht, S., and A. Cockburn. 1989. The Fate of the Forest: Developers, Destroyers, and Defenders of the Amazon. Verso, London.

Moran, E. 1981. Developing the Amazon. Indiana University Press, Bloomington.

Schmink, M., and C. H. Wood. 1984. Frontier Expansion in Amazonia. University of Florida Press, Gainesville.

Smith, N. J. H. 1982. Rainforest Corridors: The Transamazon Colonization Scheme. University of California Press, Berkeley.

Chico Mendes, who gave his life in a fight to save the rain forest and the rubber tapper's way of life, has inspired several historical accounts. Among them are:

Cowell, A. 1990. The Decade of Destruction. Holt and Co., New York.

Revkin, A. 1990. The Burning Season: The Murder of Chico Mendes and the Fight for the Amazon Rain Forest. Houghton Mifflin, Boston.

Shoumatoff, A. 1990. The World Is Burning. Little, Brown and Co., Boston.

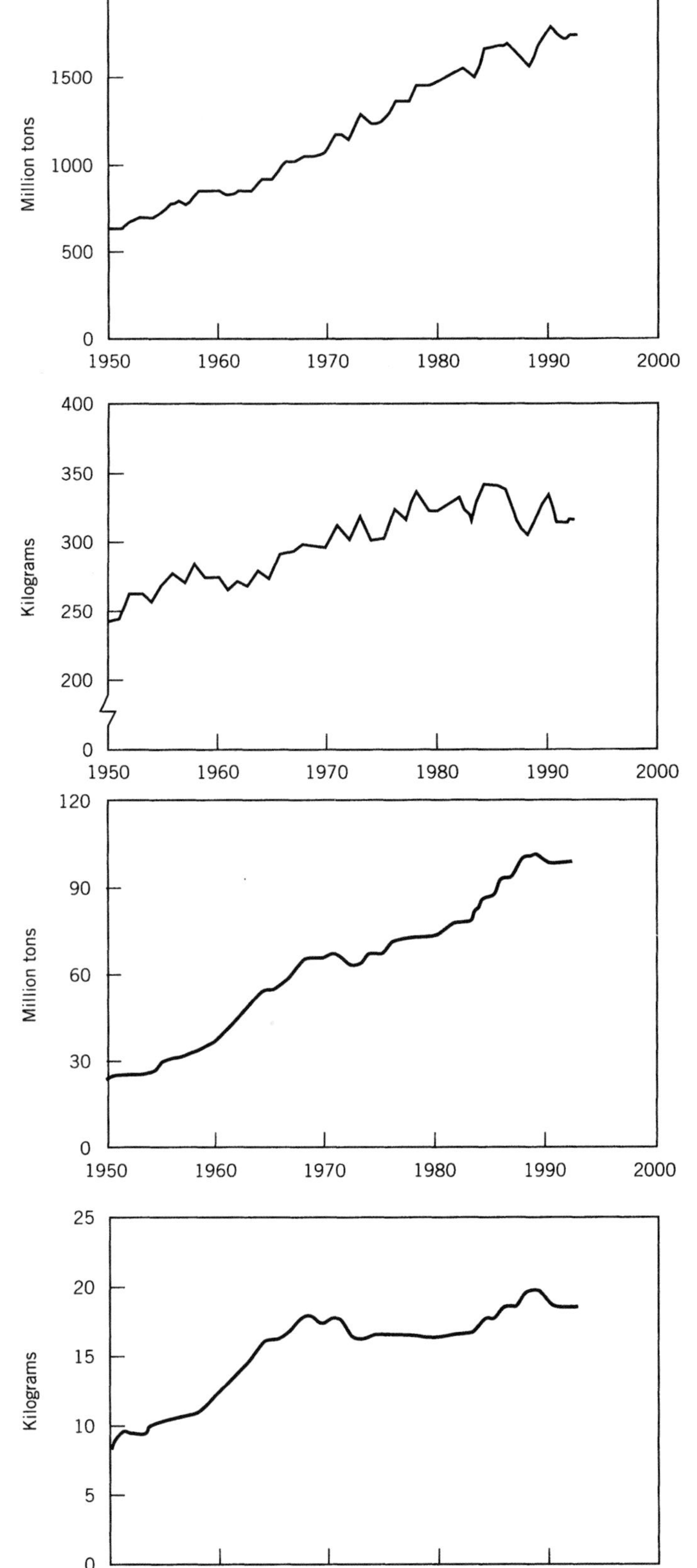

FIGURE 2.5 World grain production, 1950–92. *Source:* Brown, Kane, and Ayres (1993), p. 27.

FIGURE 2.6 World grain production per person, 1950–92. *Source:* Brown, Kane, and Ayres (1993), p. 27.

FIGURE 2.7 World fish catch, 1950–92. *Source:* Brown, Kane, and Ayres (1993), p. 33.

FIGURE 2.8 World fish catch per person, 1950–92. *Source:* Brown, Kane, and Ayres (1993), p. 33.

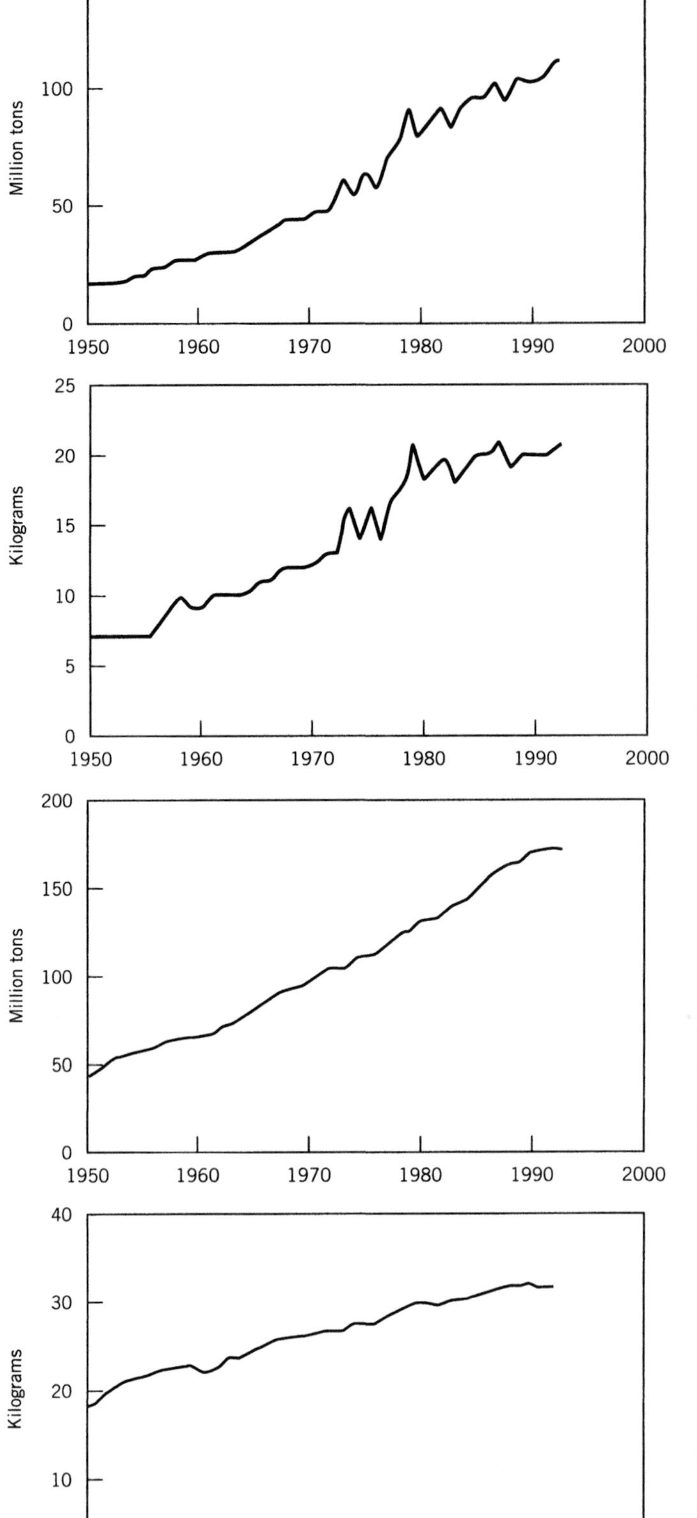

FIGURE 2.9 World soybean production, 1950–92. *Source:* Brown, Kane, and Ayres (1993), p. 29.

FIGURE 2.10 World soybean production per person, 1950–92. *Source:* Brown, Kane, and Ayres (1993), p. 29.

FIGURE 2.11 World meat production, 1950–92. *Source:* Brown, Kane, and Ayres (1993), p. 31.

FIGURE 2.12 World meat production per person, 1950–92. *Source:* Brown, Kane, and Ayres (1993), p. 31.

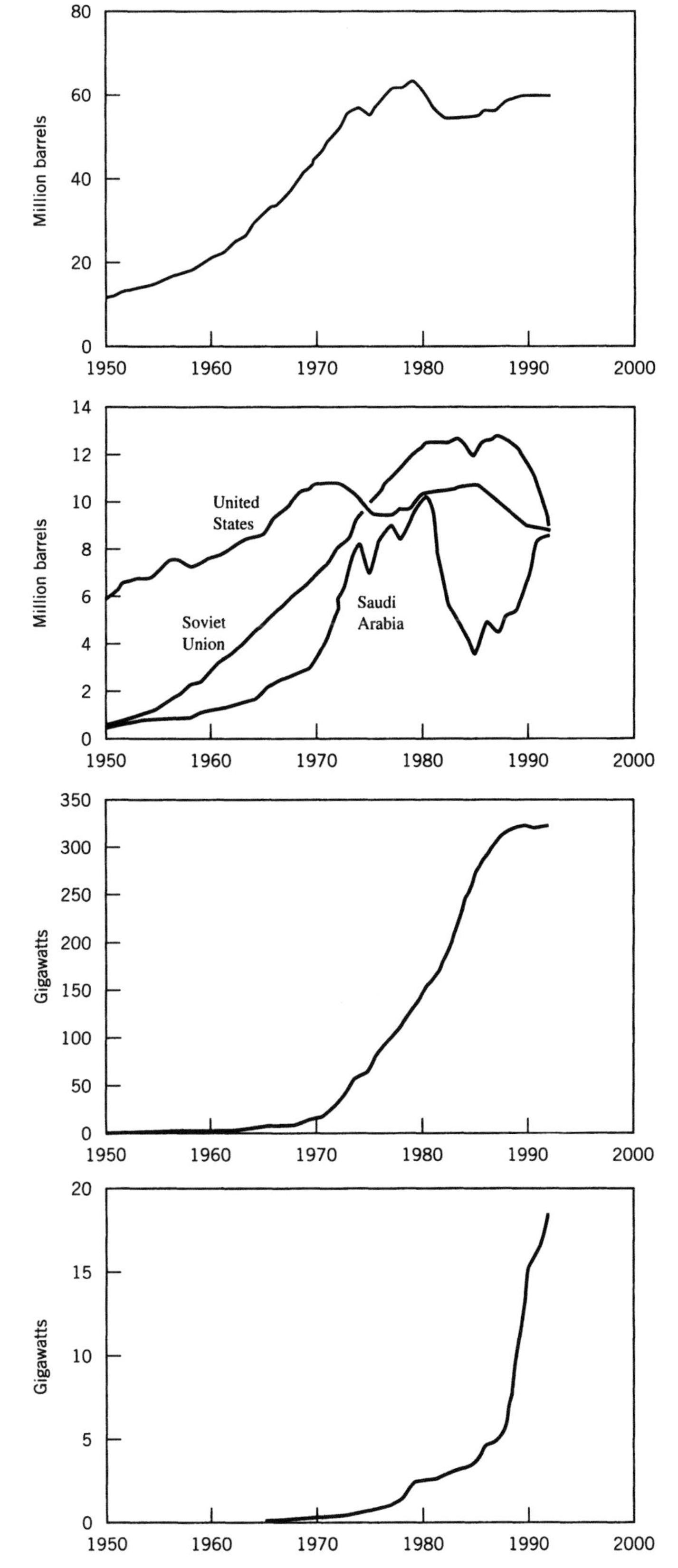

FIGURE 2.13 World oil production per day, 1950–92. *Source:* Brown, Kane, and Ayres (1993), p. 47.

FIGURE 2.14 U.S., Soviet, and Saudi Arabian oil production per day, 1950–92. *Source:* Brown, Kane, and Ayres (1993), p. 47.

FIGURE 2.15 World electrical generating capacity of nuclear power plants, 1950–92. *Source:* Brown, Kane, and Ayres (1993), p. 51.

FIGURE 2.16 Cumulative nuclear generating capacity decommissioned, 1964–92. *Source:* Brown, Kane, and Ayres (1993), p. 51.

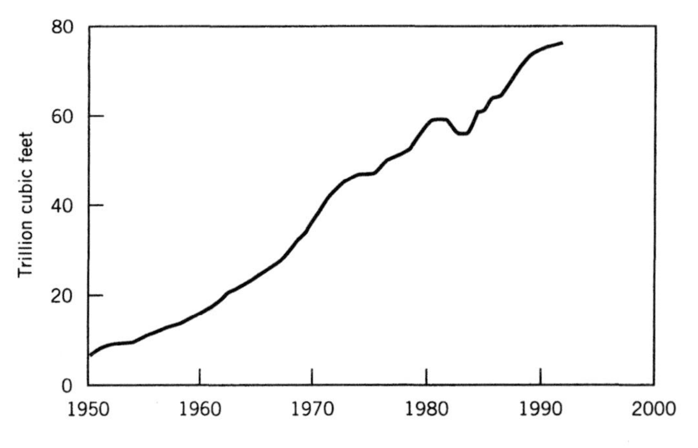

FIGURE 2.17 World natural gas production, 1950–91. *Source:* Brown, Flavin, and Kane (1992), p. 47.

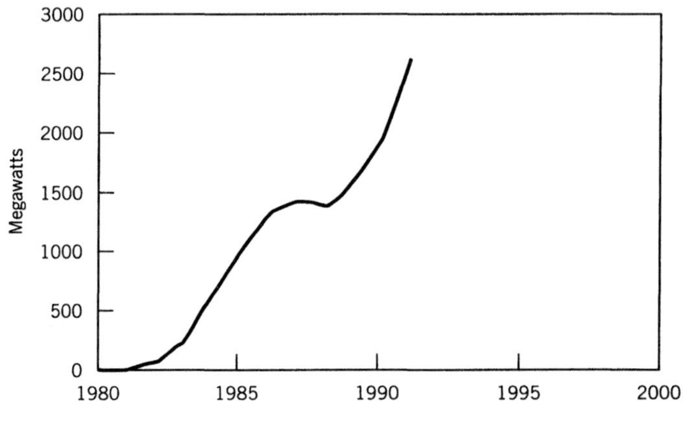

FIGURE 2.18 World wind energy generating capacity, 1981–92. *Source:* Brown, Kane, and Ayres (1993), p. 49.

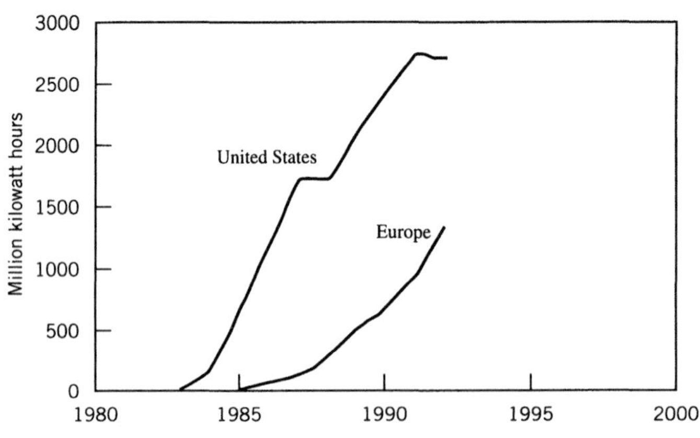

FIGURE 2.19 Wind energy generation in United States and Europe, 1981– 92. *Source:* Brown, Kane, and Ayres (1993), p. 49.

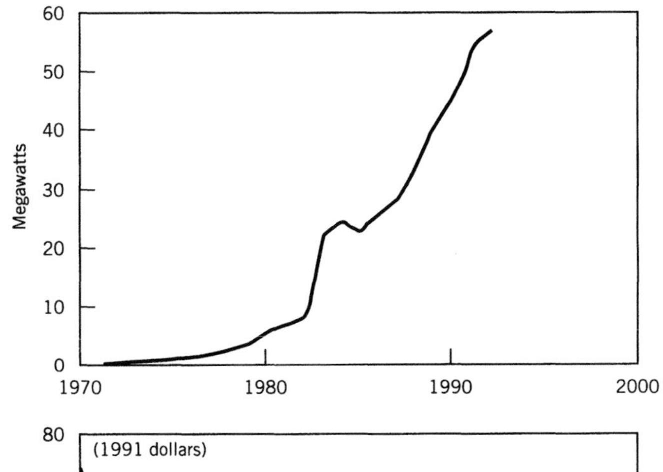

FIGURE 2.20 World photovoltaic shipments, 1971–92. *Source:* Brown, Kane, and Ayres (1993), p. 53.

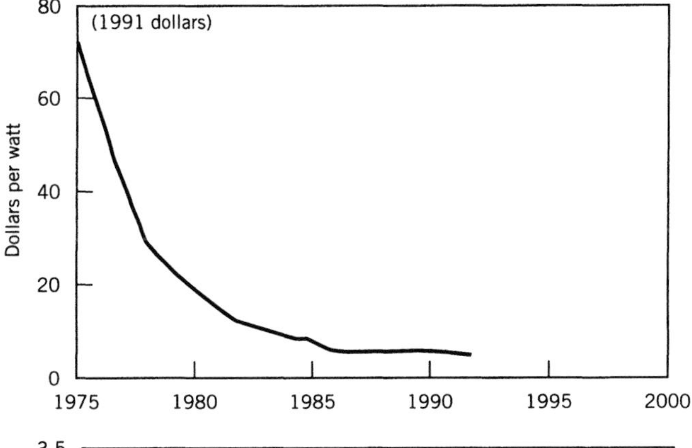

FIGURE 2.21 Average factory prices for photovoltaic modules, 1975–92. *Source:* Brown, Kane, and Ayres (1993), p. 53.

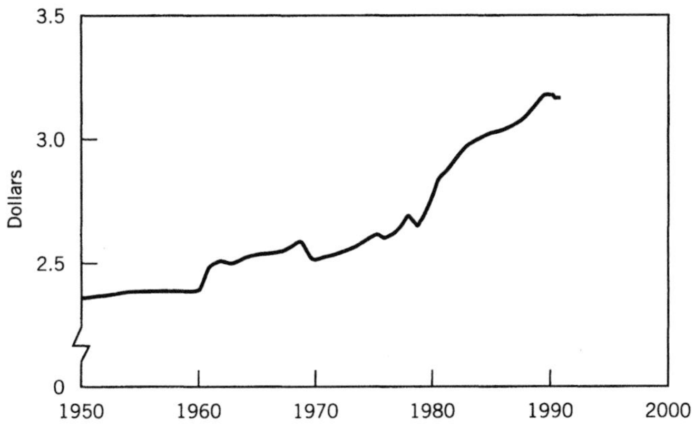

FIGURE 2.22 World economic output per kilogram of carbon emitted, 1950–91. *Source:* Brown, Kane, and Ayres (1993), p. 61.

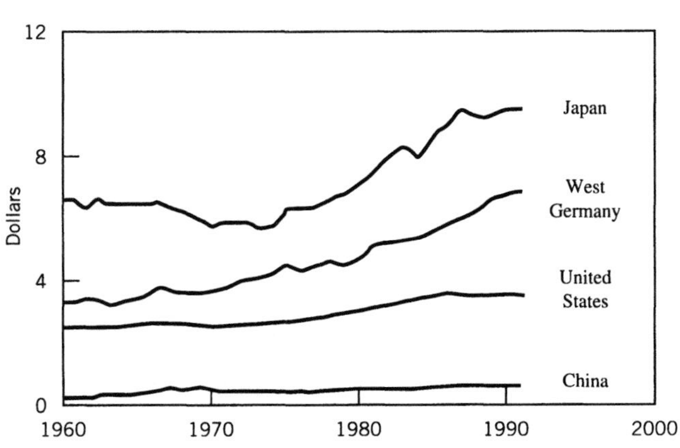

FIGURE 2.23 Economic output per kilogram of carbon emitted, Japan, West Germany, United States, and China, 1960–91. *Source:* Brown, Kane, and Ayres (1993), p. 61.

Atmosphere

As a result of the burning of fossil fuels and tropical deforestation, carbon dioxide concentration in the atmosphere continues to increase (Fig. 2.24). The increase correlates with the increase in global average temperatures over the past 30 years (Fig. 2.25).

Although there has been a recent reversal in the emissions of some air pollutants in the United States (Fig. 2.26), the long-term trend of air pollution is still sharply upward (Figs. 2.27, 2.28).

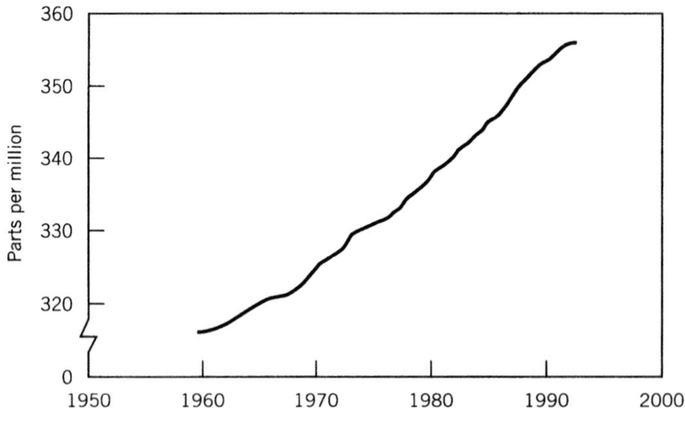

FIGURE 2.24 Atmospheric concentrations of carbon dioxide, 1959–92. *Source:* Brown, Kane, and Ayres (1993), p. 69.

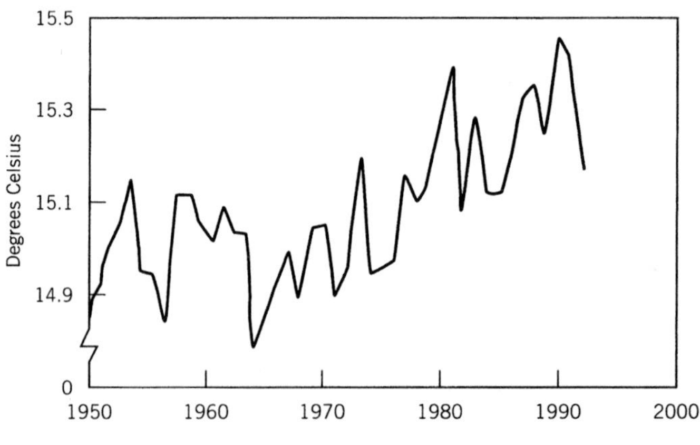

FIGURE 2.25 Global Average Temperature, 1950–92. *Source:* Brown, Kane, and Ayres (1993), p. 69.

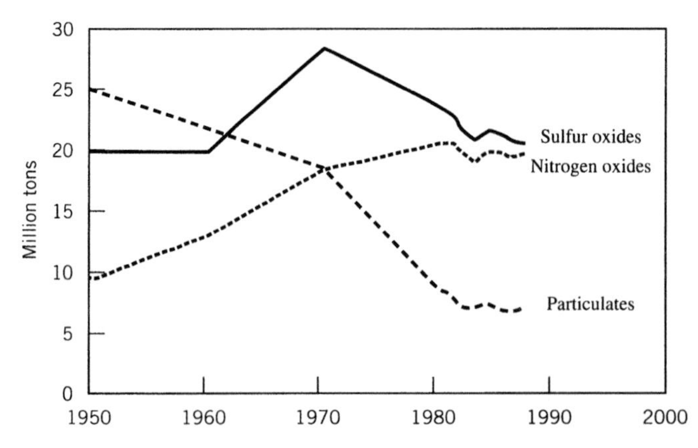

FIGURE 2.26 Emissions of selected pollutants in the United States, 1950–87. *Source:* French (1990), p. 9.

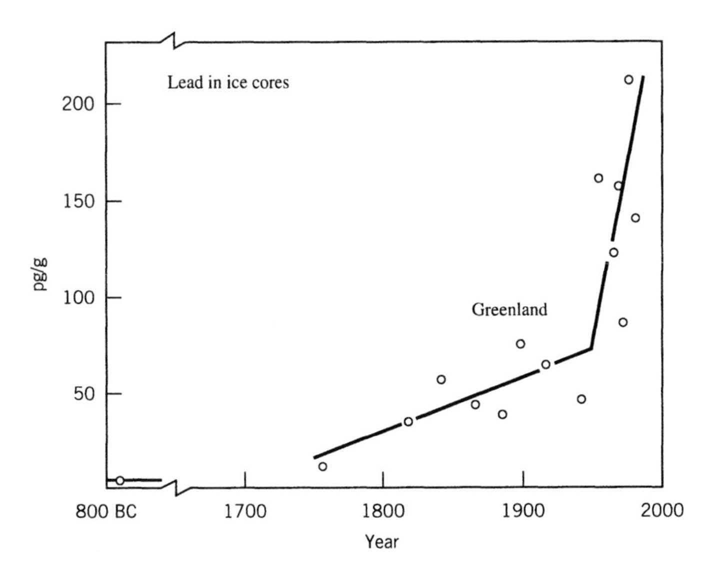

FIGURE 2.27 The amount of lead in samples of Greenland ice recovered from various distances below the surface. The units of lead concentrations in the ice are picograms (10^{-12} grams) per gram of ice, or equivalently, parts per trillion by weight. *Source:* Firor (1990), p. 12.

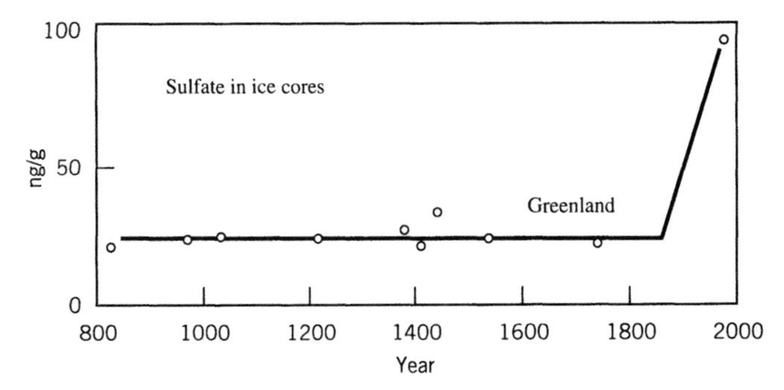

FIGURE 2.28 The concentration of sulfate ions in Greenland ice in nanograms sulfate (10^{-9} grams) per gram of ice, or parts per billion by weight. The line was drawn through the measured points to emphasize the rapid increase in recent years, but the actual time of the start of the rise, as determined by these few points, could have been any time after about 1750. *Source:* Firor (1990), p. 13.

Water

Global water use doubled between 1950 and 1970 (Fig. 2.29), and its rate of use is increasing. Two-thirds of the projected water use is for agricultural irrigation. Eighty countries, with 40 percent of the world's population, already suffer serious water shortages. There will be growing competition for water for irrigation, industry, and domestic use. River water disputes have already occurred in North America (the Rio Grande), South America (the Rio de la Plata and Paraná), South and Southeast Asia (the Mekong and the Ganges), Africa (the Nile), and the Middle East (the Jordan, Litani, and Orontes, as well as the Euphrates) (World Commission on Environment and Development 1987).

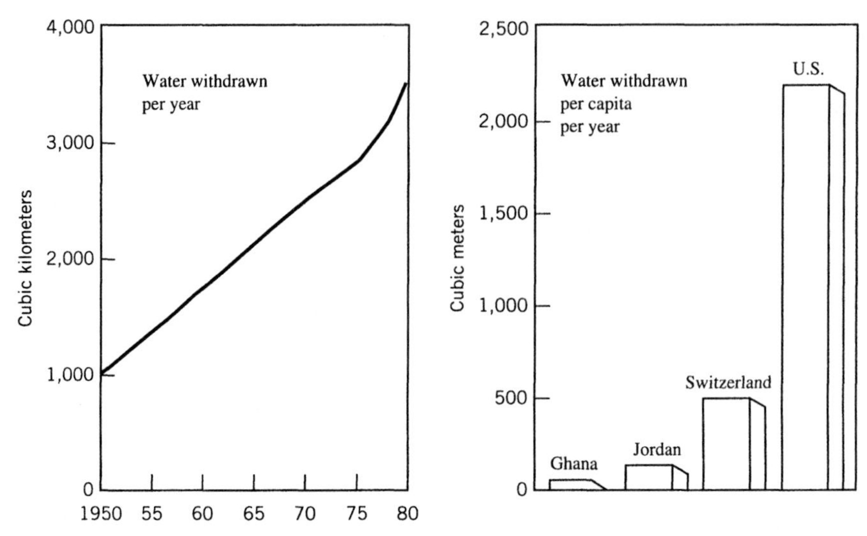

FIGURE 2.29 Global water consumption is increasing (*left*), largely in response to a growing population and increasing per capita use by agriculture and industry. Although sufficient fresh water (9,000 cubic kilometers) is currently available, sound water management is necessary to ensure an adequate supply for the future. Per capita consumption rates vary drastically (*right*); the average American, for example, consumes more than 70 times as much water as the average resident of Ghana. *Source:* Riviere (1989), p. 84.

Famine

Famines have been, and continue to be significant world phenomena (Table 2.1). What does their occurrence indicate?

Some famines are triggered by floods or droughts, but often human forces are responsible for turning local food shortages into widespread starvation. Amartya Sen, a Harvard economist, in *Poverty and Famines* (1981), claims that typically, as thousands die, there is enough food in the country to go around or enough money to import it. The poorest, most downtrodden members of society suddenly can no longer afford to buy food, usually because of unemployment or a surge in food prices. Enough food is produced that everyone in the world could be kept from starving, providing distribution problems could be worked out.

Wildlife

Species such as whitetailed deer that can easily adapt to human-modified environments have been actually increasing in the past decades. It is those species that depend on undisturbed habitat that are showing dramatic declines. For example, many migrant songbirds depend on tropical forests for winter habitat. An estimated drop of almost 50 percent of migrating birds between the 1960s and the 1980s is thought to result from tropical deforestation (Terborgh 1992).

Table 2.1 *Recent Famines with Proximate (But Not Ultimate) Cause*

DATE	LOCATION	PROXIMATE CAUSE
1846–51	Ireland	Recurrent potato famines took place. People starved as food went to England. One million died and another million emigrated.
1928–29	China	An estimated 3 million died after droughts brought a decline in agricultural productivity.
1932–34	Soviet Union	More than 5 million died during Stalin's farm collectivization program.
1943	India	Up to 3 million people died while food output declined only slightly. A war-related boom in Bengal increased urban demand. Food prices were pushed up, fueled by panic and manipulative speculation.
1958–61	China	Estimates of death range from 17 to 30 million from the failure of the Great Leap Forward, which brought about agricultural disruption.
1967–68	Nigeria	There is no reliable estimate of the famine deaths from the civil war and the blockade of rebellious Biafra. Food distribution and general economic activity were disrupted.
1973	Ethiopia	No reliable estimates of deaths. Although food availability remained the same during the drought, declines in output and unemployment in the Wollo region meant its people had no money for food.
1974	Bangladesh	No reliable estimates of deaths. Floods produced unemployment, and speculation brought higher food prices.
1975–79	Cambodia	Perhaps one million died during the Khmer Rouge's deportation of the urban population and deliberate destruction of the economy.
1983–84	Sub-Saharan Africa	Prolonged drought, and, in some cases, war took place. Ethiopia, Sudan, and Somalia failed to have relief efforts like those in democratic Botswana and Zimbabwe.

Source: *Arnold 1988; Field 1993; Harrison 1988.*

Legacy of the 1980s

In 1993, on the eve of the inauguration of Bill Clinton as the president of the United States, the Associated Press made a survey to determine how the world had changed since 1980, the beginning of the Reagan-Bush era. The results showed an increase in world population, an increase in carbon dioxide in the atmosphere, and a decrease in area of tropical forest, all of which constituted "bad" trends according to environmentalists. However, infant mortality had dropped, wheat production had increased, as had estimated world petroleum reserves, trends that technological optimists considered "good." The value of the dollar had shrunk against the yen, a trend that could be good or bad, depending on whether you were an importer or an exporter. The data did not suggest that the world was in immediate peril, or even on the brink of global famine.

The most worrisome finding was a sharp increase in the incidence of acquired immune deficiency syndrome (AIDS). Epidemics such as the plague and cholera often are mentioned in environmental literature because of the relationship between environmental conditions and these diseases. In the case of AIDS, the relationship is unclear, although we do know that a continuation of the present trend would put a tremendous burden on world resources. Not only would resources have to be redirected, but also the world's productive capacity would decrease as a result of losses in its work force.

Conclusions

Between 1950 and 1970 world production of soybeans and meat grew at a rate *faster* than the world's population. This trend shows no sign of tapering off, as would be expected as production neared the limit of capacity. Soybean and meat data give no evidence of any imminent world crisis in food resources.

Although some of the global indices such as atmospheric concentrations of carbon dioxide and availability of fresh water suggest emerging problems, there is no convincing evidence that the world is in imminent danger of collapse. The changes wrought by human practices in many cases do not seem to greatly exceed changes resulting from natural processes.

Does this mean that Simon was right and that we have nothing to worry about?

MALTHUS REPUDIATED?

Yes

Events over the past 200 years have proven Malthus's dire predictions of global starvation to be wrong. He was wrong because he saw the amount of tillable land as limiting agricultural production. He did not foresee the technological improvements that would allow greater harvests to be produced from existing agricultural land (Clark and Munn 1986). Even shortly after its first publication, Malthus became aware of the shortcomings of his argument. Walter Bagehot, the first editor of *The Economist,* quipped: "In its first form, the 'Essay on Population' was conclusive as to argument, only it was based on untrue facts; in its second form, it was based on true facts, but was inconclusive as an argument" (Heilbroner 1993, p. 25).

Simon, Sen, and the other "cornucopians" appear to be correct. There is no evidence that the swelling world population cannot be kept alive by the world's productive capacity. There is no evidence that lack of resources will stop the growth of the world economy. Technology does seem to be able to increase the supply of food and resources, to keep pace with increasing demands, at least on the global scale.

The problems that are arising certainly seem amenable to technical or regulatory solutions. For example, given the comparative wastefulness of water use in the United States (Fig. 2.29), the problems in North America should be easily solved. For some environmental problems, solutions are already being sought, as evidenced by increased efficiency of energy use and greater use of solar and wind power.

Even some early doomsayers are changing their minds. In 1992 Meadows et al. published a followup to the Club of Rome's *The Limits to Growth.* The new book, *Beyond the Limits: Confronting Global Collapse; Envisioning a Sustainable Future,* concludes that in the 20 years since the original publication, some options for sustainability have narrowed, whereas others have opened up. They maintain that, given some of the technologies and institutions invented over those 20 years, there are real possibilities for reducing the amounts of resources consumed and the pollutants generated by the human economy while increasing the quality of human life. It is even possible, they believe, to eliminate poverty while accommodating the population growth already implicit in present population age structures. However, the accommodation cannot occur if population growth goes on indefinitely and unless there is a rapid improvement in the efficiency of material and energy use and in the equity of material and energy distribution.

The authors emphasize that their three conclusions in *The Limits to Growth* are still valid, but they caution that they need to be strengthened. They rewrote the conclusions, for 1992 and beyond, as follows:

1. Human use of many essential resources and the generation of many kinds of pollutants have already surpassed physically sustainable rates. Without significant reductions in material and energy flows, the coming decades will witness an uncontrolled decline in per capita food output, energy use, and industrial production.
2. This decline is not inevitable. To avoid it, two changes are necessary. The first is a comprehensive revision of policies and practices that perpetuate growth in material consumption and in population. The second is a rapid, drastic increase in the efficiency with which materials and energy are used.
3. A sustainable society is still possible both technically and economically. It could be much more desirable than a society that tries to solve its problems by constant expansion. The transition to a sustainable society requires a careful balance between long-term and short-term goals, as well as an emphasis on sufficiency, equity, and quality of life rather than on quantity of output. It requires more than productivity and more than technology; it also requires maturity, compassion, and wisdom. "The conclusions," the authors say, "are a warning, not a dire prediction. They offer a choice, not a death sentence" (Meadows et al. 1992, pp xv–xvi).

No

"Destruction and Death" is the headline of an article in the *New York Times* of January 31, 1993 (p. E-17) that reviews the work of the Project on Environmental Change and Acute Conflict, a research group sponsored by the American Academy of Arts and Sciences and the University of Toronto (Homer-Dixon et al. 1993). Homer-Dixon et al. cite instances where scarcities of renewable resources are already contributing to dislocations and violent conflicts in many parts of the lesser developed world. These conflicts, they say, may foreshadow more violence in coming decades, particularly in poor countries where shortages of water, forests, and fertile land are already producing profound hardship. Their research has shown that land scarcity in Bangladesh, produced in part by rapid population growth, caused millions of people to migrate to India, which in turn led to brutal ethnic conflicts in the Indian states of Assam and Tripura.

Environmental damage costs China at least 15 percent of its gross national product. This burden is getting worse, mainly because of reduced crop yields owing to water, soil, and air pollution, high levels of human illness from air pollution, the loss of farmland stemming from construction and erosion, and flooding and soil-nutrient loss emanating from erosion and deforestation. The authors expect domestic strife to follow in the coming years, when huge numbers of people move from China's ecologically devastated interior to its booming coastal zone.

According to Homer-Dixon et al., a persistent insurgency in the Philippines has been given extra impetus by the desperate poverty arising from degraded forests and soils in the hilly areas of the interior. In the Middle East, severe shortages of groundwater in the Jordan River basin have reinforced the unequal distribution of water between the Israelis and Palestinians.

Homer-Dixon et al. point out that in South Africa, apartheid concentrated millions of blacks in some of the country's most ecologically sensitive areas. Wide swaths of these homelands were stripped of trees for fuel and grazed down to hard dirt. The top soil eroded, contributing to migration to cities and to the rapid growth of urban squatter settlements that seethe with violence.

Expanding population and land degradation and drought in Senegal and Mauritania helped spur a violent conflict over irrigable land in the Senegal River basin, resulting in tens of thousands of refugees. Similar factors stimulated the growth of the Maoist Shining Path movement in the southern highlands of Peru. In Haiti, the irreversible clear-cutting of forests and loss of soil have worsened the economic crisis and violent social strife, which in turn have caused an exodus of boat people.

Homer-Dixon et al. conclude that future violence influenced by scarcities will not follow traditional patterns. In the past, wars over natural resources have often been over nonrenewable resources, such as Japan's quest for oil and minerals in China and Southeast Asia before World War II. Today, by contrast, many threatened renewable resources are held in common. This makes it unlikely that they will be the object of straightforward clashes between nations.

Scarcity, they point out, produces insidious and cumulative social effects such as large migrations and economic disruption. These effects in turn lead to ethnic strife, civil war, and insurgency. These conflicts usually attract little attention in the industrialized world, but they can seriously affect the security interests of rich and poor countries alike.

Environmental degradation, first seen as mainly a problem of the rich nations and a side effect of industrial wealth, has become a survival issue for ecological and economic decline in which many of the poorest nations are trapped. Environmental degradation cannot be dealt with as a single issue isolated from population increase and the resulting social, political, and economic issues. Global habitability can be sustained only if a strategy is devised within the context of all world problems.

Maybe

Only very recently has the human impact on the environment begun to rival that of natural disasters, but in contrast to natural disasters, the human influence is continuing to increase. Relying on indices such as global averages to determine whether or not there is a crisis is like using the amount of smoke in your bedroom in the middle of the night to decide whether or not the fire downstairs is serious. Global changes that have occurred so far as a result of human actions are merely "the smoke in the bedroom."

While admitting that the data and scientific models based on current data are inconclusive, Firor (1990) believes that regardless of whether or not the data prove the planet is becoming uninhabitable, it would be foolhardy to continue those actions that are causing the trends. "The course of development we are on is a risky one," he says, "demanding knowledge and skills we may not have and leading to consequences that we may find difficult to tolerate. At the most basic level, we need to ask whether the assumption that we can manage the earth and all life upon it is leading us to the kind of world we would like to live in" (p. 106-7).

Young people born into a world in which they can't see the stars would not be disappointed because they would be accustomed to an overcast sky. They would not miss vanished species because they would never know they had existed. It may be that the real tragedy of pollution is that it won't kill us. We will gradually adjust to the changes and forget the possibilities that disappeared with the earlier world.

THE LOCAL NATURE OF MANY ENVIRONMENTAL CRISES

Using global averages as indicators of environmental problems is misleading because many environmental crises are local. Local data that by themselves would show disastrous trends are masked when they are incorporated into global averages. But the concentrations of pollution that affect people are local concentrations, not global averages.

The Data

Air Pollution

Air pollution is one of the world's most critical environmental problems. Table 2.2 shows the importance of automobile emissions to air pollution and how this pollution is concentrated in major cities. Because population is concentrated in these large cities, the effect on large numbers of people is much greater than global averages of pollution would suggest.

Water Pollution

Water pollution even more than air pollution tends to be associated with centers of population. Many cities grew primarily because they were ports or because rivers served as a source of water. Figure 2.30 shows the association between population and pollution in major rivers. As with air pollution, global averages dilute impact. The problems are concentrated where the people are concentrated.

Mining

Mineral extraction and processing often cause very serious environmental problems, yet their impact is usually local and regional (Table 2.3).

Colonization

Colonization projects such as Rondonia in the Brazilian Amazon (Figure 2.31) have little immediate negative impact on Brazil as a whole. In fact, the purpose of the colonization scheme was to reduce population pressure in other regions of the country. However, the local effects are appalling (see photo-essay, Deforestation in the Amazon).

Table 2.2 *Contribution of Road Transport to Air Pollution in Selected Cities*

Region	Year	Total Pollutants from All Sources (thousand metric tons)	Percent Attributable to Road Transport					
			CO	HC	NO_x	SO_x	Particulates	Total
Mexico City	1987	5,027	99	89	64	2	9	80
São Paulo	1981	3,150	96	83	89(a)	26	24	86
	1987	2,110	94	76	89(a)	59	22	86
Ankara	1980	690	77	73	44	3	2	57
Manila	1987	500	93	82	73	12	60	71
Kuala Lumpur	1987	435	97	95	46	1	46	79
Seoul	1983	X	15	40	60	7	35	35
Hong Kong	1987	219	X	X	75	X	44	X
Athens(b)	1976	394	97	81	51	6	18	59
Gothenburg(b)	1980	124	96	89	70	2	50	78
London	1978	1,200	97	94	65	5	46	86
Los Angeles(b)	1976	4,698(c)	99	61	71	12	X	88
	1982	3,391(c)	99	50	64	21	X	87
Munich	1974/5	213	82	96	69	12	56	73
Osaka	1982	141	100	17	60	43	24	59
Phoenix	1986	1,240(d)	87	64	77	91	1	28

[a] *Includes evaporation losses from storage and refueling.*

[b] *Percent shares apply to all transport. Motor vehicles account for 75 to 95 percent of the transport share.*

[c] *Excluding particulate matter.*

[d] *Includes 490,000 metric tons of dust from unpaved roads.*

X = not available.

Source: *From* World Resources 1992–1993, *by the World Resources Institute, Table 13.1, p. 196. Copyright © 1992 by the World Resources Institute. Reprinted by permission from Oxford University Press, Inc.*

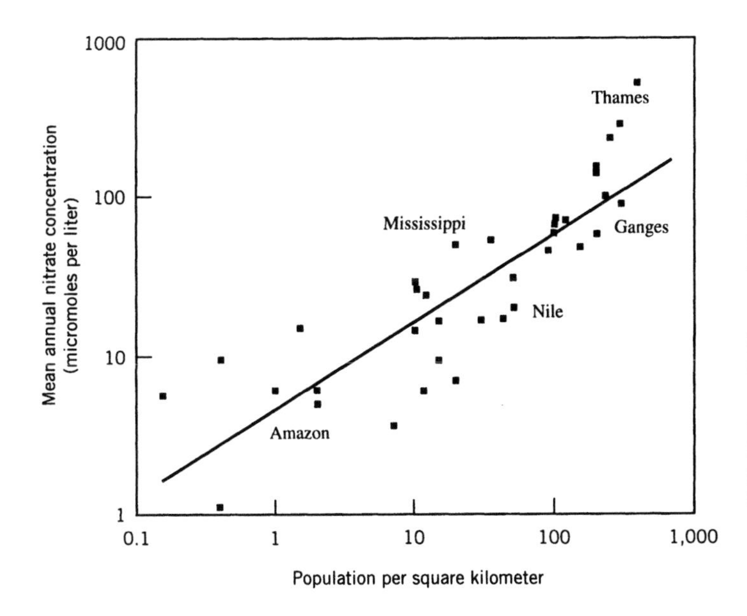

FIGURE 2.30 Pollution in 42 major rivers as a function of population in the region surrounding the rivers. *Source:* From *World Resources, 1992–1993* by the World Resources Institute. Copyright © 1992 by the World Resources Institute. Reprinted by permission from Oxford University Press, Inc., p. 176.

Table 2.3 *Selected Examples of Environmental Impacts of Minerals Extraction and Processing*

Location/Mineral	**Observation**
Ilo-Locumbo Area, Peru copper mining and smelting	The Ilo smelter emits 600,000 tons of sulfur compounds each year; nearly 40 million m^3 per year of tailings containing copper, zinc, lead, aluminum, and traces of cyanides are dumped into the sea each year, affecting marine life in a 20,000-ha area; nearly 800,000 tons of slag are also dumped each year.
Nauru, South Pacific phosphate mining	When mining is completed—in 5 to 15 years—four fifths of the 2,100-ha South Pacific island will be uninhabitable.
Pará state, Brazil Carajás iron ore project	The project's wood requirements (for smelting of iron ore) will require the cutting of enough native wood to deforest 50,000 ha of tropical forest each year during the mine's expected 250-year life.
Russia, Former Soviet Union Severonikel smelters	Two nickel smelters in the extreme northwest corner of the republic, near the Norwegian and Finnish borders, pump 300,000 tons of sulfur dioxide into the atmosphere each year, along with smaller amounts of heavy metals. Over 200,000 ha of local forests are dying, and the emissions appear to be affecting the health of local residents.
Sabah Province, Malaysia Mamut Copper Mine	Local rivers are contaminated with high levels of chromium, copper, iron, lead, manganese, and nickel. Samples of local fish have been found unfit for human consumption, and rice grown in the area is contaminated.
Amazon basin, Brazil gold mining	Hundreds of thousands of miners have flooded the area in search of gold, clogging rivers with sediment and releasing an estimated 100 tons of mercury into the ecosystem each year. Fish in some rivers contain high levels of mercury.

Source: *Brown et al.,* State of the World *(1992), Table 7.3.*

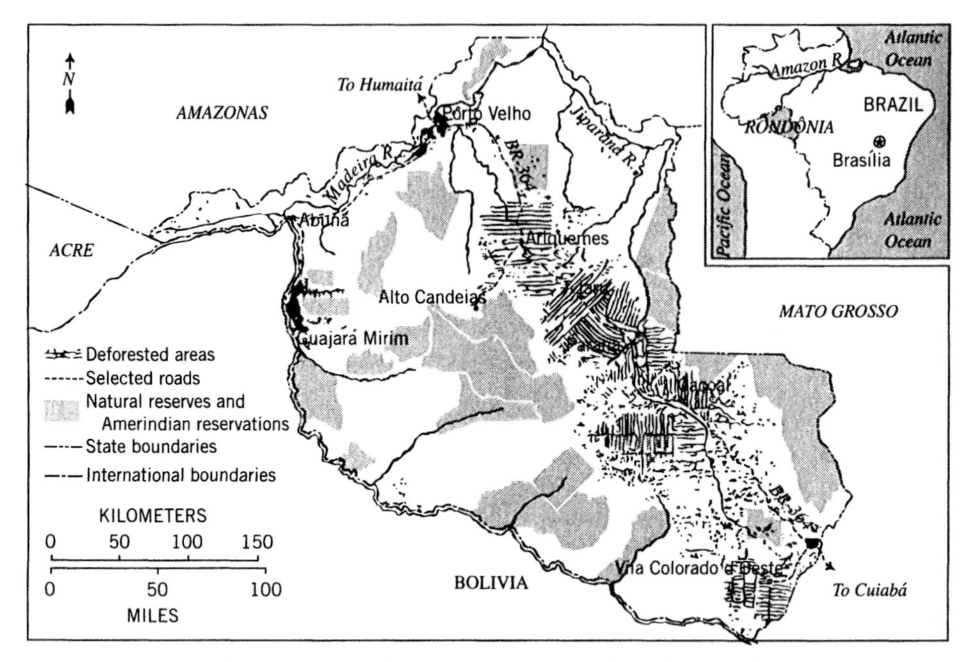

FIGURE 2.31 Deforestation in Rondonia, 1983. *Source:* Redrawn from Landsat.

Fresh Water

There is plenty of water in the world (Fig. 2.32). The problem is that the places where it is needed are often far from the places where it is available. For example, the fresh water in ice caps is not readily available to the citizens of Los Angeles, where it is needed. Another problem is that most of the water in the world (the oceans) can be used only after a relatively expensive desalinization treatment.

Environmental Accidents

We can point to many other examples of catastrophes that are local in nature. The World Commission on Environment and Development first met in October 1984 and published its report in April 1987. During that relatively brief time interval, the following disasters, were recorded:

- A leak from a pesticides factory in Bhopal, India, killed more than 2000 people and blinded and injured over 200,000 more.
- Liquified gas tanks exploded in Mexico City killing 1000 and leaving thousands more homeless.
- The Chernobyl nuclear reactor explosion sent nuclear fallout across Europe, increasing the risks of future human cancers.
- Agricultural chemicals, solvents, and mercury flowed into the Rhine River during a warehouse fire in Switzerland, killing millions of fish and threatening drinking water in the Federal Republic of Germany and the Netherlands.

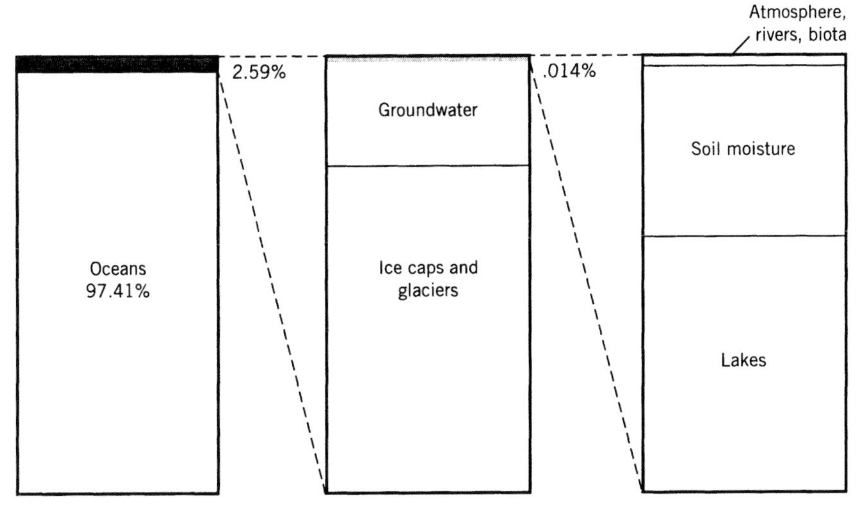

FIGURE 2.32 Distribution of water on the planet is highly uneven. Most of it (97.4%) is in the oceans: only a small fraction (2.59 percent) is on the land. Even most of the water on land is largely unavailable, because it is sequestered in the form of ice and snow or as groundwater: only a tiny amount (.014 %) of the earth's total water is readily available to human beings and other organisms. *Source:* Rivière (1989), p.84.

- An estimated 60 million people died of diarrhoeal diseases related to unsafe drinking water and malnutrition; most of the victims were children. (World Commission on Environment and Development 1987)

Global indices are an inadequate indicator of environmental problems because many of the world's environmental tragedies are local in scale. To use world market prices of metals as an index of environmental problems would be like relying on average global temperature to tell you if you needed to wear a coat. It is the conditions *right where you live* that are most important to you.

A NATIONAL ENVIRONMENTAL QUALITY INDEX

National Trends

Local environmental problems do exist, but then they always have. For example, medieval castle dwellers often used the moat as an open sewer, and the Bible talks about plagues of locusts. Inasmuch as humanity survived these crises and today seems to be better off than it was in those times, should we be so concerned with local trends today that we must burden ourselves with new laws and regulations protecting the environment? Improvement of the human condition has been and always will be interrupted by unavoidable but minor setbacks. Are these setbacks, unfortunate as they may be, important enough to stop the general trend of improvement in material well-being?

In contrast to the world of centuries past, the world of today is a global village. No one in any part of the globe remains unaffected by human activities in another part. The technology that produces an environmental crisis in one locale is often rapidly disseminated to other regions, and the scale of the technology is increasing at an unprecedented rate. Our task, then, is to try to determine whether environmental problems extend beyond the local scale.

National Wildlife is a magazine dedicated to the conservation of the North American commons, publicly owned lands such as national forests and national parks, or publicly used resources including wildlife and rivers. It is concerned with a broad range of national environmental quality concerns in the United States. Accordingly, every year the magazine publishes an Environmental Quality Index that indicates whether environmental quality is improving or deteriorating. The index takes into consideration not only environmental measures, but also governmental policies that affect environmental quality. It is useful because it is on a broader scale than the perennial local crises, yet narrower than the global scale where averages are so diluted they are of only limited significance.

The 1993 index (National Wildlife Federation Staff 1993) showed the following trends for North America.

Wildlife

The threats to endangered species are increasing. For example:

- In 1992, before his defeat in the presidential election, President Bush announced his intention to weaken the Endangered Species Act, which was coming up for reauthorization. His justification was that preservation of certain species such as the Northern Spotted Owl threatened the jobs of loggers.
- Vice President Quayle's Council on Competitiveness recommended that no action be taken to expand the use of turtle excluder devices in the shrimp fishing industry. Such devices help prevent the accidental killing of turtles when they are caught in shrimping nets.
- The U.S. Fish and Wildlife Service estimated that fewer than 30 million ducks flew from Mexico and the southern United States to Canada. This is a decline of more than 10 million since the early 1970s.

Overall, National Wildlife rated the endangered species problem as getting significantly worse. The ratings were made, however, before the inauguration of Bill Clinton who has pledged, as president, to take a stronger lead in environmental protection.

Air Pollution

In 1992 President Bush approved changes in the Clean Air Act which allowed companies to make "minor" changes in their emissions without prior notice or approval. By citing a change in production methods, a company would be able to increase pollutant emissions by as much as 245 tons a year.

California continued to present the worst and best case studies in the struggle for cleaner air. Although the number of hours per year that locations in the California basin experienced hazardous air pollution had declined 50 percent over the preceding decade, the lower average was still more than double the amount considered healthful by state standards.

Although the Environmental Protection Agency announced stricter vehicle emission inspection programs for 80 major metropolitan areas, in much of the country, cars still accounted for half of all air pollution.

In contrast to the modest gain against smog, carbon dioxide emissions remained high. The United States in 1992 remained the world's leading producer of CO_2, the principal gas implicated in potential greenhouse warming. The average American used enough energy in 1992 to emit nearly 12 tons of carbon dioxide; in comparison, the average Japanese citizen was responsible for only 2.5 tons.

Overall, the air pollution index for the United States in 1992 had worsened slightly.

Water Pollution

Water pollution has decreased in the United States since passage of the Federal Water Pollution Act in 1972. The most recent Environmental Protection Agency study found that two-thirds of all surface water in the country now meet water quality standards. However, the study also noted that contaminated runoff from farms, streets, and lawns has yet to be even partially controlled. Whereas the gross forms of pollution that cause water bodies to smell and look foul have been eliminated, contamination by substances that are neither visible nor noxious smelling continue.

Rivers in the United States are in worse shape today than they have ever been. Nearly half the rivers surveyed by the EPA were too polluted to support their intended uses for recreation, drinking water, or fisheries. About 60 percent of the impairment was traced to agricultural runoff. The most endangered waterways were the Northwest's Columbia and Snake rivers, where 200 salmon runs and 200 other native fish species are in peril. The problem is not only pollution, but also eight federal dams that block spawning runs. In Connecticut, however, an effort to restock salmon in the Salmon River may be starting to have some success. In 1965 federal and state officials began a program to restore the fish to the river, despite pollution problems and the presence of 11 large hydroelectric dams and 700 small dams. The effort finally began paying off when two Atlantic salmon were seen spawning the Salmon River in 1991.

In 1991 more than 2000 beaches were closed in 14 states because of raw sewage contamination. Florida alone discharges more than 300 million gallons of wastewater into the ocean every day. The waste often forms stagnant concentrations. "Coastal pollution is the principal culprit blamed for a spreading epidemic of blooms of harmful algae that now constitute a major planetary trend," according to oceanographer T. Smayna of the University of Rhode Island.

Not surprisingly, the index for water quality in 1992 was down slightly.

Forests

A team of NASA scientists comparing satellite images from the Brazilian Amazon and the Pacific Northwest of the United States found striking differences. The Brazilian deforestation had occurred mainly at its edges, but the U.S. forests were fragmented by clear-cuts, roads, and other development. NASA scientists warned that the Northwest forests were on the verge of losing their biological vitality. Forest Service officials discounted that analysis, insisting that clear-cut areas had been replanted but that the young trees would not show on satellite images until they were ten years old. The biologists' response was that, unfortunately, replanting could not eliminate the impact of habitat fragmentation. The ultimate outcome of fragmentation is extinction pressure, which is caused by the inbreeding of restricted animal populations and increased attacks by predators that live on forest edges.

With an increased emphasis on recreation and wildlife, the Forest Service has instituted a policy decreasing the amount of clear-cutting on national forests. Whether the policy will be enforced remains to be seen.

Overall, the National Wildlife index for forests was down slightly in 1993.

Energy

Some states, having lost patience with federal indecision about energy policy, have moved ahead with their own strategies for energy conservation. Many electric utilities and their regulators have discovered that in the long run reducing demand is far cheaper than increasing the electricity-generating capacity. As a result, New York City, for example, has cut its projected growth in electricity demand by 80 percent and expects its peak summertime demand to be less in the summer of 2008 than it was in 1991.

The federal government has sometimes opposed state-led initiatives. For example, when Maryland increased the state sales tax on gas-guzzling cars and issued a sales-tax rebate to more economical makes, federal transportation officials ordered the state to rescind the law. They argued that letting the states set different efficiency standards would unduly burden all U.S. automakers.

Overall, the index for energy was slightly worse in 1993.

Soil

Soil was the only category in which *National Wildlife*'s indices did not decrease. The staff of the Federation concluded that the overall average rate of erosion had not increased.

More farmers than ever before have begun "conservation tillage" (also known as no-till methods of soil preparation). Conservation tillage avoids plowing and retains on the field a portion of the plant's residues to reduce erosion and increase soil quality. In 1991, 28 percent of U.S. farmlands were being cultivated with conservation tillage methods.

The issue of overuse of western rangelands continued. A study of grazing lands in the Southwest deserts managed by the Bureau of Land Management concluded that practices were degrading the land and further threatening already endangered species. This study, observed Oklahoma Congressman Mike Synar, "is just more evidence of the obvious—that taxpayers are subsidizing fiscal and ecological disaster on public land." Another study seemed to support Synar's assertion. Although ranchers paid private landowners an average of $9.66 to graze one cow for a month, the U.S. government rate to ranchers was only $1.92. In addition, the government's costs for administering the program averaged $3.21 per animal-unit-month.

Recycling of Waste

The majority of the population now favors recycling even if such efforts are expensive. The major obstacle to recycling is that the present markets for recycled materials are often saturated. The aluminum recycling industry has made more progress than many in adapting to the increases in supplies. In 1991, 62 percent of aluminum beverage containers were recycled. However, it took the industry 20 years to reach this point.

Although Americans are recycling more than ever, their trash piles are also growing higher than in the past. In 1990 they threw out 196 million tons of refuse—an increase of 8 percent from 1988 and 100 percent from 1960.

Quality of Life

When they considered all factors—wildlife, air, water, forests, energy, and soil—the National Wildlife Federation staff concluded that the quality of life, as based on these indices, had decreased somewhat during 1992.

THE COMMONS AND THE PROBLEM OF LAG TIME

The Commons

By commons we mean land or resources that are used or enjoyed by everyone and yet belong to no one. Some commons such as public parks and national forests are publicly owned, and other commons such as the air that surrounds us have no ownership. In cultures that do not recognize private property, all land is a commons.

Hardin (1968) recently used the phrase *the tragedy of the commons* to encapsulate what happens to property that is commonly owned or to resources that are "free." An amateur mathematician named William Forster Lloyd sketched a scenario to illustrate the tragedy in a pamphlet in 1833. In Lloyd's example, near a farming community is some open land suitable for pasture. The farmers begin grazing their cattle on the land. The animals become fat and productive, so each farmer decides to increase the size of his herd. However, it is not long before the farmers begin to realize that if the number of cattle on the pasture keeps increasing, the cattle will overgraze the grass, causing erosion and degradation of the land.

Nevertheless, they keep on adding to the size of the herd. Why don't they stop, when they know full well that they are ruining the pasture? Because each individual farmer knows that regardless of whether or not *he* stops adding cows, other farmers will continue to increase the size of *their* herds. Then the farmer makes a calculation. What is the benefit to him if he adds one more cow to the herd? Almost the full value of that cow. What is the loss if he adds one more cow? The additional damage caused by one more animal. However, since the damages of overgrazing are shared by all the farmers, the loss to the individual farmer is only a fraction of that suffered by the community of farmers. There is the tragedy of the commons. "Each man is locked into a system that compels him to increase his herd without limit—in a world that is limited. Ruin is the destination toward which all men rush, each pursuing his best interest in a society that believes in the freedom of the commons. Freedom in a commons brings ruin to all" (Hardin 1968, p. 1244).

The pasture, if privately owned by an astute steward, would not experience overgrazing. The steward, knowing full well the consequences of overgrazing, would restrict the size of the herd long before any evidence of degradation began to appear. However, the advantage of such a dictatorial decision does not occur when the resource in question is a global or national commons. Decisions to restrict use must be made by many competing players, who know that yielding a point may work to the competitor's advantage.

The Problem of Lag Time

Imagine a farmer inspecting his crop, shortly after it begins to grow. He notices a few insects, here and there, nibbling on the edges of some leaves. What should he do? The insects could be just transients that will disappear tomorrow and do little damage today to the crop. If they are, the best thing to do is just to ignore them. On the other hand, they could be the first generation of a pest that will devour his crop. In this case, the best thing to do is to control them.

If the insects are just transients, trying to control them would be a waste of time and money. If they are potential pests, however, it would be best to act right away. The cost of controlling them immediately is small, relative to the losses that could occur if the pest developed into a plague. If the insects don't go away in a few days, the farmer's best course of action is probably to control them.

When we consider the first signs of environmental degradation on a national commons, the economic arguments develop differently. Each company that is polluting the commons realizes that if it is the first to control the pollution, it will suffer an economic disadvantage relative to other companies. Therefore, each company tends to hope that the first signs are just transient and not really indicative of a major degradation. Even when the evidence becomes strong that trends will be serious for everyone, there still is a reluctance to be the first to act.

The problem of the commons is even more serious when examined on a global scale because (1) the proportion of the biosphere occupied by the global commons is so large; and (2) the momentum that trends achieve before they can be recognized and acted upon is so great.

For example, by the time scientific evidence can conclusively prove that a dangerous buildup of greenhouse gases has taken place in the atmosphere, it may be impossible to reverse the trend before concentrations reach levels that will damage life on earth. Even if the buildup is halted, gases such as chlorofluorocarbons may remain long enough to seriously damage the ozone layer.

Global Commons

The greatest threat to the sustainability of life on earth is the degradation of the global commons. The global commons are Antarctica, the atmosphere and outer space, the oceans, and global biodiversity—the genetic base of all plants and animals. These resources are worldwide in scope and therefore must be managed as a global commons. Because of their value in sustaining all life on earth, tropical forests should also be included on a list of threatened global commons. In contrast to the oceans and the atmosphere, tropical forests are "owned," but many of those that are threatened are under the control of national governments. With the cooperation of the national government, some rain forests could be given the status of global commons, as has occurred with the rain forest in Queensland, Australia, which was recently designated as a World Heritage site.

The Atmosphere and Outer Space

Carbon dioxide was at relatively low levels prior to the latter half of the twentieth century. The earth was kept at a relatively comfortable temperature because some of the reradiated heat could penetrate the atmosphere. However, recently scientists have observed sharp increases in atmospheric carbon dioxide, as well as in methane and chlorofluorocarbons (CFCs), which have the potential to trap heat, the so-called greenhouse effect. Increases in heat-trapping gases have been attributed to the increased burning of fossil fuel, deforestation, the decay of organic matter in soils cleared of vegetation, the large number of ruminant domestic animals, the use of aerosol sprays, refrigerators, air conditioners, and other factors.

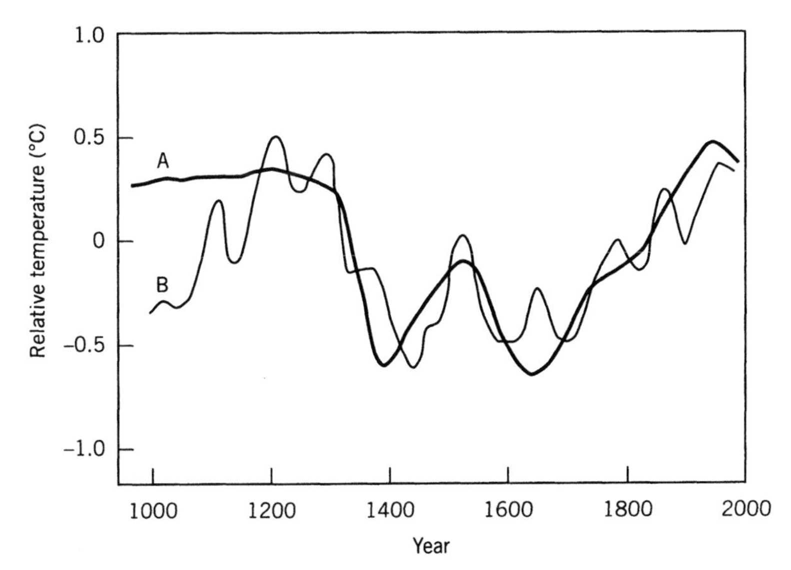

FIGURE 2.33 Estimates of temperature for Northern Hemisphere locations during the last 1000 years. Curve A was derived from various records of the severity of the winter in Europe. Curve B was derived from the width of tree rings from the West Coast of the United States. *Source:* Firor (1990), p. 66.

There has been widespread speculation that the increase in CO_2 is producing a temperature rise. However, we have no solid evidence that rising temperatures are due to a greenhouse effect. Although a rise in temperature from 1965 to 1990 could be a trend (Fig. 2.25), the change is less than 1°C. When a one-degree change is viewed as part of global trends for the past 1000 years (Fig. 2.33), it could be interpreted as a "natural" or "background" variation. The danger with such an interpretation is that if increasing carbon dioxide actually *is* the cause of a trend in increasing temperature, then the world is in trouble, judging by the long-term trajectory of CO_2 concentrations (Fig. 2.34). With a long lag time between an increase in CO_2 and a temperature response, then strong evidence for global warming will not occur until a global warming trend is well underway.

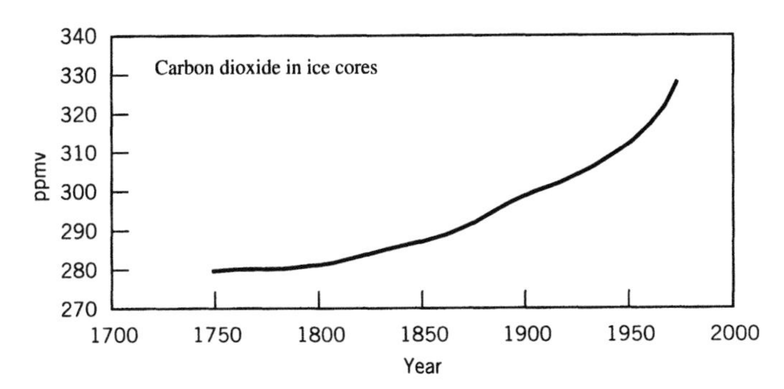

FIGURE 2.34 The concentration of carbon dioxide found in air trapped at different depths in polar ice. The concentration is in parts per million by volume. *Source:* Firor (1990), p. 51.

In his book *The Changing Atmosphere,* Firor (1990) presents an even-handed and comprehensible evaluation of changes in the atmosphere, of the significance of current changes in the atmosphere compared to changes in the past, and of the evidence that the changes could be harmful to the biosphere, including human health.

Chlorofluorocarbons are another pollutant of the global commons. Until recently, they were commonly used in aerosol sprays, refrigerants, and foams. When they are leaked or released into the air, they find their way into the stratosphere. Recent decreases in stratospheric concentrations of ozone over the Antarctic have been traced to the action of these gases. The disappearance of stratospheric ozone leads to an increase in ultraviolet radiation penetrating to earth and increases the probability of skin cancer. When this danger became apparent, a proposal to ban the use of chlorofluorocarbons as propellants was put forth in 1987 by an international agreement called the Montreal Protocol, after the location where the agreement was signed (World Resources Institute 1992). When further evidence suggested that the ozone layer in the temperate zone was also being affected, the United States led a phaseout in the production of these gases. As a result, world production has dropped sharply (Fig. 2.35).

Even if chlorofluorocarbon emissions were to stop entirely, chemical reactions causing the destruction of stratospheric ozone would continue for at least a century. The reason is that the compounds remain that long in the atmosphere and would continue to diffuse into the stratosphere from the troposphere reservoir long after emissions had ceased (Graedel and Crutzen 1989).

Acid deposition may be defined as the falling of acids and acid-forming compounds from the atmosphere to the earth's surface. Sulfur dioxides and nitrogen oxides are released into the atmosphere from large coal- and oil-fired combustion installations and are later carried to earth dissolved in rainfall or suspended in mists and fog. Often it is local, but it can have international consequences when the pollution produced by industries in one country is carried by prevailing winds across another and is "rained out" and deposited on the agricultural fields, forests, and lakes. The first evidence of the potential damage of acid-forming compounds sent into the air from industrial processes came from regions around copper smelters such as Copper Hill, Tennessee. Although evidence from

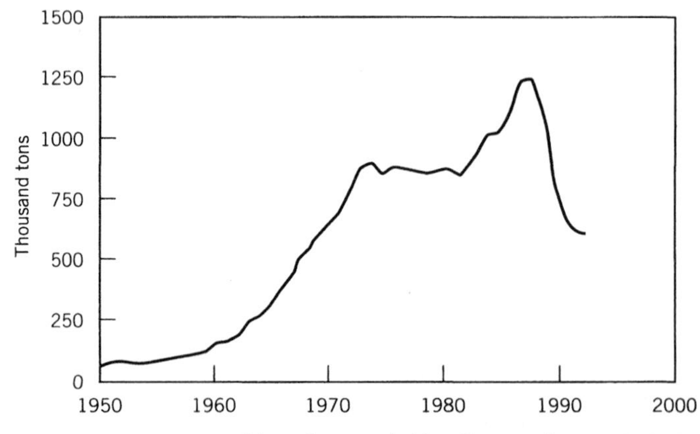

FIGURE 2.35 World production of chlorofluorocarbons, 1950–92.
Source: Brown, Kane, and Ayres (1993), p. 67.

this and other smelters around the world showed that airborne sulfur compounds could kill plants, other events were implicating sulfur in damage to people. As smokestacks were built higher in hopes of cleaning up the air in the vicinity of the emission, the sulfur dioxide lingered longer in the atmosphere before returning to earth. In the process, some of it was oxidized into sulfates and sulfuric acid, both of which harm people when inhaled. In dramatic episodes, people died by the tens or hundreds in the Meuse Valley, Belgium, in 1930, in Donora, Pennsylvania, in 1948, in London in 1952, 1953, and 1962, in New York City in 1953, and in a wide area of the eastern United States in 1966 during stagnant atmospheric spells, when sulfate accumulations in the air reached very high levels (Popkin 1986).

Space debris is yet another form of pollution in the atmospheric global commons. Orbiting satellites lose bolts, slough off paint chips, and clutter up space with debris. This space garbage hurtling around the earth can damage future satellites. The uncontrolled launching of geosynchronous satellites will result in benefits worth hundreds of millions of dollars to the users (Stone 1992), but the costs of space cleanup will be borne by future generations.

Antarctica

In 1957 during the start of the International Geophysical Year, 12 nations cooperated to establish 50 research stations in Antarctica. The success of this research effort prompted the creation of the Antarctica Treaty in 1961. This treaty turned the region into a global commons, dedicated to the peaceful pursuit of knowledge and the free exchange of scientific information. Any member country of the United Nations may now join the Antarctica Treaty System (ATS).

The continent of Antarctica provides a global physical function as an air conditioner for the earth. The ice and snow-covered terrain reflects most of the solar energy impinging on that portion of the globe, and the resulting cold air and water are important in maintaining global circulation systems. The conservation of Antarctica is also important because more than half of the earth's above-ground fresh water is bound up in Antarctic ice (Rivière 1989).

Antarctica is also important biologically. It hosts a large population of wildlife, including squid, fish, penguins, other seabirds, seals, and whales. Antarctic food chains are heavily dependent on krill—small shrimplike creatures that are being overfished. The abundance of krill is due in part to the upwelling of the Antarctic Ocean, which supplies nutrients to the phytoplankton at the base of the oceanic food chain.

Despite the unprecedented arrangements for sharing Antarctica as a common global laboratory, serious problems have emerged. Many scientific laboratories have left their trash around the stations. Thousands of discarded oil drums have been carried away by the sea ice. Seabeds next to some stations have been transformed into huge graveyards for thousands of tons of scrap metal, beer cans, and worn-out vehicles. As a result, entire bays are polluted. Levels of dioxin and hydrogen chloride in incinerators built by the U.S. National Science Foundation exceeded standards for large municipal incinerators. Remedial measures began to be implemented in 1991, when the 26 countries that do Antarctic research signed an environmental protocol to the original treaty aimed at minimizing the impact of human activity (Mervis 1993).

ATS members also have fortified the original treaty by including agreements on protection of the wildlife, waste disposal, and mineral development. Many environmentalists oppose the agreement on mineral development because, they argue, the Antarctic environment is very susceptible to accidents and pollution, and mining for minerals should be banned completely.

The Oceans

All the oceans of the world have been affected by pollution and overexploitation. Floating oil released by accidental spillages or by illegal cleaning of tankers is one of the most devastating forms of marine pollution. Although major oil spills that threaten coasts receive much publicity, those on the open oceans often go unreported. Another important cause of pollution to the oceans is waste disposal. Because the oceans are so huge, it has been assumed that they have an almost infinite capacity to absorb wastes. However, coastlines act as giant filters, and although trash may drift freely in the open seas, it accumulates on the coasts. There it poses a hazard to fisheries and is a nuisance to the tourist industry. The threat may be direct as in the case of hospital wastes dumped at sea or indirect as in the case of pollutants such as fertilizers which result in algal blooms. Such blooms are detrimental both to fisheries and to the tourist industry.

Overfishing also is a problem in the ocean commons. In the last decades, modern fishing technology, including drift nets, has exceeded what the United Nations Food and Agriculture Organization considers the maximum yield sustainable in the long term. Numerous commercial fisheries have collapsed in the last few decades, destroying the livelihoods of fishermen and their communities (Weber 1993).

Biodiversity

Another global tragedy of the commons is the loss of biodiversity—that is, the increasing rate of extinction and the increasing rate of destruction of natural ecosystems which is causing these extinctions. Loss of species means the loss of genetic material that has the potential to provide humans with new pharmaceuticals and with alternative sources of food and fiber. Loss of ecosystems represents loss of certain services of nature such as prevention of soil loss, purification of air and water, and regulation of water flow. Although the estimates as to the rate of species extinction (Table 2.4) and rates of tropical deforestation (Table 2.5) may vary, there is little doubt that the rate is unprecedented in the history of humankind.

QUANTITY VERSUS QUALITY AS A GOAL FOR GLOBAL DEVELOPMENT

How Much?

In areas where pollution is not bad and resources are reasonably priced in relation to income, many people believe that the world is a better place in which to live than it used to be, especially people who have lived through the Great Depression and World War II. Of course, they would agree that it certainly would be better if the occasional small-scale but violent conflicts could be eliminated. Nevertheless, physical well-being as measured by quantity of goods consumed has certainly increased among the people of almost all developed countries and among many of the peoples of lesser developed countries as well.

Table 2.4 *Various Estimates of the Rate at Which Species Are Becoming Extinct in the Tropics*

ESTIMATE	BASIS OF ESTIMATE	SOURCE
1 species/day to 1 species/hour between 1970s and 2000	Unknown	Myers, 1979
33–50% of all species between the 1970s and 2000	A concave relationship between percent of forest area loss and percent of species loss	Lovejoy, 1980
A million species or more by end of this century	If present land-use trends continue	National Research Council, 1980
As high as 20% of all species	Unknown	Lovejoy, 1981
50% of species by the year 2000 or by the beginning of next century	Different assumptions and an exponential function	Ehrlich and Ehrlich, 1981
Several hundred thousand species in just a few decades	Unknown	Myers, 1982
25–30% of all species, or from 500,000 to several million by end of this century	Unknown	Myers, 1983
500,000–600,000 species by the end of this century	Unknown	Oldfield, 1984
0.75 million species by the end of this century	All tropical forests will disappear and half their species will become extinct.	Raven, Missouri Botanical Gardens, personal communication to WRI and IIED, 1986
33% or more of all species in the twenty-first century	Present rates of forest loss will continue.	Simberloff, 1983
20–25% of existing species by the next quarter of century	Present trends will continue.	Norton, 1986
15% of all plant species and 2% of all plant families by the end of this century	Forest regression will proceed as predicted until 2000 and then stop completely.	Simberloff, 1986

Source: *Lugo 1988, Table 6.1, p. 59.*

The available evidence supports the view of Julian Simon and the other technological optimists that the world could easily support many more inhabitants than it does at present, if local quarrels could be eliminated.

How Good?

Despite the lack of data showing that resources are not now generally limiting quantitative increases in agricultural production and material goods, all is not fine with the world. Many environmental problems are local, and data showing environmental deterioration are often significant only when restricted to centers of population. Other environmental problems may be occurring in the global or national commons, in situations where there

Table 2.5 *Preliminary Estimates of Tropical Forest Area and Rate of Deforestation for 87 Tropical Countries, 1981–90 (thousand hectares)*

Regions/Subregions	Number of Countries Studied	Total Land Area	Forest Area 1980	Forest Area 1990	Area Deforested Annually 1981–90	Annual Rate of Change 1981–90 (percent)
Total	*87*	*4,815,700*	*1,884,100*	*1,714,800*	*16,900*	*−0.9*
Latin America	32	1,675,700	923,000	839,900	8,300	−0.9
Central America and Mexico	7	245,300	77,000	53,500	1,400	−1.8
Caribbean Subregion	18	69,500	48,800	47,100	200	−0.4
Tropical South America	7	1,360,800	797,100	729,300	6,800	−0.8
Asia	15	896,600	310,800	274,900	3,600	−1.2
South Asia	6	445,600	70,600	66,200	400	−0.6
Continental Southeast Asia	5	192,900	83,200	69,700	1,300	−1.6
Insular Southeast Asia	4	258,100	157,000	138,900	1,800	−1.2
Africa	40	2,243,400	650,300	600,100	5,000	−0.8
West Sahelian Africa	8	528,000	41,900	38,000	400	−0.9
East Sahelian Africa	6	489,600	92,300	85,300	700	−0.8
West Africa	8	203,200	55,200	43,400	1,200	−2.1
Central Africa	7	406,400	230,100	215,400	1,500	−0.6
Tropical Southern Africa	10	557,900	217,700	206,300	1,100	−0.5
Insular Africa	1	58,200	13,200	11,700	200	−1.2

Source: *From* World Resources 1992–1993, *by the World Resources Institute, Table 8.2, p. 119. Copyright © 1992 by The World Resources Institute. Reprinted by permission from Oxford University Press, Inc.*

is a long lag time between the start of a trend and the emergence of statistical significance of the trends.

Anyone who breathes the air in big cities or who explores the wilderness to study wildlife knows something is wrong. The discrepancy between the evidence trotted out by advocates of unlimited growth and the evidence of our own senses occurs because each addresses a different question. The first ask, "What is the maximum number of people that can be supported by the world's resources?," whereas the second asks, "What is the optimum number to achieve a desirable quality of life?"

People living on isolated frontiers welcome an increase in *number of people,* for on the frontier, a minimum number of people is necessary to make life sustainable. But increases cannot go on indefinitely. Eventually, the benefits of increased neighbors are outweighed by the costs of the pollution and environmental degradation associated with the needs of swelling populations. Eventually, the benefits of mutual aid conferred by a growing community are outweighed by the stress produced by overcrowding.

There is some optimum level of population on the planet Earth, beyond which the benefits of additional population are outweighed by the costs of the addition, notably, crowding stress, drain on resources, and production of pollution. There is no law or scientific principle that establishes the optimum level of material goods and of population. Everyone has a threshold at which he or she feels that further increases bring more pain than comfort. By the time one experiences the threshold, however, it is already too late. By the time we perceive that we are passing the optimum level of population, resource exploitation, and pollution accumulation, the momentum will carry them well beyond the optimum.

Anyone who argues that what the world needs is continued quantitative economic growth, including population increases in order to create the demand that will drive that growth, should be required to spend a year where growth has run amok and virtually destroyed the livability of the area. They should be made to breathe the air in Mexico City; to fight the traffic jams in Bangkok; to dodge the street thieves in Rio de Janeiro; to confront the homeless in New York; to rent a modest apartment in Tokyo; to try and experience nature close to any big city.

"Certainly we don't want *that* kind of growth," the advocates of unlimited growth might argue. "What we want is the kind of growth that improves the *quality* of life for individuals."

The conservationist would ask in turn: "And what exactly would characterize such an improvement in quality? More material goods such as automobiles?"

"No," the thoughtful economist might reply, "Rather, air that is clean enough not to cause respiratory illness, commuting that does not start the day with high stress, the freedom to go outside without the fear of crime, the pleasure of seeing the homeless cared for, and to reward the conscientious worker with a decent place to live."

"Now we are getting some place," the conservationist would say. "Let's talk about how we can achieve *that* kind of growth—or perhaps we should call it qualitative development, instead of growth."

The problem of how the world can achieve development that will provide its inhabitants with an improved standard of living, when the standard includes a healthy environment and not merely more material goods, is the subject of the rest of this textbook.

"What's the use of development, if man is dehumanized?"

(Bishop Guadencio Rosales, 60-year-old leader of the antilogging movement in Mindanao, The Philippines (reported in The Wall Street Journal, January 18, 1993, p. 1).

SUGGESTED READINGS

For useful statistics on the world's environment and resources:

Brown, L. R., C. Flavin, and H. Kane. 1992. Vital Signs. The Trends That Are Shaping Our Future. W. W. Norton, New York.

Tolba, M. K. 1992. The World Environment: Two Decades of Challenge. Chapman and Hall, London.

World Resources Institute. 1992. World Resources 1992–93. A Report by the World Resources Institute, in collaboration with the United Nations Environment Programme and the United Nations Development Program. Oxford University Press, New York.

For the *State of the World.* This is an annual publication by the Worldwatch Institute. For 1993, the publication citation is:

Brown, L. R., Project Director. 1993. State of the World. A Worldwatch Institute Report on Progress Toward a Sustainable Society. W. W. Norton, New York.

For the use of modeling to predict the future of the earth:

Meadows, D. H., D. L. Meadows, and J. Randers. 1992. Beyond the Limits. Confronting Global Collapse; Envisioning a Sustainable Future. Chelsea Green Publishing Co., Post Mills, Vt.

For overviews of global environmental problems:

Managing Planet Earth. Scientific American, September 1989. (Entire Issue.)

For an analytical approach to global environmental problems:

Clark, W. C., and R. E. Munn, eds. 1986. Sustainable Development of the Biosphere. Cambridge University Press, Cambridge.

Vitousek, P. M. 1992. Global Environmental Change: An Introduction. Annual Review of Ecology and Systematics 23.

For a popularized presentation of the global climate problem:

Firor, J. 1990. The Changing Atmosphere: A Global Challenge. Yale University Press, New Haven, Conn.

For a recent popularized Malthusian view:

Kennedy, P. 1993. Preparing for the Twenty-First Century. Random House, New York.

For the view that environmental conditions have been declining due to overexploitation of resources in chapters written by traditional resource managers:

Jeske, W. E., ed. 1981. Economics, Ethics, Ecology: Roots of Productive Conservation. Soil Conservation Society of America, Ankeny, Iowa.

For a head-in-the-sand approach to the coming century:

Bailey, Ron. 1993. Eco-Scam: The False Prophets of Ecological Apocalypse. St. Martin's Press, New York.

CHAPTER

ENVIRONMENTAL ETHICS

CHAPTER CONTENTS

CHAPTER OVERVIEW

Some conservationists have argued that adoption of an environmental ethic is crucial for our long-term survival. However, debates about whether we are part of nature or above nature have paralyzed progress on formulating an ethic. Other conservationists have urged that we move forward and deal with environmental deterioration as best we can, even though we lack a universally acceptable ethic.

PRINCIPLE

One of our most important responsibilities is to pass on to future generations an environment whose health, beauty, and economic potential are not threatened.[1]

[1]From the July 1990 G-7 Declaration of the Economic Summit (El Serafy, 1991, p. 68).

IS THERE AN ENVIRONMENTAL ETHIC?

Environmental ethics is a branch of philosophy concerned with the moral relations between humans and the natural world. It examines questions such as

Does nature have any value, apart from those commodities it supplies to economic man?

Do species have the right to exist? Or, conversely, do humans have the right to drive a species to extinction in the course of their pursuit of material well-being?

Do humans have any right to cause animals to suffer, or to kill them?

Do humans have the right to kill any living thing?

Do animals and plants have rights?

Do humans have a responsibility toward nature?

Does this generation have a responsibility toward future generations?

Although conservation deals with some of the same questions that concern environmental ethicists, conservation is not synonymous with environmental ethics. For environmental ethicists, philosophy is an end in itself. In contrast, many conservationists are interested in environmental ethics primarily to justify doing what they already believe to be important. In other words, they seek an ideological basis for their actions.

Two philosophical questions are of particular interest to Conservationists: (1) Do humans have a responsibility toward nature? (2) Does this generation have a responsibility toward future generations?

Do Humans Have a Responsibility Toward Nature?

Some philosophical writers believe that people do indeed have a responsibility toward nature. Taylor, for example, asserts that "in addition to and independently of whatever moral obligations we might have toward our fellow humans, we also have duties that are owed to wild living things in their own right. Our duties toward the Earth's non-human forms of life are grounded on their status as entities possessing inherent worth" (Taylor 1986, p. 13).

Not all philosophers agree with Taylor. In fact, some believe that the idea of an environmental ethic is alien to the philosophy of modern science. The conception of the physical world, as viewed by Descartes was of a purely mechanical, purposeless system, unfolding according to impersonal laws that it was the aim of the scientist to discover. This viewpoint holds that the individual's ability to stand back and ask questions about nature is unique among living creatures, and marks humans as separate from nature. Humans have purposes and desires, but nature is purposeless, and therefore has no value, no inherent worth (Matthews 1989). Descartes aspired to a "practical philosophy by means of which, knowing the force and the action of fire, water, the stars, heavens, and all the other bodies that environ us, as distinctly as we know the different crafts of our artisans, we can in the same way employ them in all those uses to which they are adapted, *and thus render ourselves the masters and possessors of nature*" (cited in Passmore, 1974, p. 20; emphasis in original).

This philosophy was absorbed into the ideology of modern Western societies, communist as well as capitalist. It found expression in a metaphysics for which the human being is the sole finite agent and nature is a vast system of machines for humans to use and modify as they please. An extreme statement of this view is that:

> *. . . there is no basis in the history of Western civilization for the idea that there should be a new ethic that will help protect the natural environment from destruction. Preservationist-orientated environmentalists are nothing more than antiscientific nature mystics who have abandoned the "analytical, critical approach" which has always been the "peculiar glory" of Western civilization, and have frivolously justified their irrational positions in terms of strange Oriental religions. Preservationist attitudes and actions, in addition, are entirely inconsistent with man's fundamental mission on Earth—civilizing the world—and if they are permitted to flourish, it may well mean the end of Western civilization and the end of man. There is only one good way to relate to nature—in terms of its possible usefulness to human beings in terms of economic utilitarianism. (Cited and criticized by Hargrove, 1989, p. 77)*

Does This Generation Have a Responsibility Toward Future Generations?

The American Indians answered this question nicely in a proverb that today's environmentalists have adopted as a slogan: "We do not inherit the earth from our parents. We borrow it from our children." Philosophical writers sympathetic with this view believe that it is highly presumptuous for our generation to imagine that its wants and its political causes might conceivably justify our jeopardizing not just our inheritance, but our inheritors as well—our sons and grandsons and the myriad unborn generations whose hopes and achievements we cannot know. That this generation thinks its own transient conflicts are weightier than the infinity of the human future demonstrates truly colossal arrogance (anonymous, cited in Partridge 1981). To the believers, the commitment is held as an imponderable—a conviction, transmitted in the cultural heritage, that is so covertly and extensively accepted that it never occurs to them to submit the issue to analysis. It is a deeply held conviction that is a nonissue (Wagner 1981).

To others, it is not so obvious. Shrader-Frechette (1981) raises several issues. Because it is not possible, she says, for present persons to enter into an explicit, legal bargain with future individuals, social contracts between current and forthcoming persons are a chronological impossibility. It is also impossible because present individuals can benefit future ones, but not vice versa. Furthermore, it can be asked, would recognition of the rights of future generations diminish the extent to which the needs of current individuals (especially the poor or socially disenfranchised) were served?

Still others, on the basis of their religion, feel that intergenerational responsibility is pointless. A recent U.S. secretary of the interior, James Watt, was accused of permitting a despoliation of the environment on the basis that it didn't matter, since he believed that the apocalypse would soon be coming. Specifically, Wolf (1981) wrote: "Based on a very literal interpretation of Scripture, this [Watt's] view holds that the Earth is merely a temporary way station on the road to eternal life. It is unimportant except as a place of testing to get into heaven . . . . The earth was put here by the Lord for His people to subdue and to use for profitable purposes on their way to the hereafter. James Watt is steeped in this latter tradition. It forms the core of his life and values" (p. 65).

Bratton (1983) tempers the criticism of Watt by saying that this characterization was not consistent with all of the secretary's remarks. Watt, she said, had a higher view of creation than just as a "temporary way station." Nevertheless, the point here is that we cannot assume that everyone accepts the idea that there is an intergenerational responsibility to care for the environment.

WESTERN VERSUS EASTERN VIEWS ON THE INDIVIDUAL AND NATURE

Nature and Christianity

According to one school of thought, Christianity and the Judeo-Christian tradition are at the heart of both the past and present environmental degradation. The main points of this attack (Callicot and Ames 1989) are that in the Christian view God transcends nature; humans exclusively were created in the image of God and thus are segregated from the rest of nature; God gave them dominance over nature; and God commanded them to subdue nature and multiply themselves.

White (1967, p. 1207) emphasizes that "we shall continue to have a worsening ecologic crisis until we reject the Christian axiom that nature has no reason for existence save to serve man." White also includes science and technology in his indictment. He claims that:

> *. . . modern technology is at least partly to be explained as an Occidental, voluntarist realization of the Christian dogma of man's transcendence of, and rightful mastery over, nature. But as we now recognize, science and technology—hitherto quite separate activities—joined to give mankind powers which, to judge by many of the ecologic effects, are out of control. If so, Christianity bears a huge burden of guilt. (p. 1206)*

McHarg's (1992) criticisms of Christianity are equally harsh:

> *The Biblical creation story of the first chapter of Genesis, the source of the most generally accepted description of man's role and powers, not only fails to correspond to reality as we observe it, but its insistence upon dominion and subjugation of nature encourages the most exploitative and destructive instincts in man, rather than those that are deferential and creative. Indeed, if one seeks license for those who would increase radioactivity, create canals and harbors with atomic bombs, employ poisons without constraint, or give consent to the bulldozer mentality, there could be no better injunction than this text. Here can be found the sanction and injunction to conquer nature—the enemy, the threat to Jehovah. (p. 26)*

Most Christians believe that this kind of criticism is unwarranted. Walter Cook, professor of forest resources at the University of Georgia, defended Christianity against such an attack during a seminar by saying: "God said to subdue the earth, not beat it to death; He meant for us to tame it, farm it, and use its resources, not ravage it, poison it, and waste its resources."

The debate about whether God meant for human beings to subdue nature or to preserve nature has been carried out in many formats. One has been in *American Forests,* the journal of the American Forestry Association. On one side, Isherwood (1982) said:

> *Our forbears [sic] thought it was their theological duty to subdue the American wilderness. Wild places were worse than useless; they were an embodiment of some evil. Trees were weeds, wild animals unclean and expendable. Cleared land was not only a physical*

> *need but a moral need. They were simply carrying out the scripture of Genesis, "fill the earth and subdue it" (p. 34).*

In reply, Selle (1983), speaking on behalf of God in denying that Christianity was responsible for environmental degradation, said:

> *Scriptural theology has too long absorbed the blame for a polluted and abused environment. If some misguided souls have adopted a philosophy such as described above, they are feeding on their own ignorance and not on my clear teachings . . . . I intended my creation to be used with love and respect. Yes, I want it to be* used *for the good of humanity and for the glory of the Heavenly Father to be* used *for shelter, food, clothing, medicines, refreshment, and inspiration . . . . The means to a discerning land ethic is the same means toward a happy life; justice, wisdom, sharing, mercy, love. (p. 6)*

Passmore (1974) shifts the burden away from Christianity and toward the Greek tradition on which it is based. He maintains that the critics of Western civilization are justified in their historical diagnosis that in accordance with a strong Western tradition humans are free to deal with nature as they please, because it exists only for their sake. But they are incorrect, he says, in tracing this attitude back to Genesis. Genesis and the Old Testament certainly tell humans that they are or have the right to be masters of the earth and all it contains. But at the same time Genesis insists that the world was good before people were created and that it exists to glorify God rather than to serve humankind. It is a result of the Greek influence, Passmore argues, that Christian theology was led to think of nature as nothing but a system of resources that human beings could use or waste at will, with no moral implications.

The basis for the human-centered/nature-centered debate is understandable in the context of the development of civilization. In the early history, there were few people, and the forces of nature were a threat to their continued existence. It was important for human survival that nature be conquered. Therefore, the command to subdue the wilderness was a key component of the world's best known guide to survival, the Christian Bible.

Through technology, however, humans have now subdued nature, and their continued existence is no longer threatened by nature. *The human-centered view has helped people survive the forces of nature until now. Now we need a nature-centered view in order to help nature survive human forces.*

Nature and Asian Traditions

Criticism of Christianity for supposedly giving human beings the right to exploit, and thereby pollute and degrade nature has given rise to an interest in Eastern religions. In these religions humans are seen as part of nature, not above it.

The ancient Eastern cultures were the sources of respect for and religious veneration of the natural world. The Indian philosophy of Jainism, for example, proposed that humans not kill or harm any living creature. Early Buddhists and Hindus professed a feeling of compassion and a code of ethical conduct for all that was alive (see photo-essay, Social Forestry in Thailand, Chapter 6). Similarly, China and Tibet produced philosophies that honored life other than the human's, and promulgated elaborate dietary rules in this interest.

In the Far East the relationship between humans and nature was marked by respect, bordering on love. The human was understood to be part of nature. And wilderness, in

Eastern thought, did not have an unholy or evil connotation but was venerated as the symbol and even the very essence of the deity. Chinese Taoists postulated an infinite and benign force in the natural world. Far from avoiding wild places, the ancient Chinese sought them out in the hope that they would be able to sense more clearly the unity and rhythm that they believed pervaded the universe. In linking God and the wilderness, instead of contrasting them as did the Western faiths, Shintoism and Taoism fostered love of wilderness rather than hatred (Nash 1967).

Differences in Asian and Western philosophy have been important in shaping national character. The jewel of the Western tradition is the insistence on the uniqueness of the individual and the preoccupation with justice and compassion. The Western assumption of superiority has been achieved at the expense of nature, whereas the Asian harmony of man and nature has been achieved at the expense of the individuality of man (McHarg 1992).

Passmore (1974) rejects the appropriateness of Asian philosophies for Western societies: "Although it is one thing to ask Western societies to be more prudent, it is quite another thing to suggest that they can solve their ecological problems only if they abandon the analytic, critical approach which has been their particular glory and go in search of a new ethics, a new metaphysics, a new religion" (p. 3).

Passmore's concern about losing the particular glory of the West is probably moot. The change is in the other direction. In much of the Orient, the West's "particular glory" is rapidly replacing the Asian tradition. A visit to Tokyo, Seoul, Bangkok, Singapore, and other Asian centers reveals that the Asian respect for and religious veneration of the natural world is losing out to the Western style of development in its most excessive and unrestrained form. Industrialization and population pressure have made appreciation of the natural world an occasional pleasure at best in the urban and industrialized environment of most of Japan, Korea, and other industrialized Asian countries.

Deep Ecology

The deep ecology movement pioneered by Arne Naess (1973, 1984) vigorously rejects an anthropocentric view of the world in favor of what he calls "the relational, total-field image." The movement advocates a "biocentric equality" (Golley 1987, —that is, the notion that humans are part of nature, and as such, have no more right to exploit other species as those other species have to exploit humans. In this respect, deep ecology is similar to Asian traditions of thought such as Buddhism, whose fundamental teaching is that humans are part of nature, not above it; humans have no right to kill other creatures, except as it becomes necessary to feed themselves.

Naess contrasts his deep ecology movement with the shallow ecology movement, which he criticizes as being concerned only with the health and affluence of people in the *developed* countries.

Although many conservationists are sympathetic with the deep ecology philosophy, most believe that there is little possibility that it will be adopted in Western countries. Reforming the existing system from within offers a more realistic hope of achieving at least some success.

In addition, there are the environmental extremists, whose tactics such as "spiking" old-growth stands of timber is disdained even by deep ecologists. ("Spiking" a tree is driving an iron pin deep into a tree, so that when a chain saw hits the spike, the blade will shatter. Such tactics have seriously injured loggers.) Nevertheless, extremism in environmentalism has the same advantage as extremism in any field: it makes any less extreme position appear more reasonable.

Puritan Ethics

Puritanism has had a large, but generally unrecognized, role in the American conservation movement. "Waste not, want not" is an old American folk saying that reflects the Puritan philosophy. It is a guide to economy of action, for individuals, communities, and nations. It has found new life in modern conservation.

The Puritans who migrated to the North American continent in the seventeenth century sought a life of purity: churches that were pure, and communities bound together in unity, harmony, and purity. In their quest for purity, the Puritans tried to follow a God-given pattern of life which permitted no deviations from the path laid down for them in their interpretation of the Holy Scriptures, and taught to them by their parents and their properly appointed ministers. The Puritan pattern of purification is probably best presented as a quest for *simplification* (Staples 1988).

The belief that people don't need fancy houses and fancy clothes and that materialism is sinful is a Puritanical ethic that runs deep in the North American character. Lack of ostentation was a characteristic of Puritan society. Today, it is a point of pride among conservationists.

Despite the rampant consumerism that has plagued the United States, "materialistic" still is a pejorative term in many segments of American society. Acquisition of "status symbols" is considered to be a crass substitute for qualities that truly should confer status in society, qualities such as selflessness, public service, and artistic and intellectual expression. Puritanism also is mirrored in preferences for certain recreational activities. Challenges in nature, such as hiking, mountain climbing, and kayaking, are considered to be character-building and thus morally superior to materialistic activities such as shopping and television watching. Hobbies that require considerable skill and that depend on an understanding of nature are considered worthier than those that merely exploit mechanical conquest of nature. Thus sailing is better than power-boating, surfing is better than jet-skiing, fly fishing in natural streams is better than fishing in sonar-equipped power boats, and horseback riding is better than ATV (off road, all terrain vehicle) driving. And nondestructive activities such as nature photography are more desirable than destructive activities such as hunting for endangered species. (Hunting for species such as deer when they are ecologically destructive because of overabundance is another matter.)

The Quakers and Amish provide a strain similar to that of the Puritans. These religious groups settled in rural areas and even today will not use the trappings of modern society. Many do not use cars and trucks, and restrict their farm implements to those powered by draft animals. They fertilize their fields with manure and organic residue, and they refrain from chemical herbicides and pesticides. These groups, while formerly considered anachronistic by much of mainstream America, are now viewed as having not only a sustainable agriculture but also a sustainable way of life.

MALTHUS AND ETHICS

Neo-Malthusian Theory

In his "Essay on the Principle of Population" Malthus predicted that unlimited population growth would result in disaster. It was impossible, he said, for agricultural production to sustain growing populations indefinitely. War, epidemics, and famine were "positive" checks on population growth, whereas abortion and infanticide were "preventative" checks. "Moral restraint" from sexual intercourse was another check added in a later edition (Mellos 1988).

Since the time of Malthus, the world population has increased by a factor of almost six. Although wars and famine have taken a toll, they have scarcely made a dent in the exponentially rising number of people.

Although Malthus was wrong about how fast the food supply could grow, the idea that an expanding world population would result in increasing stress on natural resources and environment has been reformulated in terms of neo-Malthusian theory. It is not a question of whether food per se can be supplied for an expanding population. The question, rather, is how the short-term production of food, fiber, and other material goods demanded by the growing population will affect the long-term state of the world. Specifically, increases in agricultural production will cause increases in pollution owing to agricultural chemicals and soil erosion. Increased forest harvesting will eliminate much of the biological and genetic diversity on which future agriculture and forestry must be based. Increased air pollution from automobiles and industry results in dramatic increases in health problems. Increased generation of material goods greatly increases the problem of waste disposal.

As noted in Chapter 2, *The Limits to Growth* (Meadows et al., 1992) proposed a model for relating the dynamics of various forms of growth to the limits of growth as determined by resource limitations. It examined the relationship between population, capital, production, and pollution in light of the planet's capacities to remain habitable. The model suggested that the exponential growth of population generates an exponential demand for food, but the supply of food is dependent on land, water, fertilizer, and agricultural machinery, which depend in turn on capital growth and nonrenewable resources of which there is a finite stock. The study concluded that the factors of growth are interrelated such that exponential growth as a whole becomes a threat to the planet's survival.

The study has received a great deal of criticism on the basis of its estimates of given factors and its mode of measuring estimates. Yet there is little disagreement that the growth of population does trigger off patterns of demand and consumption of natural resources which expand exponentially and in the process generate an exponential production of pollution (Mellos 1988).

The Ethics of Population Control

In 1969 *The Population Bomb* aroused a firestorm of controversy. In that book, Ehrlich warned that the exponentially increasing population of the world would cause an increase in poverty, hunger, and environmental deterioration, as billions of people devastated the remaining resources. He argued that one suggested solution—increased agricultural production—would be inadequate. As he reasoned: "A cancer is an uncontrolled multiplication of cells; the population explosion is an uncontrolled multiplication of people. Treating only the symptoms of cancer may make the victim more comfortable at first, but eventually he dies—often horribly. A similar fate awaits a world with a population explosion if only the symptoms are treated" (p. 148).

As possible solutions, Ehrlich suggested a change in tax laws, so that taxes would increase instead of decrease with additional children, as well as a national policy encouraging contraception, laws permitting abortion, and distribution of information and birth control devices. He stopped short of recommending compulsory birth control, but acknowledged that many of his colleagues felt that it might be necessary.

Ehrlich recognized that an official government program for birth control would interfere with the people's supposed right to have as many children as they wanted. He saw the need for an ethic that recognized the dangers of overpopulation and that held that humans had a moral duty to limit themselves. He did not believe that Christianity was capable of incorporating that ethic:

> *Somehow we've got to change from a growth-oriented exploitative system to one focused on stability and conservation. Our entire system of orienting to nature must undergo a revolution. And that revolution is going to be extremely difficult to pull off, since the attitudes of Western culture toward nature are deeply rooted in Judeo-Christian tradition. Unlike people in many other cultures, we see man's basic role as that of dominating nature, rather than as living in harmony with it. (p. 151)*

Lifeboat Ethics

"Lifeboat ethics" refers to ethics that would be adopted when lifeboat space was insufficient to accommodate all passengers from a sinking ship—a situation such as occurred when the *Titanic* sank in the North Atlantic. Hardin (1974) uses the term as a metaphor for the ethics that would occur among inhabitants of a rich country—the "lifeboat"—that was surrounded by hordes of people from poor countries, all trying to scramble aboard.

The rich nations of the world may be thought of as lifeboats with moderate numbers of persons on board, while the poor countries are severely crowded lifeboats. The poor, says Hardin, continuously fall out of their lifeboats, swim for a while, and hope to be admitted to one of the rich lifeboats. The ethical dilemma is whether the passengers on the affluent, less crowded boats should help the swimmers or allow them to come aboard. Like the land of every nation, each lifeboat has only a limited carrying capacity.

Hardin examines this ethical dilemma from the viewpoint of those in the rich lifeboat—"our lifeboat" he calls it. There are many more poor people outside our boat than there are rich inside, and it is loaded to capacity if we consider the "safety factor" necessary to ensure the well-being of its passengers. If one follows Christian or Marxist ethics, we are bound to admit everyone to the boat since the needs of those inside and outside the boat are the same. Doing so swamps the boat, however. "Complete justice, complete catastrophe."

Hardin's proposed solution is to protect the survival of those on board, preserve the safety factor, and admit no more people to the rich boats. He justifies this solution by explaining that if rich nations gave resources to poor ones or admitted many of their people through generous immigration policies, then disaster would eventually strike the wealthy countries. The prolific passengers would swamp the lifeboat. Furthermore, Hardin states that if the goods of the earth are assumed to be part of a commons open to all, then "the tragedy of the commons" will occur. People will attempt to safeguard their own interests and in doing so will bring ruin to all.

Even the noble idea of a "world food bank" says Hardin, to which nations would contribute according to their needs, would lead to the tragedy of the commons. Although special interest groups such as farmers and pesticide manufacturers gain through such policy, all U.S. taxpayers must pay for it. Thus, he maintains, this allegedly humanitarian scheme really hides a powerful special interest lobby for extracting money from taxpayers. Moreover, he claims, if poor countries are "bailed out" whenever there is an emergency, they will not learn from their experience. As Hardin explains, if their emergencies are not met by outside help, their populations will drop back to the carrying capacity of the land. Relief from the outside ultimately hurts poor nations because it escalates population. The input of food thus acts, he says, as the pawl of a ratchet, "preventing the populations from retracting" (p. 564).

Hardin's words sound cold and inhuman, for they show no pity and recommend no help to the starving millions at our door. But is the view that it is more ethical to save a small population for a long time more inhumane than the attempt to help everyone, only to have total disaster shortly thereafter?

How Many People Do We Want?

Chapter 2 on environmental trends presents evidence that the limits of growth may be determined more by the earth's ability to absorb waste than by its ability to provide resources. As a result, the question for conservationists shifts from "How many people can we have?" to "How many people do we want?"

An overnight train-trip across central China in a hard-bed sleeper gave me an idea of how many I *don't* want. At the station, people jammed the platform, arguing, spitting, gargling. To get aboard the train, I fought the swarming crowd, flowing with the surge, falling against those in front, being elbowed by those behind. The triple-stacked flat boards that were intended as beds for one person were crowded with three. The bathrooms were filled with a stench so bad I had to hyperventilate and hold my breath to get near them. At station stops, throngs of vendors thrust at me musty-looking morsels in wrinkled paper. As the sun rose, the heavy cloud of smoke from the wood and charcoal stoves in the houses that blanketed the hillsides gave the landscape an incinerated aspect.

I would much prefer not to live in the midst of so many people. Am I immoral or am I merely selfish for having such a preference?

SCIENCE AND ETHICS

Does Science Have an Ethic?

Science has its own internal ethic governing the behavior of scientists. It is unethical for scientists to plagiarize, to falsify data, or to use grant funds for purposes other than for which they were granted. For those who breach the code of ethics, the penalties can be exclusion from further grants, failure to receive promotion, or even dismissal.

Scientists face a broader ethical issue, however: Are there (or if there are not, then should there be) any ethical guidelines governing what a scientist should study, when the results of his or her study could have a significant impact on society outside the scientific world?

Descartes' dualism of mind and nature, the philosophical foundation for early modern science, would proscribe any ethical considerations concerning the subject matter of scientific study. Nevertheless, scientists often are faced with ethical dilemmas. An important case was that which faced physicists when they realized the implications of their work on the atomic structure of matter. In 1939, when President Roosevelt was contemplating ordering work on an atomic bomb, F. W. Aston, a British scientist, wrote to Alexander Sachs, a Roosevelt adviser, saying "There are those about us who say that such research should be stopped by law, alleging that man's destructive powers are already large enough" (Robinson, 1956, p. 16). As World War II drew to a close, some of the scientists who helped develop the atomic bomb became worried about the social and political implications of atomic energy and about the use of the atomic bomb. They felt that the release of "the forces that power the sun" (as one of them put it to a Senate committee) meant a qualitative change in weaponry and in international politics (Strickland 1968).

The Green Revolution

Crop scientists have seldom felt a conflict between their applied goal—to increase crop production to better feed the increasing population of the world—and the science they use to achieve that goal. In the years after World War II, the world's population began to increase dramatically, owing in part to new medicines that were made available to many parts of the world. Agronomists quickly responded to the new demand with research on increasing crop production through increased fertilization, pest and weed control chemicals, hybrid grains that responded to the chemicals, and increased machinery to cultivate larger areas. This so-called green revolution greatly benefited some regions. A classic example was the rapid rise in wheat production in northwest India in the late 1960s.

Considerable polarization developed between those with positive and optimistic views about the Green Revolution and those with negative and pessimistic outlooks. The optimists have included biological and agricultural scientists involved in creating the new technologies. In the early days of the Green Revolution, some of them saw an enormous potential for increased production. They were fired with enthusiasm and faith, excited at the way in which the new dwarf wheats and rices shifted yield potentials to new high levels. Attention was concentrated on geographical areas that were well endowed with irrigation water and infrastructure, most notably the Punjab and Haryana in India, where the new seed–fertilizer–water technology was exploited very quickly. The spectacular trebling of wheat production there during the 1960s encouraged optimism. As the Green Revolution spread to other crops, some saw the prospect of banishing hunger from the world.

Those who took negative and pessimistic views included social scientists concerned about the political economy and about the question of who gained and who lost from the Green Revolution. Many studies showed that the new technologies were captured by the rural elites and benefited those in the more favored regions. The new high-yielding varieties of foodgrains, planted, fertilized, irrigated, and protected by pesticides, were usually found on the fields of the larger and more prosperous farmers. As a result of the new technology, major social and economic consequences arose. They included an increase in the number and proportion of landless households, a growing concentration of land and assets in fewer hands, and a widening disparity between the rich and poor households. In their negative assessments, some social scientists believed the Green Revolution had sharpened social tensions, and some spoke of it as turning red (Chambers 1984).

Has the Green Revolution been unethical? Are the scientists who plunge blindly ahead with "increasing productivity" unethical, if their labor results in population increases that thereby increase human dependence on limited agricultural soil, nutrients, and water, and on petroleum energy that surely will someday run out and finally create greater human misery?

They can be seen as unethical only if, in fact, the fruit of their labor *does* result in greater human misery. Because we cannot know the answer until the future, the ethics of increasing crop production so that more people can be fed cannot be resolved at the present time. If life does get better sometime in the future as a result of having more people in the world, then those who took steps to increase population will be cheered. However, if continuous increases in population result in disaster, then they will be blamed.

It is impossible to know the answer. All that can be stated now is that before they plunge headlong into some new technology for further subduing nature or further increasing the number of people on earth, scientists should seriously ask themselves what the result might be if their work is as successful as they hope. It is important for them to contemplate the consequences of their success.

Biotechnology

The Green Revolution is only one of many agricultural revolutions that has occurred during human history (Grigg 1984). Biotechnology is the latest innovation that envisions itself as the savior of starving millions. At a recent meeting of crop scientists, biotechnologists reported on proposals for expanding world food production. One of the proposals was to introduce genes into crops such as wheat in order to enable them to grow in regions for which they previously were not adapted (Moffat 1992). "It is now just a matter of time," stated one biotechnologist, "to get improved cold hardiness. When that happens, the world's breadbasket could include the Canadian prairies and much of Siberia—the largest, harshest agricultural region in the world" (p. 1347).

The plant breeders surely felt good about carrying out research with such a noble goal—feeding the increasing population of the world. But did they not consider their ethical dilemma?

1. We already have evidence for the results of increasing production without controlling population: increased suffering and starvation.
2. The far northern ecosystems are far more fragile and more prone to disaster than the good soils and better climates where the first revolution took place. It will take less of a natural mishap to disrupt these systems and to throw the increased millions of people dependent on them into critical conditions.
3. The harsh ecosystems that are destroyed in the process are the last refuges of wild nature and the wild genes that have the capability to survive in that habitat.

The scientists and their funding agencies, investors, lobbyists, and corporate boards do not have some hidden agenda or harmful intent. Rather, they are so fixated on their own narrow goals that they fail to see the broader ramifications and potentially harmful consequences of their endeavors (Fox 1992). Is it ethical to plunge blindly ahead without at least stopping and considering the implications of a biotechnological breakthrough?

Vice President Al Gore in his book, *Earth in the Balance,* comments eloquently on the technological rush to control the natural world:

> *Our seemingly compulsive need to control the natural world may have derived from a feeling of helplessness in the face of our deep and ancient fear of "Nature red in tooth and claw," but this compulsion has driven us to the edge of disaster, for we have become so successful at controlling nature that we have lost our connection to it. And we must also recognize that a new fear is now deepening our addiction; even as we revel in our success at controlling nature, we have become increasingly frightened of the consequences, and that fear only drives us to ride this destructive cycle harder and faster. (p. 225)*

Even if biotechnologists are able to engineer a kind of grain that can grow in Siberia, by doing so they are tightening up the ratchet one more notch, increasing the potential for disaster should the increasingly delicate balance between technology and nature be disrupted by natural disasters, or by a new and virulent virus that evolved to attack even the most cleverly crafted genes.

Gore comments on the presumed power of biotechnology to conquer nature, or to use biotechnology to enable humankind to adapt to an increasingly polluted earth:

> *Some even imagine that genetic engineering will soon magnify our power to adapt even our physical form. We might decide to extend our dominion of nature into the human gene pool, not just to cure terrible diseases, but to take from God and nature the selection of genetic variety and robustness that gives our species its resilience and aligns us with the natural rhythms in the web of life. Once again, we might dare to exercise godlike powers unaccompanied by godlike wisdom.*
>
> *Our willingness to adapt is an important part of the problem. Do we have so much faith in our own adaptability that we will risk destroying the integrity of the entire global ecological system? If we try to adapt to the changes we are causing rather than prevent them in the first place, have we made an appropriate choice? Can we understand how much destruction this choice might finally cause? (p. 240)*

Gaia

Lovelock, in his book *Gaia: A New Look at Life on Earth* (1979), examines from a different perspective the problem of humankind's attempt to subjugate and dominate the earth. "Gaia" is Greek for the concept of Mother Earth. Lovelock uses the term as shorthand for the hypothesis that the biosphere is a self-regulating entity with the capacity to keep our planet healthy by controlling the chemical and physical environment. He explains that Gaia is a complex entity involving the earth's biosphere, atmosphere, oceans, and soil. The totality constitutes a feedback or cybernetic system that seeks an optimal physical and chemical environment for life on this planet. The maintenance of relatively constant conditions by active control may be described by the term *homeostasis.*

The Gaia hypothesis suggests that the biotic community plays a major role in keeping the chemical composition of the oceans and atmosphere relatively constant. The contrary hypothesis is that purely abiotic geological processes produced conditions favorable for life and that organisms merely adapted to these conditions. The question then is: Did physical conditions change first and life evolve to those conditions, or did both evolve together? As evidence for the hypothesis that the changes coevolved, Lovelock compared the atmosphere of the earth with that of the planets Mars and Venus, where it is unlikely that life exists. Earth's low carbon dioxide and high oxygen and nitrogen atmosphere is completely opposite from the conditions on those planets. Photosynthesis, which evolved soon after the first appearance of life on earth, removes carbon dioxide from and adds oxygen to the atmosphere. The accumulation of fossil fuels is evidence that photosynthetic activity in the geologic past often exceeded the reverse gaseous exchange of respiration. Therefore, it is logical to conclude that the biotic community is responsible for the buildup of oxygen and the reduction in carbon dioxide over time.

Recent models of the evolution of the earth's atmosphere have indeed suggested that atmospheric oxygen levels rose naturally, but not immediately, as a consequence of photosynthesis and the evolution of life as it now exists on earth (Kasting 1993).

Critics of the Gaia hypothesis argue that for the atmosphere and life on earth to be a cybernetic (feedback) system, a control system and a set point must exist, neither of which is evident in the global cycles of gases. Without a control, they argue, there can be no feedback system; that is, coevolution could not have occurred. However, Odum (1989) has pointed out that natural systems do not have set points; they are not absolutely homeostatic. Control at the biosphere level is not accomplished by external, goal-oriented thermostats or other mechanical feedback devices. Rather, control is internal and diffuse, involving many feedback loops. (For examples, see the section on feedback systems in Chapter 6.) In fact, natural systems are constantly changing.

It is the constant change that allows the system to survive, always in a new and slightly different form. The world, like an ecosystem, is not in homeostasis but in dynamic equilibrium. Lovelock did not likely use the term *homeostasis* as meaning static or stable at a set point. Rather, he may have meant it to mean fluctuating, but within the physical and chemical limits necessary to sustain life.

Lovelock points out that human activities, especially pollution, have the potential to disrupt the system beyond the bounds within which the regulating mechanisms can function. He concludes, "From a Gaian viewpoint, all attempts to rationalize a subjugated biosphere with man in charge are as doomed to failure as the similar concept of benevolent colonialism. They all assume that man is the possessor of this planet. The Gaia hypothesis implies that the stable state of our planet includes man as part of, or partner in, a very democratic entity" (p. 145).

Nature Knows Best

Barry Commoner, in *The Closing Circle* (1971), formulated a law of ecology which held essentially that "nature knows best." It implied that any major synthetic change in a natural system is likely to be detrimental to that system.

The book provoked a tremendous amount of criticism. Some critics interpreted it as advocating a back to nature movement, a kind of "therapeutic nihilism" that is clearly impractical. However, Commoner did not really say this. He seems to have meant that whenever human beings have a need to modify nature, for purposes of agriculture, forest management, and so on, the system can be managed in a variety of ways. The technique that disturbs the system least is often the most desirable. In other words, whenever possible, the individual should try to work with and not against nature. For example, the packaging for foods and milk in supermarkets should be made of biodegradable materials, not undecomposable plastics. In this way, nature will get rid of the trash, and people will not be burdened with practices such as incineration.

THE EVOLUTION OF ETHICS

As noted earlier, three phases have characterized the development of human ethics: the relation between individuals; the relation between the individual and society; and the relation to land and to the animals and plants that grow on it.

Man's Relation to Man

The relation between individuals began in antiquity and was formalized in Western culture by the ten commandments in Exodus 20. Three of these commandments have been formalized into laws in almost all societies:

- Thou shalt not kill.
- Thou shalt not commit adultery.
- Thou shalt not steal.

The Individual's Relation to Society

Until the 1700s, development of an ethic between individual and society developed slowly. Then, technological change prompted the need for a revolution in attitudes about society's responsibility toward the individual. Bronowski (1973), in his *Ascent of Man,* summarized the changes:

> *The idea that science (and technology) is a social enterprise is modern, and it begins at the Industrial Revolution. We are surprised that we cannot trace a social sense further back, because we nurse the illusion that the Industrial Revolution ended a golden age. The Industrial Revolution is a long train of changes starting about 1760. It is not alone: it forms one of a triad of revolutions, of which the other two were the American Revolution that started in 1775, and the French Revolution that started in 1789. It may seem strange to put into the same packet an industrial revolution and two political revolutions. But the fact is that they were all social revolutions. The Industrial Revolution is simply the English way of making those social changes. (p. 259)*

Changes in the social contract between individual and society accelerated in the 1800s; most notably, slavery was abolished, and labor unions were founded. In the early 1900s, women's suffrage was finally granted. The economic crisis of the 1930s gave rise to Franklin D. Roosevelt's "New Deal" and the start of Social Security. In the years after World War II, a plethora of social benefits quickly became accepted as individual rights within a modern society. Examples of society's "duties" toward its people include health insurance and benefits, pensions, unemployment compensation, and food stamps. The Civil Rights Act, which ensured the blacks' equal access to public services, marks a recent advance in the contract between society and its citizens.

Every change in the relationship between society and the individual has been met with resistance from those who stood to lose as a result of the change. Thus factory owners often opposed labor laws because they decreased profit. Social Security was opposed because it meant increased taxes. Nevertheless, most of the changes resulting from the increased responsibility of society for its citizens have been for the better of society as a whole.

The Individual's Relation to Nature

The same forces that have prompted the need for a social ethic have given rise to the need for an environmental ethic. The forces are those of science and technology which multiply human power over nature without any moral or ethical consideration of the effect of that power on nature. There is need for an ethic and for laws to enforce that ethic regarding people's treatment of the environment, in the same way as there was a need for social ethics and laws.

In "The Land Ethic" (1949), Leopold pointed out that the range of recipients of our moral attention had grown slowly but steadily, from members of other tribes or language groups, to prisoners of war, to people with different colored skins, and even to women. Nothing but habit, Leopold implied, would prevent us from making the evolutionary move toward incorporating in our ethics the land and animals and plants that live on it: "The extension of ethics to this third element in human environments is, if I read the evidence correctly, an evolutionary possibility and ecological necessity" (p. 203).

Leopold urged a new *standard* for ethics. In judging the very meaning of "right" and "wrong," he said, we should put the living land at the center: "A thing is right when it tends to preserve the integrity, stability, and beauty of the biotic community. It is wrong when it tends otherwise" (pp. 224–5).

In this regard, Leopold offers a holistic, biocentric ethic, in contrast to the mainly atomistic, anthropocentric ethics familiar in all the Western traditions. It is extremely important to develop an alternative viewpoint because exclusive attention to what seems to be good for *Homo sapiens* in the short term has proven ruinous and promises to inflict even worse environmental damage in the future. To lead us out of this anthropocentric morass, it might seem that *a land ethic* of holistic biocentrism could be an important guide.

The Land Ethic Versus the Social Ethic

Unfortunately, the land ethic cannot serve as a guide to modern people. It leaves unresolved the fundamental conflict between social ethics and environmental ethics posed by Hardin and other proponents of "lifeboat ethics."

> *One of the earth's great problems both today and as far as we can see into the future, is human overpopulation. However we may decide to define Leopold's "integrity, stability, and beauty of the biotic community," it almost certainly would be enhanced by many fewer people burdening the land. Therefore, anything we could do to exterminate excess people—especially where they are congregated in large, unsanitary, destabilizing slums—would be morally "right"! To refrain from such extermination would be "wrong"! "Culling" individuals, if held short of extinction, is a good thing biologically, as long as the species is plentiful; and the human species is obviously too plentiful and getting more so. We have here what could be used as a justification for mass murder, in particular to support policies of deliberate extermination by the wealthy few in the global North against the teeming global South. Is this an ethic, or a potential excuse for ruthless genocide?*
>
> *Taken as a guide for human culture, the Land Ethic—despite the best intentions of its supporters—would lead us toward classical fascism, the submergence of the individual person in the glorification of the collectivity, race, tribe, or nation. Leopold's view could easily become an excuse for radical misanthropy. (Ferré 1993, pp. 444–445; reprinted by permission of Blackwell Publishers).*

Rolston's *Environmental Ethics* (1988) emphasizes the contrast between the land ethic and the social ethic. Although he is aware that the tenderheartedness we cultivate for dealings among human beings is unsupported and unsupportable in nature, he is also aware that the predacious standards of biotic health in nature are morally outrageous when imported into human culture. He states the contrast very clearly:

> *Nature proceeds with a recklessness that is indifferent to life; this results in senseless cruelty and is repugnant to our moral sensitivities. Life is wrested from her creatures by continual struggle, usually soon lost; those few who survive to maturity only face eventual collapse in disease and death. With what indifference nature casts forth her creatures to slaughter! Everything is condemned to live by attacking or competing with other life. There is no altruistic consideration of others, no justice. (p. 39)*

Despite our tender human sympathies for an innocent fawn, Rolston says, we must accept that a hungry cougar will make a meal of it, if it can; and even if we have a chance to intervene to save the fawn, we should not. "There is no human duty to eradicate the sufferings of creation" (p. 56).

If, however, the cougar were to threaten to eat a lost child, we certainly would be morally compelled to act to save the child. However, on the basis of sheerly biological principles, there would be no difference whether a hungry predator were to eat a wandering fawn or a lost child. We should *not,* on the basis of the land ethic, save the fawn; but our ethical intuitions strongly urge us that we *should* save the child.

The decision on whether or not to act is more difficult when the prey eaten by the cougar is neither a fawn nor a child but a sheep. A sheep farmer would naturally act to stop the predation. But what if the cougar were extremely endangered and down to the last dozen or so as is the Eastern panther in Florida. Is one of the last dozen panthers in the wild worth less than one (lousy) sheep?

ENVIRONMENTAL ETHICS AND ECONOMIC DEVELOPMENT

Conflicts between ranchers who lose stock and government programs to preserve or reintroduce predators have been defused by arrangements whereby ranchers are compensated for loss by predators. Other conflicts are not as easily resolved. Examples are as follows.

Cutting down a woodland to build a medical center.
Destroying a freshwater ecosystem in establishing a resort by the shore of a lake.
Replacing a stretch of cactus desert with a suburban housing development.
Filling and dredging a tidal wetland to construct a marina and yacht club.
Bulldozing a meadow full of wildflowers to make way for a shopping mall.
Removing the side of a mountain in a stripmining operation.
Plowing up a prairie to plant fields of wheat and corn.

Ranchers, businesspeople, economists, and politicians can feel threatened by the implications of a land ethic that is antithetical to capitalism, at least as capitalism has existed during the Industrial Age, when business empires were built on exploitation. An environmental ethic, which states that pollution and population should be controlled and that resources should be nurtured, not exploited, conflicts with the traditional model of economic growth, which depends on ever-increasing industrial production, exploitation of resources and labor, and of population to consume the material goods of industry (Larkin 1981).

Arguments that nature should be preserved because it has an inherent value are sometimes difficult to make within a culture that glorifies a frontier ethic and holds sacred laissez-faire capitalism. Businesspeople who see environmental regulations as interfering with their assumed right to exploit nature ridicule such statements as coming from "butterfly collectors," "nature lovers," or "hippies." Even the argument that humans ultimately depend on the integrity of nature meets resistance: Humans have always come up with the technical means to conquer or outwit nature; they can do it again.

Thus, given that there are often fundamental and irreconcilable conflicts between a land ethic and a social ethic, how are we to proceed? How can the fundamental conflict be resolved, asks Golley (1993a), when others view the environment in a fundamentally different way than we do? "How does one find a compromise with people who do not divide the world into property and who do not convert nature into a resource that can be exploited to extinction on a theory of economic convertibility?"

One hundred years ago, Karl Marx recognized the exploitative nature of capitalism with regard to labor, and developed his theories of communism based on a system that theoretically did not exploit people. The fact that communism has failed does not negate the fact that in its early days capitalism was extremely exploitative of human labor and was rightly condemned for being so. In the twentieth century, the labor union movement and other "socialistic ideas" such as Social Security and Medicare have gone a long way toward counteracting the social abuses of the capitalistic system.

In the same way that communism has "threatened" capitalism, environmentalism is seen by some as the most recent threat to capitalism. Environmental regulations such as the Endangered Species Act are perceived as threatening the "right" of loggers to clear-cut national forests. Federal air pollution emission standards are seen as "interfering" with the free enterprise system. But just as socialistic ideas gradually were accepted and improved the status of the workers within capitalistic society, so environmental ideas can be expected to gradually become more widely accepted.

WHAT TO DO WHILE WAITING FOR AN ENVIRONMENTAL ETHIC

There may be a wait before a widely accepted ethic and a new set of values become available to help resolve environmental policy and management problems. If we wait until such time, the damage may be irreparable. Instead of endless debate, Carpenter (1987) suggests that we set aside the arguments and move ahead with immediate action. He titled his essay, "What to Do While Waiting for an Environmental Ethic." His suggestions are either to ignore the split and move ahead, or reformulate the split in terms of conventional politics.

Ignore the Split and Move Ahead

"The environmental movement is a notably pragmatic affair" says Hays (1993). "While philosophers have often found environmental matters to be grist for their mill, they have remained apart from the main scene of action and often appear to be somewhat irrelevant to it" (p. 1822).

For his part, Norton (1991) believes that the issues of controversy within the environmental movement are real but that at the same time they are artificial because the dominant course of environmental action is practical, pragmatic, and focused on solving problems. After reviewing the traditionally defined tension between the biocentric view as epitomized by John Muir and the use-oriented tradition of conservation as exemplified by Pinchot, he argues that environmentalists often waste much time and energy and foster unnecessary divisions by continuing to fight that battle. It is far more important, he points out, that environmentalists focus heavily on practical situations in which those who are philosophically at odds work out solutions to problems on which they fundamentally agree. To illustrate the point, Norton cites examples in which the two sides have joined to confront issues such as the pressures of growth, pollution control, biodiversity, and land-use policy.

Even though environmentalists have no generally agreed-on ethic, the general public increasingly recognizes that there are environmental problems that must be remedied. For example, in a public survey carried out in Puerto Rico in 1992 to determine awareness of environmental problems, 75 percent of the 800 persons interviewed agreed that urban development, lack of education about natural resources, and lack of reforestation have contributed to the deterioration of these resources. The specific problems included water resources, air quality, green belts, garbage, urban development, energy resources, coastal zones, population growth, toxic wastes, and preservation of historic sites (*El Nuevo Dia,* San Juan, Puerto Rico, December 26, 1992, p. 8). Although loss of biological diversity was not identified in the poll as a discrete category, the inclusion of problems in green belts and coastal zones suggests that the public is aware of threats to biological resources. Lack of a formal ethic does not preclude the public from pressuring their government to take action.

Reformulate the Split in Terms of Conventional Politics

The humankind versus nature debate is virtually moot with regard to the solution of environmental problems. Conservationists and politicians of all stripes are beginning to realize that people cannot exist without nature and that nature cannot exist without people. To argue which is more important is pointless. Solutions to environmental problems must rise above mere "environmentalism" and encompass all of the social and economic issues with which the environmental problems are intertwined. We cannot solve environmental problems in isolation. Solutions to environmental problems must be dealt with as part of an overall package of solutions for the world's social and economic, as well as environmental, problems.

Reformulating conservation issues in terms of practical politics is important. "Man centered," the philosophy of conservationists like Gifford Pinchot, becomes "individual responsibility" in the rhetoric of right-wing politics. The individual takes responsibility and the individual receives the full consequences of his or her action, whether the consequences are good or bad. This sounds good to political conservatives.

"Nature centered," the philosophy of preservation-oriented conservationists is, in the rhetoric of left-wing politics, "action for the common good." Only through government acting on behalf of the global or local commons, can environmental problems—which are basically problems of the commons—be solved. This sounds good to liberals.

CONCLUSION

Conservation is a guide to action designed to ensure that *the present generation cares for the environment in such a way that future generations can enjoy a land with undiminished productive capacity, and a landscape with undiminished beauty and undiminished diversity of structure and species.*

If such action constitutes ethical behavior, then conservation *is* a matter of ethics.

SUGGESTED READINGS

For a western view of man as part of nature:

Callicott, J. B. 1987. The Land Aesthetic. Pp 157–171 **in** *J. B. Callicott, ed. Companion to a Sand County Almanac.* University of Wisconsin Press, Madison.

Leopold, A. 1949. The Land Ethic. Pp. 201–226 **in** *A Sand County Almanac, and Sketches Here and There.* Oxford University Press, New York (reprinted in 1987).

For a critique of the land ethic perspective:

Rolston, H., III, 1988. Environmental Ethics: Duties to and Values in the Natural World. Temple University Press, Philadelphia.

For a modern, Western view of the world, including humankind, as an integrated, cybernetic system:

Lovelock, J. E. 1979. Gaia: A New Look at Life on Earth. Oxford University Press, Oxford.

For an overview of Asian thought regarding the individual as part of nature:

Callicott, J. B., and R. T. Ames. 1989. Nature in Asian Traditions of Thought: Essays in Environmental Philosophy. State University of New York Press, Albany.

For a presentation of the view that environmental ethics is inconsistent with Western philosophy and Western traditions:

Passmore, J. 1974. Man's Responsibility for Nature: Ecological Problems and Western Traditions. Duckworth, London.

For a rebuttal to Passmore:

Hargrove, E. C. 1989. Foundations of Environmental Ethics. Prentice Hall, Englewood Cliffs, N.J.

For the view that the Man above Nature/Man in Nature controversy is debilitating:

Norton, B. G. 1991. Toward Unity Among Environmentalists. Oxford University Press, New York.

CHAPTER

ECOLOGICAL ECONOMICS

CHAPTER CONTENTS

CHAPTER OVERVIEW

Resources are not managed sustainably and the environment is becoming polluted because the economic system does not account for the total value and utility of most resources, or for much of the value and utility of a clean environment. The total utility value of resources and of clean surroundings is not reflected in the market price. As a result, they are not used efficiently. The situation is particularly critical in less developed countries where whole resource segments such as forests and fisheries are being sold off at prices much lower than replacement costs, and as a consequence are being destroyed and lost forever.

Only when the utilitarian value of a resource is reflected in the economy is the resource used in a sustainable manner. Only when the value of a clean environment is incorporated in the economy is pollution abated.

PRINCIPLE

The market system of a frontier economy puts little or no value on the services of nature as a supplier of natural resources and as a sink for pollution. This leads to inefficient use of resources and a degradation in nature's capability to provide services. As an economy matures and sustainability becomes a social goal, the total utility value of resources and services of nature must be incorporated into the economic system.

THE MARKET MODEL AND VALUATION

In the sense that the practice of bartering and trade is very old, the marketplace is very old. Barter was based on the value of the goods traded, but value was simply a metaphysical concept that had a very ambiguous but emotive meaning (Robinson 1962). Only very recently in the history of civilization has there emerged a model that attempts to explain how the marketplace works and to define value in operational terms. Development of the model began in 1776 with Adam Smith's *Wealth of Nations.* Over 100 years of effort by economists, sociologists, philosophers, mathematicians, and political scientists resulted in a market model that essentially culminated with Alfred Marshall's book *Elements of Economics* (1892). In this work, the model is developed through graphical analysis of demand and supply curves and differential calculus in order to handle marginal changes and equilibrium. In equilibrium, under pure competition, and with perfect knowledge, the utility-maximizing economic man would equate the price he paid for the last unit of a commodity purchased with the utility he assigned to that commodity. Under these conditions market price equals value.

A question that still remained, involved the mechanism by which the market coordinated everything and finally came to a specific price and a determinant allocation. Smith called this mechanism the *invisible hand* (1776), whereas Marshall termed it *equilibrium* (1892).

The market model, including its language, vocabulary, and concepts, is neoclassical economics. In the Western world, the model is universally taught in schools and is the internal language of commerce and industry. It is the language used by the media in referring to business, and it is adopted by governments whenever questions of value, allocation, or distribution are considered.

The neoclassical market model is generally considered to be complete, robust, and transmittable. It is complete in that the vocabulary and set of concepts are internally consistent, and the assumptions, logical propositions, and conclusions are free of contradiction, paradox, and ambiguity. The model is robust because it is applicable to most marketplace phenomena. Finally, the model is transmittable because it can be communicated with relative ease (Farnworth et al 1981).

The Marketplace

The marketplace declares price, allocates resources, and distributes products. It also processes information about preferences, resources, and technology. According to the model, the marketplace represents an auction between buyers and sellers that leads to an efficient use of society's resources. The model concludes that the price of an item will accurately represent how society values that item and that the distribution of products to consumers will mirror how society values the contribution of each individual to the common production.

The fundamental event in the real-world marketplace is the freely engaged, two-person trade. It is freely engaged in that both parties are volunteers and free from coercion or lack of alternatives. Trade occurs if, and only if, two people or groups feel better off as a result of the transaction. The buyer must be willing and able to buy, and the seller willing and able to sell, and the two must come to an agreement. If essential resources for a commodity suddenly become scarce or expensive, the commodity will become less available than formerly and prices will rise, whereupon consumers will be likely to substitute alternatives. If a commodity is viewed as less desirable than formerly, that is, if it is judged to have reduced utility, consumers will buy less, profits from producing it will decline, and eventually resources engaged in its production will be released into other occupations.

In this sense, the marketplace is an information-processing system. Inputs are values (individual preferences, tastes, or utilities), resource availabilities, costs, and the technological means to transform resources into products. Outputs of the system are market prices (value), allocation of resources (land, labor, capital, and entrepreneurship) into the production of a particular set of commodities, and the distribution of those commodities among the resource suppliers and others. The market processes information about agreements to declare values and allocate resources.

The marketplace may or may not approximate the assumptions of the market model. If conditions closely match the assumptions of the market model, then the results of market activity will be more or less efficient and value will be correctly determined. But the assumptions of the market model cannot be met for many goods and services because the market model requires perfect knowledge by all parties, perfect competition among producers, unrelenting maximization by all decision makers, nonsatiation of the market, and perfect separability of all utility and production functions. The market model can thus be only an approximation of reality, and the results of marketplace activity can only approximate perfect economic efficiency. Market price can only approximate society's true valuation, and market allocation of resources can only approximate the efficient quantity of many goods.

Market and Value

Adam Smith (1776) made profound intuitive distinctions between value and price, and about the problem of maintaining equilibrium. His first insight concerned the distinction between what he termed *value in use* and *value in exchange.* The word "value," he observed, has two different meanings, sometimes expressing the utility of a particular object and sometimes the power of purchasing other goods that the possession of that object conveys. The one may be called value in use and the other value in exchange. The things that have the greatest value in use frequently have little or no value in exchange. Few things are more useful than water, but it will not purchase very much. Scarcely anything can be had in exchange for it. In contrast, those things that have the greatest value in exchange have frequently little or no value in use. A diamond has hardly any value in use except for decoration, but a very great quantity of other goods may be had in exchange for it. (Today, diamonds have industrial uses, but the quality useful for industry is different from the quality imparting value in a jewelry store.)

Clearly, value in exchange is market price, whether in terms of a standard commodity such as gold or money or in terms of all other commodities in a barter system. Value in exchange clarifies how producers are capable of determining the price which, if received, will induce them to sell an item or service. If the value in exchange of an item is greater than the cost of producing or providing one more of those items (the marginal cost), this will provide the incentive for the supplier or the manufacturer to produce that additional item.

Value in use reflects the total utility of something for satisfying physical or psychological needs or desires. Value in use of something is unrelated to its marginal cost. At the margin, diamonds are worth much and water is worth little. But for total utility, diamonds are worth little and water is worth much. Prices measure marginal utility. Welfare is determined by total utility.

Table 4.1 *Determination of Value Through the Interaction of Exclusive and Nonexclusive Uses, and Exhaustible and Inexhaustible Goods*

PROPERTY RIGHTS ASSOCIATED WITH USE	NATURE OF THE GOOD	
	Exhaustible	*Inexhaustible*
Exclusive	This traditional market good is provided by the private sector at efficient prices and quantities if a perfectly competitive market exists. *Value is determined in private markets.*(e.g., a loaf of bread)	Good may be provided by market or by the public sector depending on the specific case. If it is the public sector, a "second best" solution of efficient pricing must be used. If provided by private markets, then pricing is the revenue device but not the rationing device. *Value must be imputed and depends on the specific good.* (e.g., timber in a national forest[a])
Nonexclusive	Institutional framework does not recognize ownership or the technology to establish ownership, and control does not exist. *Value must be imputed.* (e.g., public water supply)	Market provision is not possible. The good must be provided by the public sector. *Value must be imputed.* (e.g., clean air)[a]

[a] *Timber in national forests and clean air have traditionally been presumed to be "renewable" natural resources, and therefore "inexhaustible." It is now clear that timber, if intensively and carelessly logged, is nonrenewable and therefore exhaustible. Clean air is a sink for pollutants, and if a meteorological inversion hangs over a city, clean air is exhaustible.*

Source: *Farnworth et al. 1983, Table 1, p. 13. Reprinted by permission of Gordon and Breach, copyright © 1983.*

Private Goods versus Public Goods

Although the marketplace works well for privately owned goods, it is inefficient for public goods. Private goods are characterized as being exclusive and exhaustible. *Exclusiveness* is an attribute of property rights associated with use. By paying a market price, the purchaser may enjoy exclusive use of that good or resource and may exclude others from its use if he or she chooses. *Exhaustible* means that the total quantity is finite and may be totally used or consumed within a time period. For example, anyone who purchases a loaf of bread has the exclusive right to benefit from the bread, and once the day's supply is sold, there is no more. In such a case, competitive markets can establish value and allocate resources efficiently.

Public goods and services do not have the attributes of exclusiveness and exhaustibility. Examples of public goods and services include national defense, aesthetic environments, the national highway system, and benefits from clean air and water. The benefits an individual derives from public goods do not exclude another person from benefiting, nor do they exhaust the supply. Valuation of public goods and services is therefore the summation of the willingness of all individuals to pay for a specific quantity of the public good. Once the public good or service is available, additional users may be added at zero marginal cost (up to the point of congestion). Unfortunately, the zero marginal cost (the cost of each additional user) can lead to the problem of the *commons*—the condition of too many users taking advantage of the zero marginal cost, resulting in the degrading of public goods and services. Public parks are an example. In some cases, free access has led to overcrowding and deterioration of facilities.

Ownership rights associated with use of a good or service, as well as the exhaustible or inexhaustible nature of the supply of a good or service, determines whether the market system transmits the correct price, or whether the market system fails and values must be imputed. Table 4.1 illustrates how the interactions between exclusive and nonexclusive property rights, and exhaustibility and inexhaustibility determine value. Value is determined by private markets only for the exclusive-exhaustible case; all other cases require an imputed nonmarket value for efficient resource utilization. When the private markets do not provide correct prices, private decision makers who are motivated by self-interest will inefficiently utilize resources. Without correctly imputed values for public goods, governmental decision makers also cannot allocate resources efficiently.

MARKET AND ACCOUNTING FAILURES

Since Marshall (1892) brought together the necessary elements of the market model, other scholars such as Pigou (1920) and Wicksell (1935) have pointed out its limitations.

Monopolies

Under ideal market-model conditions of competition, self-interest is channeled to create results in the best public interest. In such instances, market values accurately reflect society's values. However, under monopoly and its variations such as cartels and oligopolies, self-interest does not lead to the greatest public good, and so the market fails.

When a monopoly forms, collective action by government is necessary to protect the people. Government may enact antitrust laws that constrain self-interest by keeping businesses small and competitive, or it can charter a monopolist and regulate pricing by providing guidelines. For instance, the price of electricity is determined not by the marketplace but politically by public utilities' commissions that weigh the costs of production and the needs of the stockholders and consumers.

Monopolies can occur in the ownership of natural resources. The Organization of Petroleum Exporting Countries (OPEC) is an example of a resource monopoly. Because monopolies of natural resources, like all monopolies, result in prices that are higher than would occur under a perfectly free market, they can be considered antithetical to optimum functioning of the global economy.

When monopolies or cartels agree to reduce the rate of resource extraction or production, buyers are forced to pay a higher price. This can affect the environment in several ways. The so-called global oil crisis in 1973 caused by an agreed-upon limitation in the oil supply by OPEC members resulted in sharply higher petroleum prices. The higher prices benefited the environment by encouraging conservation within consuming countries. The high prices harmed the environment by stimulating exploration for new oil in non-OPEC countries, including environmentally damaging drilling offshore and in rain forests.

Table 4.2 *Total Estimated Environmental and Social Costs from Pesticides in the United States*

IMPACT	COST ($ MILLION/YEAR)
Public health impacts	787
Domestic animal deaths	30
Loss of natural enemies	520
Cost of pesticide resistance	1400
Honeybee and pollination losses	320
Crop losses	942
Fishery losses	24
Bird losses	2100
Groundwater contamination	1800
Government regulations to prevent damage	200
Total	8123

Source: *Pimentel et al. 1992, Table 4, p. 757. Reprinted by permission of the American Institute of Biological Sciences, copyright © 1992.*

Externalities

Externalities are market imperfections that occur whenever impacts from the production or resource extraction process are not compensated for through markets. In classical economics, whenever the production activities of a seller or the consumption activities of a buyer have impacts on third parties, an externality occurs. Although externalities can be positive, often they are negative and convey a cost that is not recognized in the market transaction.

Table 4.3 *Estimated Annual Economic Costs of Human Pesticide Poisoning and Other Pesticide-Related Illnesses in the United States*

EFFECTS	COST ($ MILLION/YEAR)
Hospitalization after:	
poisonings: 2380 × 2.84 days @ $1000/day	6.759
Outpatient treatment after:	
poisonings: 27,000 × $630	17.010
Lost work due to poisonings:	
4680 workers × 4.7 days × $80/day	1.760
Treatment of pesticide-induced cancers:	
< 10,000 cases × $70,700/case	707.000
Fatalities:	
27 accidental fatalities × $2 million	54.000
Total	786.529

Source: *Pimentel et al. 1992, Table 1, p. 752. Reprinted by permission of the American Institute of Biological Sciences, copyright © 1992.*

Pigou (1920) gave the classic example of an externality. A steel factory produces and sells steel, and both the buyers and sellers of steel benefit from the transaction. But the steel factory sends smoke into the local city. Because of the mill, the residents of this city may have significantly higher laundry and health-care costs and a lower standard of living than the people in a cleaner, neighboring city. These diffuse costs are not included in the price of the product. The smoke was considered external to the economic costs of steel production. If the mill were forced to clean up its emissions, its costs would be higher and its products less competitive. Such a step, manufacturers believed, would result in the mill closing and loss of jobs and income. However, if *all* polluting industries are required to internalize the costs of pollution, the competitive disadvantage is eliminated.

Use of pesticides is another economic activity that produces undesirable externalities. The environmental and social costs of pesticide use in the United States have been estimated to be approximately $8 billion each year (Table 4.2). Users of pesticides in agriculture pay directly for only approximately $3 billion of these costs, which includes problems arising from pesticide resistance and destruction of natural enemies. Society indirectly pays this $3 billion through higher costs of produce. Society also pays the remaining $5 billion, $787 million of which covers public health costs (Table 4.3).

In tropical forests of developing countries, the costs of renewing natural resources such as forests, wildlife, and fisheries are usually considered an externality. Aside from the cost of a "concession" from the government, the only costs that a logging firm considers are those of extraction such as chainsaws, trucks, and loggers. Sometimes the government or an international development agency even pays part of the extraction costs such as roads. It is usually assumed that there is an inexhaustible supply of trees and wildlife and that nature will replace itself. This assumption is no longer valid. Many species of plants and animals in the tropics are disappearing because they have been harvested beyond their capacity to regenerate.

Table 4.4 *Resource Classification Scheme That Takes into Account the Degradability of the Resource Through Use, and the Substitutability of the Resource.*[a]

Resource Type	Substitutability	Not Necessarily Degraded or Dispersed in Use	Necessarily Degraded or Dispersed in Use
Nonrenewable (at current use rates)	Essential	Stratospheric ozone, tropical forests, biodiversity	Time or opportunity
	Substitutable	Diamonds and gold for repository	Fossil fuels, some minerals
Renewable	Essential	Decomposers, pollinators	Solar energy, fresh water, some agricultural soils
	Substitutable	Animals for power, vaccines; trees for cooling	Wood, food products

[a] *The examples are representative but are not exhaustive.*

Source: *Daily and Ehrlich 1992, Table 1, p. 764. Reprinted by permission of the American Institute of Biological Sciences, copyright © 1992.*

Because of the danger associated with the failure to recognize the vulnerability of resources such as tropical forests, it is important to reconsider their classification as a "renewable" resource. Table 4.4 presents a recent resource classification scheme that recognizes the nonrenewability of some resources that were formerly considered renewable.

The Concept of Capital

Capital or wealth in classical economics is the stock of material objects owned by human beings at an instant of time. It includes the inventory of all consumer goods as well as producer goods. Factories, aluminum ingots, shoes, automobiles and Barbie dolls are capital.

Factories and apartment houses are examples of capital that produce income for their owners. Factories and apartment houses eventually get old and must either be repaired or replaced. Because of the costs of repair that result from depreciation of capital, a certain amount of money can be deducted annually from the income tax owing to "capital depreciation." The money saved can be used to maintain the capital.

Nonrenewable resources such as coal, iron, and oil are used up once they are taken out of the ground. For tax purposes, in many economies they are treated as though they were capital, in the sense that governments allow owners of oil wells a depletion allowance. Just as an old factory has accumulated more allowance for depreciation than a new factory, an old oil well will have been allowed a greater depletion allowance than a new well. Governments have recognized that both manufactured capital and nonrenewable resources eventually will disappear, and therefore some allowance should be made to compensate the owners, to encourage them to build new factories, or to explore for new oil.

Natural resources such as forests and fisheries have traditionally been considered renewable in the sense that harvesting them will not destroy them. The widespread notion has been that they will replace themselves. This is true as long as the amount harvested is small in relation to the total stocks. However, recent increases in population and resource-harvesting rates are resulting in the overharvesting of many forests, fisheries, and other "renewable" resources, with the result that they no longer are renewable. The reasons why this has happened can vary. There may be insufficient trees or fish for a viable population. Or the environment may be polluted or eroded to the point where biota cannot survive.

The failure of modern economics to count such resources as forests and fisheries as capital has resulted in their depletion, with no formal recognition of any economic loss. A country can become impoverished through loss of its resources, yet officially it is no worse off.

Capital depreciation allowances were instituted long ago to help make industry more sustainable. The same sort of recognition must be afforded resources if they are to be sustainable. Forests and fisheries are just as crucial to material well-being as are factories, and they should be recognized as capital that will be depleted if they are not cared for.

The Problem of the Commons

The lack of property rights often drives many commonly owned natural resources to exhaustion or to extinction. For example, hunters for passenger pigeons in the American East and bison in the American west drove the pigeon to extinction and the bison to near extinction. Because no one "owned" the animals, they were free goods, and because each hunter assumed that if he did not shoot the animals, someone else would, the species were overhunted.

Although the exploiters of a common resource stock should have an incentive to conserve that resource, rarely do they practice sustainable management. Hardin's (1968) "tragedy of the commons" described in Chapter 2 illustrates the problem for publicly owned lands treated as a nonexclusive resource. Economist H. S. Gordon (1954) made the same argument for fisheries: As long as fish can be caught profitably, he said, fishermen will do so, and this activity will lead to severe overfishing. The more valuable the fish and the more technologically efficient the fishing vessels, the more severe will be the overfishing. Gordon predicted that only when the fish stock had been so reduced that revenues from fishing barely covered operating costs would an equilibrium be reached. The collapse of the North Atlantic cod, haddock, and flounder fisheries (Holmes, 1994) has proven him correct.

Biodiversity is a global commons (Chapter 2) that suffers from the world's most irredeemable threat—extinction. Many biologists consider its loss to be the greatest threat to the sustainability of resource production systems, because it is the source of new genetic material for foods and pharmaceuticals. The threat to the biodiversity commons comes from humans' increased ability to alter the environment.

The problem is especially serious in the tropics. Before the advent of large-scale clear-cutting with heavy machinery, ecosystem regeneration was not threatened. Clearings were small enough in relation to remaining forest that the insect pollinators, bird and mammal seed-dispersers, and fungal nutrient recyclers could find nearby refuge and recolonize once the disturbance ceased.

In contrast, the scale and intensity of modern deforestation is eliminating seed sources and forcing genetic stock to extinction. For example, much of the eastern portion of the Amazon basin has been deforested and will probably never regain its original complement of species because the areas cleared have been too large. When the only remaining seed trees are hundreds of miles away, a forest cannot regenerate. When there are no birds and animals to disperse the seeds, even remaining trees are ineffective. And even when by chance a seed reaches a cleared area, its chances of germinating and surviving are very slim, because the microclimate is so different from what it requires and the predators such as ants are in great abundance (Nepstad et al. 1990).

The Problem of Scale

The free market system does one thing very well. It solves the problem of allocating goods and services by providing necessary information and incentive. What it does not do is solve the problem of optimal scale (Daly 1991b).

A dominant paradigm in conventional economics is defining the ideal rate of economic growth. The rate of growth of consumption approaches the ideal level when full employment is nearly reached, profits are being maximized by resource holders, and the rate of technical progress is steady. The economy is considered capable of growth into the indefinite future, through technological innovations. However, the free market model does not count the environmental and social costs that result from unrestrained growth (Hannon 1992). Conventional economics does not address these problems of scale.

The term *scale* is shorthand for the physical scale or size of the human presence in the ecosystem, as measured by population times per capita resource use. Daly (1984) illustrated the point by making an analogy with loading a ship. The objective is to maximize the load carried by the ship. If all the weight is placed in one corner, the boat will quickly turn over and sink. The weight has to be spread out evenly, and to do this, we invent a price system. If one corner of the ship becomes lower in the water than other corners, the price for putting another kilogram of cargo in this corner increases. This in an internal optimizing rule for allocating space among alternative uses. It is an allocative mechanism only. It does not consider how much weight the ship can carry before sinking. It is like a very useful but unintelligent computer that keeps on allocating space until the optimally loaded ship sinks to the bottom of the ocean.

Simply saying that the end purpose of the economy is to create utility and to organize the economy accordingly is to ignore the fact that, ultimately, a closed system such as the earth sets limits or boundaries as to what can be done by way of achieving that utility (Pearce and Turner 1990). With the age of colonialism and the frontier coming to an end, economics must learn to deal with the problem of scale.

That the environment has a finite limit to support a population has long been recognized by biologists. The limit is called *carrying capacity,* and it has been defined as the maximum number of individuals of a particular species that can be supported by a given area of habitat over a given period of time (Miller 1990).

The earth's carrying capacity for humans depends on the assumptions used in the calculation. Humans, with the aid of technology, are capable of extracting more resources from the earth than other species, and therefore assumptions must be made about the rate of technological innovation. Other assumptions concern the social system, which can range from minimum subsistence to luxurious consumption. Assumptions about how long energy resources can last, and what level of pollution and crowding is acceptable also enter the equation (Daily and Ehrlich 1992). Although the carrying capacity of the earth for human beings will vary according to the kind of existence desired, *Homo sapiens* can claim no exemption from limitations of earth's finite resources.

The Discount Rate

A common criterion used by investors when they are deciding where to put their money is the interest rate. If a bank offered a higher interest rate on a deposit than did a government bond, the investor might choose the bank. When an investor wants to put money into developing a business or a resource, the same rationale is used. However, the return on a business or resource investment works differently than interest return. On a bank deposit, interest is paid at regular periodic intervals, often monthly or quarterly. On an investment in a business, returns do not begin until the business begins to make money. On an investment in a resource such as a forest plantation, the investment does not pay off until the forest is harvested.

When deciding whether or not to invest in a timber plantation, an investor will look at something called its net present value (NPV). Net present value expresses the amount of money that a long-term investment would be worth today if that money could be had now instead of in the future. The usual way to calculate the NPV of a potential investment such as a forestry scheme is to devalue future profits by a discount rate. The discount rate is used to calculate the present value of a good, discounted because we must wait a period of time to get it. For example, if we are thinking about investing in a tree plantation, how much must that plantation be worth in the future to entice us to invest today? If we consider a discount rate of 10 percent ($i = 0.10$), this means that if we think it will be worth \$1000 next year, that would be equivalent to \$909.09 today [$(909.09)\ (1 + i) = \$1000$] (Clark 1991). Discounted annually, \$1000 in 10 years time is equivalent to 385.54 today, and \$1000 in 100 years time is equivalent to 7 cents today.

Saying that $1000 in proceeds from the sale of timber 10 years from now is equivalent to $385.54 today is like saying that if we invest $385.54 today at a 10 percent interest rate (compounded annually), in 10 years, we should get back $1000.

A low rate of return is often the reason for the lack of long-term investment necessary to regenerate many natural resources. Low return is due, in part, to the long time necessary for production of resources such as forests, coupled with the exponential effect of a compounded discount or interest rate.

Another reason for lack of investment in resources such as tree plantations or sustainably managed natural forests is that there is competition in the marketplace from timber taken from locations where no investment has been made, such as forests in the Amazon or Alaska. As this "virgin" timber disappears, prices will increase, and returns on forest plantations will increase. If governments and development agencies could be persuaded to stop subsidizing roads into undeveloped areas, this would have the effect of taking virgin forests out of the market. Then the price of wood would rise faster, and it would become more profitable to invest in plantations or to manage forests sustainably.

Because a high discount rate has a similar effect as a high interest rate, development projects are often chosen or managed to maximize the discount rate. In resource development projects, a high discount rate is bad for the environment because it speeds the rate of depletion of nonrenewable resources and shortens the turnover and fallow periods in the exploitation of renewables. It shifts the allocation of capital and labor toward projects that exploit natural resources more intensively (Daly 1991b). High discount rates often are applied in resource projects to make the investment appear more economically competitive. The sole use of a high discount rate as a criterion for resource development is an accounting method that encourages depletion of the natural capital and thereby foregoes sustainability.

Table 4.5 *Example of Net Present Value (NPV) Calculations Illustrating the Effect of Discount Rate*

	Value at a Future Time			Present Value					
	(DR[a] = 0% Year[a])			(DR = 3% Year[a])			(DR = 10% Year[a])		
Year	Cost	Benefit	Gain	Cost	Benefit	Gain	Cost	Benefit	Gain
Destructive Exploitation									
1	50.00	130.00	80.00	50.00	130.00	80.00	50.00	130.00	80.00
2	0.00	0.00	0.00	0.00	0.00	0.00	0.00	0.00	0.00
3	0.00	0.00	0.00	0.00	0.00	0.00	0.00	0.00	0.00
100	0.00	0.00	0.00	0.00	0.00	0.00	0.00	0.00	0.00
	Total = 80.00			Total (NPV) = 80.00			Total (NPV) = 80.00		
Sustainable Management									
1	10.00	13.00	3.00	10.00	13.00	3.00	10.00	13.00	3.00
2	10.00	13.00	3.00	9.71	12.62	2.91	9.09	11.82	2.73
3	10.00	13.00	3.00	9.43	12.25	2.83	8.26	10.74	2.48
100	10.00	13.00	3.00	0.54	0.70	0.16	0.00	0.00	0.00
	Total = 300.00			Total (NPV) = 97.64			Total (NPV) = 33.03		

[a] *DR, discount rate.*

Source: *Fernside (1989), Table 1, p. 64. Reprinted by permission of Elsevior Science Publishers. Copyright © 1989.*

Low discount rates favor projects that are more environmentally benign. Arguments for using a lower discount rate when judging resource projects such as forest plantations in competition with other investments include: adjusting present value calculations to correct for expected increases in the value of forestry products relative to other commodities; increasing the weight given to future costs such as insect control; using shadow prices (discussed later in this chapter) in the calculations to reflect the social and ecological benefits of forests; and assigning additional weight to irreversible costs such as species extinction (Fearnside 1989). Such devices are seldom considered in economic decisions.

The assumption that high discount rates maximize profits is not necessarily correct for resource management systems. Table 4.5 illustrates how a natural forest, when managed sustainably, can result in a greater net present value with a 3 percent discount rate than with a 10 percent discount rate. If the discount rate is 10 percent, this encourages immediate clear-cutting, which yields a net present value of 80 but with no benefit after the first year. In contrast, sustainable management yields a net present value of only 33. However, if the forest is managed sustainably for 100 years and the discount rate is 3 percent, net present value is 97.64. The key to the difference is that, with sustainable management and a low discount rate, only a few trees are harvested each year or periodically, and the remaining trees lose their value less quickly, or even increase their value because of growth (Fearnside 1989).

The choice of a discount rate indicates the extent to which we feel an obligation to future generations. A positive discount rate has been compared to taxation on future generations without representation. A zero discount rate provides for intergenerational equity in resource conservation decisions. It amounts to weighing the interests of future generations as equal to those of the present. Many conservation-minded economists may actually recommend a negative discount rate. This would weigh the interests of the future more heavily than those of the present. This may be especially important when the commitment of scarce resources has irreversible implications that may deprive future generations of the option to use those resources later.

Nonsubstitutability

Conventional economic analysis assumes three basic factors of economic activity: land, labor, and manufactured capital. These factors are considered partially substitutable. The production of timber, for example, requires inputs of both land and labor. However, it is possible to grow the same amount of timber using more land and less labor, and vice versa. The degree of substitutability among inputs depends on the technology of production, which changes over time.

Natural capital is not included in this scheme because conventional analysis usually assumes that manufactured capital can be substituted for natural capital, or at least that there is partial sustainability. For example, "fiber-board" (glued-together residues of sawmills) can in some cases be substituted for lumber sawn from trees. However, as the services of nature become increasingly stressed, it is clear that in many cases substitutability does not exist. There is no known human-made capital that can substitute for the services of a forest in protecting soil and preventing erosion into a hydroelectric reservoir; for the services of nature as the *ultimate* sink for pollution (although garbage and pollution can be condensed and temporarily stored in synthetic structures); or for the genetic diversity that has given rise to all of our food crops and many of our pharmaceuticals. The failure to recognize the nonsubstitutability of many natural resources is a weakness of conventional economic analysis.

Table 4.6 Some Market and Nonmarket Values of Goods and Services of Tropical Moist Forests

	Nonmarket	
Market	Attributable or Assignable Values	Intangible or Non-assignable Values
Lumber: logs, plywood, veneer Fiber: paper, fodder, clothing and shelter Fuel: firewood, methanol Chemicals: oils, resins, esters, phenols Pharmaceuticals: quinine, nicotine, caffeine, alkaloids Exotic flora and fauna: house plants, pets Research plants and animals	Maintenance of global air quality: removal of particulate and gaseous material Maintenance of tropical water quality: erosion control, flood control, regional water quality Recreation: hunting, tourism, filming, aesthetics Genetic stocks New food plants and animals New chemicals Potential biological control agents New germ plasm to reinvigorate food or fiber stocks (disease control or yield improvement Fixation of solar energy to support production systems	Maintenance of global carbon balance Maintenance of atmospheric stability Habitat for native people Intrinsic worth of species, culture, and ecosystems Natural laboratory for evolution and natural selection Maintenance of diversity: backup systems for ecosystem services and global life support system Cycling of essential nutrients Absorption and breakdown of pollutants

Source: *Farnworth et al. (1983), Table 2, p. 14. Reprinted by permission of Gordon and Breach, copyright © 1983.*

THE PROBLEM OF VALUATING RESOURCES AND THE SERVICES OF NATURE

The market and its present accounting system have failed to adequately serve humanity because the market price for most natural resources, including a pollution-free environment, does not reflect their utilitarian value. Their value in use is not reflected in their value in exchange. The price that is paid for them is not indicative of their value for satisfying physical or psychological needs or human desires. This section deals with the problem of inadequate evaluation of resources and the services of nature.

Market Values

Some resources have a market value. These are values in exchange, the price paid for a natural resource such as a hardwood log in the free market. Ecosystems such as forests yield many products that have a market value. Products of a tropical moist forest that have market values are listed in the first column of Table 4.6.

Although these products have a value in the market, the price reflects only the cost of extracting them. For logs, the costs would include chainsaws and the wages of timber cutters. Usually, the price does not reflect the cost of managing the forest in a sustainable manner or replacing the forest if it is clear-cut. Because the price does not include replacement costs, the forest, once damaged, is not replaced.

Attributable Values

Economists have come to recognize that certain functions of nature, although they have no market value, are necessary for the functioning of the market economy. Certain utilitarian values of nature are so important to the economy that a monetary value must be imputed to these resources and services in order to protect them. The attributable values of tropical moist forests are listed in the center column of Table 4.6.

A well-recognized utilitarian value of nature is the prevention of siltation of reservoirs by intact forests. The value of a forest in protecting a watershed can be attributed by calculating the costs associated with dredging a reservoir where the protective forest has been destroyed. If the cost of dredging the reservoir is greater than the market price for the timber in the forest, it makes economic sense to ensure that the trees are left in place.

Clean air and clean water also can be considered attributable values. The costs of not having clean air are the medical costs associated with people suffering from conditions caused by air pollution. To calculate these costs, the doctor and hospital bills for all problems related to air pollution such as asthma could be tallied, as well as the costs to the employer to cover the employee's sick time. Like watershed protection, clean air and water are public goods, and their value must be attributed outside the marketplace.

The cost of pollution in a fishery such as oysters in the Chesapeake Bay has been calculated in terms of dollars lost to the oystermen when they experience lower harvests because of pollution. But a clean bay is worth much more than merely the price of the oysters that were not harvested. People prefer living and boating on a clean bay, not a smelly, dirty one. A house overlooking a clean bay has a higher value than a house overlooking a dirty bay. Determining the difference in price is one way of attributing the value of a clean bay.

Intangible Values

Although the value of some nonmarket resources can be attributed, there are others for which it is impossible even to impute a value. These resources are said to have intangible values. Examples of intangible values for tropical forests are given in the last column of Table 4.6.

Some intangible values, such as maintaining the global carbon dioxide balance, can be approximated by estimating the cost to society of a rising sea level, but such estimates are so rough and would take place so far in the economic future that they have little meaning. Other intangible values such as the value of a native culture may not have an attributable value at all. Such values can be preserved only if it is considered "ethical" to do so.

We can point to many examples of government intervention into private markets to assert values that are not adequately defined by the marketplace and to reallocate resources. The abolition of slavery and the promotion of labor unions are two notable examples. Through public works projects and public utilities, in the 1930s the U.S. government created the Tennessee Valley Authority (TVA) and promoted rural electrification. The government has also chartered and delimited the operation of natural monopolies and has enacted environmental legislation. The point here is that the existence of values that are undeclared by the marketplace, but that have been at least partially corrected by political negotiation is not a revolutionary concept. Rather, it has long been accepted within the framework of countries that are basically capitalistic. However, in such cases the information processing is not automatic, as in the marketplace. Market and accounting failures are corrected through negotiations to achieve agreement.

Case Studies

A number of reports have been issued that calculate the total value, that is, the value in utility of various ecosystems. Two examples are as follows.

Tropical Forests

De Groot (1992; see Table 4.7) has estimated the socioeconomic values of environmental functions provided by tropical moist forest ecosystems, if they were to be managed on a sustainable basis. The classification system differs from the market/nonmarket value system in that he calls market values "production functions," and his nonmarket values are divided among regulation functions, carrier functions, and information functions.

Myers (1992), in his plea to conserve tropical forests, has discussed little-recognized values of species, values that are in addition to the prices of the highly prized timbers such as mahoganies, rosewoods, and teak. Some tropical fruits such as coffee, bananas, cacao (for chocolate), and avocados have an important market in developed countries. Such fruits cannot continue flourishing without season-by-season infusions of new genetic material from their wild relatives. In fact, all modern crops, because they are refined products of selective breeding, constantly require new genetic material from "support germ plasm" in order to maintain their productivity, enhance their nutritive content, improve their taste, and resist diseases and pests as well as environmental stresses.

Many entirely new foods, discovered in the wild of rain forests, are being used by indigenous tribes but are unknown to Western society. Myers' references have yielded a list of at least 1650 plants of tropical forests that have vegetablelike materials that can be used by humans. Other products derived from rain forest species include some that can be used for natural sweeteners and control of agricultural pests. Forest animals are also a potential source of protein. The capybara, a large rodent that inhabits the wetlands of South America, provides excellent meat and can produce well in many areas of the wet tropics that are unsuitable for cattle. In some areas along the Orinoco, capybara ranching is already becoming important.

Perhaps the best publicized natural chemical products of tropical forests are the pharmaceuticals, especially vincristine and vinblastine, developed by Eli Lilly and Company as a followup to clues provided by tribal herbalists and other traditional healers in Madagascar and Jamaica. The *Vinca* alkaloids derive from a dainty, almost unnoticeable plant, the rosy periwinkle. The alkaloids are used against Hodgkin's disease, leukemia, and breast, cervical, and testicular cancers.

Myers cites estimates that researchers should discover at least 8000 more species that have some form of activity against laboratory cancers. Extrapolating from past discoveries, he predicts that three should eventually rank as "superstar" drugs.

Some biocompound extracts from tropical plants can be used directly as drugs. A prime example is rauwolfia, a material from the so-called snakeroot plant of the monsoon forest in India. Rauwolfia was used for more than 4000 years by Hindu seers to

Table 4.7 Socioeconomic Value of Environmental Functions Provided by Tropical Moist Forest Ecosystems (based on maximum sustainable use levels)[a]

	Types of Values					
Environmental Functions	1 Conservation Value	2 Existence Value	3 + 4 Social Values[b]	5 Consumptive Use Value	6 Productive Use Value	7 Value to Employment
Regulation Functions	>> 11	++	++		++	
Buffering of CO_2	++		+		*	
Climate regulation	++		+		*	
Watershed protection	++				*	
Erosion prevention	++				*	
Storage/recycle human waste	+		+		*	
Bio-energy fixation	(20000)[c]	+			*	
Biological control	11	+	+		*	
Migration habitat	++	++	++		*	
Maintenance of biological diversity	(800)[c]	++	++		*	
Carrier Functions	>> 2	++	++		> 41.0	++
Habitat for indigenous people	++	++	++		*	
Cultivation					15.0	+
Recreation			+		26.0	++
Nature conservation	2	+++	++		*	+
Production Functions			++	++	471.0	++
Food/nutrition			++	+	300.0	++
Genetic resources			++		40.0	+
Medicinal resources			++	+	100.0	+
Raw materials for manufacture				+		++
Rubber/latex					16.0	
Timber					15.0	

Table 4.7 Continued

	Types of Values					
Environmental Functions	**1 Conservation Value**	**2 Existence Value**	**3 + 4 Social Values[b]**	**5 Consumptive Use Value**	**6 Productive Use Value**	**7 Value to Employment**
Production Functions continued						
other					0.3	
Biochemicals					*	+
Fodder and fertilizer				+	*	+
Ornamental resources				+	*	+
Information Functions	++		++	++	>4.0	+
Aesthetic/Spiritual/Historical information	++		++	++	*	
Cultural/artistic inspiration			+	+	0.8	+
Educational & scientific information				+	3.2	+
Total annual value	>>13	++	++	++	>>516.0	

[a] *Values are expressed qualitatively (++) or in US$/ha/year.*

[b] *Social values consist of the importance of environmental functions to human health and the option value placed on a safe future.*

[c] *Figures given within brackets were not used in calculating the total value because the calculation was too speculative.*

* *These functions do contribute to economic productivity, either directly or indirectly, but no market or shadow price could be determined due to lack of information and/or shortcomings of the market mechanism.*

Source: *De Groot,* Function of Nature, *(1992), Table 4.1.5, p. 192. Reprinted by permission of Wolters–Noordhoff Groningen, The Netherlands, copyright © 1992.*

treat nervous disorders, mental illnesses, dysentery, cholera, and fever. Today it is used as the base of a tranquilizer that has sales in the United States of $30 million per year.

Other valuable products of tropical forests include fibers and canes for wickerwork furniture made of rattan; kapok as an insulating and soundproofing material and as stuffing for life jackets; and essential oils in mouthwash, deodorant, and in beverages and medicines. Extractable oils are found in detergents, candles, emollients, lubricants, cellophane, and explosives. Exudates such as gums, resins, and latexes are used on envelopes and stamps, in varnish, and for chewing gum.

Marshlands

Another valuation of natural resources and the services of nature is provided by Reisner (1991) in his description of the marshlands of southern Louisiana.

One hundred and fifty years ago, there were over 6 million acres of Louisiana coastal wetlands. These marshes and swamps formed an interface between the fresh water carried down the Mississippi River and the salt water of the Gulf of Mexico. Under natural conditions, the marshes were a buffer between the two and were in dynamic equilibrium. In the region south of New Orleans, the main river channel was shallow, and constantly shifting. Ordinarily, the water flow would slow down, flatten out, and filter through the vast areas of marshes, depositing the silt and building up the marsh.

Occasionally, a big storm would blow through, churn up the mud, and wash it out to sea. Some marshland would disappear, but it would immediately begin to build up again, sometimes in the same place, sometimes in different places.

Then the dredging of canals and the stabilization of the channels began in order to accommodate shipping. The Mississippi no longer flowed over its banks. Its channel was dug out, and levees were constructed to keep the flow concentrated and in one place. The result was fine for shipping because it provided a deep, constant channel, but it robbed the surrounding marshes of their supply of silt. Thus, when the hurricanes hit the coast, the silt and clay were washed away, but in between such storms, previously the time of replenishment, replenishment decreased, in some cases dramatically.

In 1913 it was estimated that the marshlands of Louisiana were disappearing at the rate of about 7 mi^2 per year. By 1946 the figure had reached 16 mi^2, and in 1967 28 mi^2. Although the present official value is 39 miles for 1991, other estimates put it as high as 50.

Oliver Houck, a professor of environmental law at Tulane University in New Orleans, has calculated what the loss means in term of attributable values (Reisner 1991). Fifteen years ago, the state's fishery was worth $3.2 billion annually, after processing. The overwhelming bulk of the catch is estuarine-dependent species, such as menhaden, Atlantic croaker, sea trout, spot, drum, blue crabs, brown and white shrimp, and oysters. These species spawn and mature in the marsh, and when it is gone, the value of this catch will be gone.

Furs and hides taken mainly from the marshes were worth $24 million in 1976. Sport fishing is worth $100 million or more. Sport hunting for ducks is worth perhaps half that but will decrease because the marshes constitute a critical part of the Mississippi flyway. By Houck's equation, Louisiana's coastal wetlands were worth $10.5 billion a year in 1978 and should be worth half again as much now. To that, Houck adds the value of the marsh to protect communities along the coast from the advancing Gulf. To replace this function of the marsh, the Corps of Engineers has proposed building ring levees and extended river bulwarks that would cost more than a billion dollars. "That, in effect, is the value of the marshes as flood protection, as storm buffer, " says Houck (p. 289).

Houck asks whether the value of the dredged canal for shipping isn't worth more than the value of the marshes as buffers. He concludes that, for a while at least, it is. However, it is not an either/or choice. It may be possible to have a shipping channel and marshes. Recently, the Corps of Engineers and the state of Louisiana have endorsed a project to siphon some of the water from the Mississippi over the levees so that the silt carried by the water can begin to replace the marsh. It will cost money, but it will be easier to justify when the marshes are valued as an investment in a capital asset, just as the levees are valued capital assets.

ACCOUNTING AND THE ENVIRONMENT

If resources are to be sustainable and if our environment is to be livable, sustainability and livability must be accounted for and incorporated into the economy. This section discusses problems inherent to our current accounting systems and outlines ideas that have been proposed to include values and services of nature in our accounting systems.

GNP and GDP

Economists refer to a nation's total consumption by the general term of *national output* or *net output.* The net output of a modern economy is somewhat arbitrarily but officially defined as the amount of personal and government consumption (cars, food, highways, defense, etc.), plus the amount of net export of goods and services, plus the amount of new capital formed (investment for expansion and replacement), plus any changes in the inventory of goods, all in a given time period. When taken as an annual sum, this amount is called the gross national product or GNP. The GNP is a flow, not a stock, and it is measured in dollars per year.

There is an alternative definition of the same number: the sum of all salaries and wages, plus the sum of all taxes paid, plus the sum of all profits, dividends, and interest payments, minus the total depreciation of capital stocks. When these items are in equivalent monetary units, they can be added together to give the same dollar value as the GNP. This list is sometimes referred to as the *value added,* or *net input* to the economy (Hannon 1992).

GNP is part of the basis on which economic and social welfare policy decisions are made at the national level (Costanza, Daly, and Bartholomew 1991). If the GNP is high, inflation fears aside, the economy is thought of as being "healthier" than when the GNP is low. Gross national product, as well as other related measures of national economic performance such as gross domestic product (GDP) (Repetto 1992), have become extremely important policy objectives, political issues, and benchmarks of the general welfare. If the quarterly GDP drops even marginally, a recession is declared and the administration's competence is questioned. Throughout the world, the GNP and GDP are primary measures of economic performance.

Yet GNP as presently defined ignores the contribution of nature to production, often leading to peculiar results. For example, a standing forest provides real economic services for people: by conserving soil, cleaning air and water, providing habitat for wildlife, and supporting recreational activities. But as GNP is currently figured, only the value of harvested products such as timber is calculated in the total. And in many countries, the loss of the timber is not even treated as a debit. Thus, deforestation is encouraged because the activity appears, on the books, to increase economic indicators. Such accounting is analogous to a system that treats merchandise sales from a department store as a credit on the books but does not include the decrease in inventory as a debit on the books.

An example of the problem of an economic indicator that counts environmentally destructive activities as a plus in the national scheme of bookkeeping is the billions of dollars that the Exxon oil company has spent on the cleanup of the oil spill at Valdez, Alaska. This spending actually *improved* the nation's apparent economic performance because cleaning up oil spills creates jobs and consumes resources, all of which add to the GNP. GNP adds up all production without differentiating between costs and benefits; therefore, it is not a very good measure of economic welfare (Costanza, Daly, and Bartholomew 1991).

The System of National Accounts (SNA)

Scarcity of natural resources was of little concern to nineteenth-century neoclassical economists, who laid the basis for the current economics. In nineteenth-century Europe, food grains and raw materials were easily available from the United States, Australia, Russia, and the colonies, while steamships and railroads lowered transport costs.

The current system of national accounts was developed by Keynes and his contemporaries during the 1930s, when the world was preoccupied with the Great Depression. The Keynesian model (on which the United Nations' System of National Accounts is based) focuses primarily on consumption, savings, and investment (Repetto 1992). This system assumed that, unlike conventional capital assets such as factories and machines, no investment needed to be made in natural resources. At that time, scarcity of natural resources and environmental degradation were not major factors causing unemployment and business failures. Natural resources were viewed as being so abundant that they had no marginal value—that is, an increase in demand would not affect the price, because the supply was presumed to be "infinite."

In recent years, however, the SNA's failure to recognize depletion of natural resources has caused severe problems in many developing nations. Because the SNA assumes that natural resources are so abundant that they have no marginal value, their depletion is not registered.

This approach is fundamentally inconsistent. If a country's balance sheets at two different times indicate that an asset such as a forest has been depleted, then the income and product accounts for the intervening years should show a charge for the depreciation. This follows from the most fundamental identity of accounting: the difference in stocks between two temporal points equals the net flow in the intervening period. For example, the difference between a company's net worth at the start and end of the year equals the savings or loss during the year.

The U.N. System of National Accounts violates this procedure with respect to natural resource assets. This bias gives false signals to policymakers. It reinforces the illusion that a dichotomy exists between the economy and the environment and so leads policymakers to ignore or destroy the environment in the name of economic growth. It confuses the depletion of valuable assets with the generation of income. The result can be illusory gains in income and permanent losses in wealth.

There is nothing wrong with drawing on natural resources to finance economic growth. However, a reasonable accounting representation should recognize that one kind of asset has been exchanged for another. The accounting should be the same as that of a farmer who decides to cut and sell the timber in his woods to raise money for a new barn. His accounts would reflect the acquisition of a new income-producing asset, the barn, and the loss of an old one, the woodlot. Were the farmer to ignore the cutting of his woodlot as a loss of an asset and use the proceeds from his timber sale to finance a winter vacation, he would be poorer on his return and be unable to afford the barn. But if his calculations followed the SNA, he would be able to take the vacation and still afford the barn (Repetto 1992).

Because the United States is becoming more and more dependent on resources from abroad, the failure to account for resource depletion will have less of an effect and influence over all economic signals in the United States than in developing countries. There are localized exceptions: the market value of old-growth forests in the Pacific Northwest does not reflect their resource value, and consequently they are being sold at prices much lower than replacement costs. It is in the developing countries that the failure to account for natural resource assets can lead to economic disaster. "A country can cut down its forests, erode its soils, pollute its aquifers, and hunt its wildlife and fisheries to extinction, but its measured income is not affected as these assets disappear. Impoverishment is taken for progress" (Repetto 1992, p. 94).

This skewed system of accounting is having devastating results. For example, Costa Rica has been using up its natural capital at an extremely rapid rate. From 1970 to 1989, the accumulated depreciation in the value of its forests and soils, and its fisheries (shown in Fig. 4.1), exceeded $4.1 billion in 1984 prices (Repetto 1992). Relative to the size of the economy, the annual loss is huge. This loss does not even include losses of other values of the forest such as future potential for attracting the tourist trade.

In the Philippines, annual losses resulting from deforestation averaged 3.3 percent of the gross domestic product between 1970 and 1987. Thus, as a result of this factor alone, the GDP was really 3.3 percent lower than the government said it was. A 3.3 difference

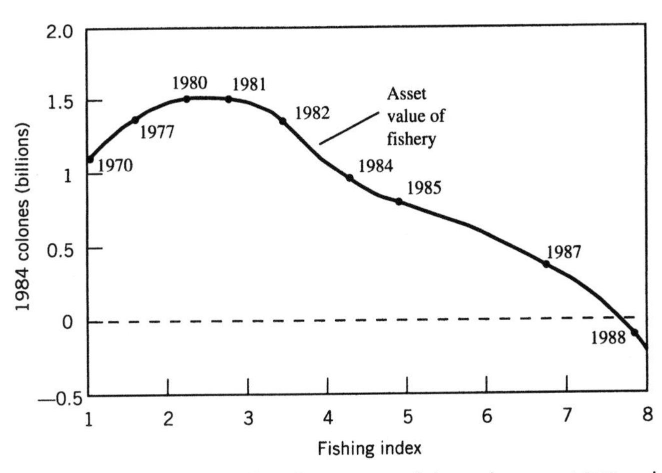

FIGURE 4.1 Asset value of Costa Rican fisheries between 1970 and 1988. Fishing index is a measure of effort expended by the industry. After 1980, the greater effort depleted the fishery. *Source:* Repetto "Accounting for Environmental Assets." Copyright © 1992 by *Scientific American, Inc.* All rights reserved.

in the GDP is a tremendous difference. Presidential elections can be won and lost depending on whether an economy is perceived to be gaining or losing 3.3 percent per year. But loss owing to deforestation was not the only unaccounted-for asset in the Philippines. In 1988 erosion of soil as a result of bad farming practices and bad logging practices caused tremendous losses. Farming losses were 2.5 percent of the GDP, and silting of reservoirs resulted in a loss of an additional 5 percent. Furthermore, when the silt washed out to sea, it damaged the coral reefs that support small-scale fisheries. Although the nation's accounts showed a mounting external debt, the accounts gave no sign of the destruction of the resources, destruction that makes repaying the debt more and more difficult (Repetto 1992).

Although changing the SNA could begin to affect governmental economic decisions, it would not directly influence the way individual companies keep financial accounts and make business decisions. However, it would have a major effect on the standard yardstick used to measure progress. In time, it would affect public policy leading to a revision of corporate accounting procedures.

Accounting for Resources and Services of Nature

Shadow Pricing

The idea that nature has value to humans that is not counted in free market prices is not new to economists. Resource economists are accustomed to working with what is sometimes called *shadow pricing.* It is frequently used, for example, in evaluating parks for recreation such as hunting and fishing. The benefits of wildlife, parkland, fish streams, and other recreational resources can be measured by travel cost as revealed by evidence of willingness to pay. This approach uses visits to sites for recreation, education, or other purposes to develop an econometric model that can be used to estimate net economic benefits. Benefits also can be measured by a contingent valuation method. Here, interviewers try to assess people's expressed preferences by asking people directly how much they value wildlife, fish, wilderness and so on, rather than by inference as under the travel cost method. One approach is to interview the hunters to determine how much they would be willing to pay for the privilege of hunting in a public forest.

Contingent valuation can also be used to estimate the value placed on wildlife by nonvisitors (Willis and Benson 1988). One can get an idea of how this works just by asking oneself, "How much would I pay, just to know that whales will not become extinct, even if I never get to see a whale?"

Problems such as air pollution are sometimes evaluated by estimating "compensation costs." People are asked about their willingness to accept compensation for the loss of environmental quality. For example, some people might be willing to live in a region with polluted air if it meant a job with a higher salary. The difference in salary would be the compensation cost for that person.

Benefit-Cost Analysis

In the 1930s when the amount of spending on public works was rapidly increasing to combat the effects of the Great Depression, benefit-cost analysis (BCA) was often used to evaluate government-sponsored projects. It resulted from a need to remove decisions on public works and development projects from subjective political influences and to give them a more objective evaluation.

In the United States, BCA originally began with the River and Harbor Act of 1902, which required the Army's Corps of Engineers to evaluate federal expenditures for navigation in a way that identified both commercial benefits and their cost. However, it was not until 1936 and the Flood Control Act that BCA became an important part of government decision making. This legislation was passed "for flood control purposes if the benefits to whomsoever they may accrue are in excess of the estimated costs" (Flood Control Act of 1936, cited in Campen 1986, p. 16).

As a result of this act, water projects aimed at providing flood control and other benefits such as hydroelectric power, recreation, and irrigation were studied to determine whether they met this standard. Because public works projects often benefit a relatively small interest group at the expense of all taxpayers, BCA seemed to be a mechanism for removing the evaluation and choice of alternative proposed public expenditures from the realm of politics and entrusting it as much as possible to expert practitioners with no vested interest in the outcome. Because the BCA combined all benefits and costs together to give a single summary measure, it was supposedly "scientific" and "quantitative." When benefits are exactly equal to costs, the benefit-cost ratio is 1.0. For a project to be justified, the BCA has to be greater than 1.0.

Several federal agencies, including the Interior Department's Bureau of Reclamation, the Agriculture Department's Soil Conservation Service, and the Tennessee Valley Authority, adopted the BCA. Unfortunately, each agency evolved mutually inconsistent sets of standards and procedures for carrying out analyses. Practice developed on the basis of tradition and in response to political and bureaucratic interests, with scant reference to economic theory. Each agency's primary concern was apparently to maximize the extent of its own involvement in water projects. Benefit-cost studies were used to justify projects that the agencies had already decided to undertake rather than to provide serious, critical analyses of the merits of the project (Campen 1986).

The problem arose from the assumptions necessary for the BCA. An illustration is provided by the BCA in the Bureau of Reclamation's 1967 document supporting the proposed Nebraska Mid-State project, the main feature of which was the diversion of water from the Platte River for irrigation of cropland (Hanke and Walker 1974, cited in Campen 1986, p. 53).

The document claimed a benefit-cost ratio of 1.24. That ratio was calculated by

1. Using an artificially low discount rate of 3.1 percent.
2. Assuming a project life of 100 years rather than the more common and supportable 50-year assumption.
3. Counting positive wildlife and fish benefits, when the project would have resulted in the Platte drying up over half of the time.
4. Valuing increased farm output at support prices that incorporated a substantial federal subsidy.
5. Including secondary benefits that would simply be transfers of economic activity out of other regions and into the Platte region.
6. Ignoring the costs resulting from the failure of other farms that would be displaced by increased agricultural output in Nebraska.

Besides the assumptions and selective data inclusion that can bias the outcome of benefit-cost analyses, BCAs have another weakness—namely, they usually do not take into consideration *who* gets the benefits and *who* pays the costs. This is a particular concern in developing countries. The costs of development projects in less developed countries are often borne by the local populations (those whom the project is presumably to help)—and the benefits accrue to outside investors, businesspeople, and government officials. For example, the *costs* of a big hydroelectric dam in the Amazon region are borne by those living where the dam is to be built, in that

1. Native indigenous cultures are destroyed.
2. There is social displacement, often to urban slums.
3. Hunting and fishing grounds are destroyed.
4. Stagnant water behind the dam increases the incidence of disease.
5. Species may be driven to extinction.

The dam *benefits*

1. Those who use the electricity, often residents or industrial complexes in distant cities.
2. The contractors who build the dam, usually with headquarters in large metropolitan areas or overseas.
3. Landowners, who acquired land in and around the dam site or in nearby villages in anticipation of increased land prices.

In short, BCA often tends to ignore the interests of those who are most directly affected by the proposed project.

Yet another weakness of BCA is that if often fails to consider the nonmarket values of resources (Table 4.6) that would be destroyed as a result of a proposed project.

Because BCA has been flawed does not mean that it must be abandoned, however. Abandonment could make the process even more political, an alternative that is even less desirable. Rather, BCA could be improved by changing the system of evaluation, just as the SNA must be changed in order to reflect the value of the resources affected.

For example, if an international lending bank proposes to finance a road to open up a tropical rain forest, the BCA should be required to evaluate the following:

1. What is the value of the undisturbed forest? (Table 4.7).
2. For what period of time is that value sustainable?
3. Who benefits from the extractable but nondestructive products of the forest (such as rubber and Brazil nuts), and where do they live?
4. How many people receive employment and benefits from harvest of the products?
5. What is the value of the timber to be harvested?
6. How long will lumbering be sustainable?
7. If ranching follows logging, as in many tropical lowlands, what is the value of the products from the proposed pasture.
8. For how long is the pasture sustainable, and what is the cost of maintaining sustainability?
9. Who makes the profit from the pasture, and where do they live?
10. How many people will be hired as cowboys, and so on, to run the ranch, and how much will they earn?

Benefit-cost analysis also should include opportunity costs that are lost by the proposed action. The concept of opportunity costs applies to the monetary or other advantage surrendered for something in order to acquire it in competition with other potential uses. An opportunity cost of building a pulp mill in a forest is the loss of utility of that site for recreation.

Opportunity costs of the land often are externalities for resource extractors on public properties. Loggers on public lands usually do not have to pay the costs of opportunities forgone when a forest is clear-cut for timber. If the highest value alternative use of a forest is as a national park, then the opportunity cost of logging the forest includes the value of the national park. The full cost of logging should include this cost as well as the opportunity cost of the labor and equipment used in the logging operations. If the benefits of logging (price times output) do not exceed the social opportunity costs, then logging should not take place.

Failure of governments to charge opportunity costs to resource extractors on public lands often makes it economically unattractive for private entrepreneurs to develop privately owned resources. Because loggers on most public lands throughout the world usually do not have to pay opportunity costs, they can undercut private landowners who wish to start a forest plantation or to reforest a cutover area. Private owners must pay an opportunity cost of the land; that is, they must compete on the open market and buy or rent the land at prevailing rates.

If resource extractors such as loggers and ranchers in national forests were required to pay the fees equivalent to those demanded by private owners of equivalent lands, the price of wood and beef might be higher. However, higher prices would ensure a sustainability of supply because they would better motivate private entrepreneurs to get into the business.

Environmental Impact Analysis

The National Environmental Policy Act was passed in 1969. It required that any project undertaken by a U.S. government agency be evaluated in order to estimate its environmental impact before the project was started (Miller 1990). The act also required that international projects financed by American institutions have an environmental assessment.

Many early environmental impact statements were mainly lists of wild species that occurred on the tract to be developed. Some considered alternative approaches of achieving the same goals (for example, a gas-fired power plant instead of a nuclear power plant). However, identification of a more environmentally acceptable alternative did not mean that the alternative *had* to be substituted for the original.

In general, most early environmental impact statements still failed to remedy the shortcomings of the traditional BCAs. For example, lists of species and landscape description were of little help in deciding whether a project should be approved. If a dam was to be built, obviously all the species living in the area to be flooded would be killed or driven out. Listing the species did little to help decide whether or not the project should be done.

After the passage of the Endangered Species Act in 1973, the presence of an endangered species in an area to be flooded could halt a major construction project. However, the finding of an endangered species and the subsequent prohibition against continuing construction sometimes did more harm than good. For example, in 1976 conservationists filed suit against the Tennessee Valley Authority to stop construction of the $143 million

Tellico dam on the Little Tennessee River in Tennessee. The reason was that the area to be flooded was the only known breeding habitat of an endangered fish, the snail darter. There was much public sentiment against halting construction of the dam because of the presence of a seemingly insignificant fish. Eventually, because of political lobbying, the Tellico dam was exempted from the Endangered Species Act, and construction was completed (Bennett 1992).

The snail darter episode was unfortunate because it caused the conservationists who filed the suit to seem frivolous and perhaps gave conservation a negative image. Many who were involved in the controversy were more concerned with the conservation of the ecosystem of the Little Tennessee River and its valley than with the snail darter. The director of the U.S. Fish and Wildlife Service described the habitat of the darter as "only in the swifter portions of shoals over clean gravel substrate in cool, low-turbidity water." A law that protected the water-purification function of fast-flowing river ecosystems might have been more publicly acceptable than the act that protects snail darters.

In order for conservationists to protect ecosystems and their functions, they are sometimes forced to invoke the law that protects species. If a species is protected, its habitat cannot be destroyed. But as the case of the snail darter demonstrates, fighting to save a species if the species seems insignificant can result in a backlash. It would be much more desirable if there were a law that protected endangered ecosystems or endangered ecosystem functions. Such a law would not only protect the values and functions of nature, but also could protect endangered species. A first step toward such an endangered ecosystem function act is the Clean Water Act, which preserves wetlands because of their functional importance.

Another weakness of the environmental impact statements has been that they considered one project at a time. This has led to "the tyranny of the small decisions" (W. E. Odum 1982, p. 728). For example, the ecological integrity of the Florida Everglades has suffered not from a single adverse decision, but from a multitude of tiny ones, including a series of independent choices to add one more drainage canal, one more roadway, one more retirement village, and one more well to provide Miami with drinking water. No one decision reduced the annual surface flow of water into the Everglades National Park so much that the effects of droughts intensified, or that unnaturally hot, destructive fires occurred. Yet all of these things have happened.

The decision to construct nuclear power plants near the shoreline of Lake Michigan, so that water from the lake could be used as coolant, is another example. An impact statement was prepared for each plant, and the findings were that each plant, considered individually, would cause no significant change in the lake. However, no assessment was done of the cumulative effects of all the plants.

Some impact assessments have done a very thorough job in considering all environmental aspects. For example, the environmental assessment of the Tucurui Hydroproject on the Rio Tocantins (Goodland 1978), a branch of the Amazon in Brazil, predicted

- The probable increase in malaria as a result of the project.
- The occurrence of schistosomiasis, leishmaniasis, and onchoceriasis.
- Social disruption following the forced removal of Indians and colonists living in the area.
- Flooding of archaeological sites.
- Loss of flora and fauna.
- Invasion of the reservoir by waterweeds.

- Oxygen deficiency in the water caused by rotting vegetation.
- Logs interfering with the turbines.
- The occurrence of deforestation and disturbance associated with dam building.

Despite the projected problems (all of which eventually materialized), construction of the project proceeded. With partial funding from the World Bank, the dam was completed in the early 1980s.

Accounting for Resources and the Services of Nature: New Directions

In any calculation of sustainable net income for a country, depletion of natural resources must be subtracted from GNP. Figure 4.2 illustrates the effect of incorporating depletion in Indonesia's national accounts. As in most economies, the gross national product (GNP) or gross domestic product (GDP) is the sum of all economic activity within the country. Net national product (NNP) or net domestic product (NDP) is GNP minus depreciation of natural capital. The growth of gross domestic product in Indonesia is compared with the growth of net domestic product derived by subtracting estimates of net natural resource depreciation for only three resources: petroleum, timber, and soils. Clearly, conventionally measured gross domestic product substantially overstates net income because consumption of natural resource capital is not considered (Repetto et al. 1989).

Other important macroeconomic estimates are distorted even more. Figure 4.3 compares, for Indonesia, estimates of gross domestic investment (GDI), which does not consider depreciation of natural resource capital, and net domestic investment (NDI), which does. Such distortions of statistics that are central to economic planning result in a misrepresentation that can be disastrous. It is as though the driver of a car were to judge the sustainability of his journey by the speedometer instead of by the gas gauge. He would drive along at 65 mph, blissfully unaware of any impending problem, until the moment that the gas tank went dry.

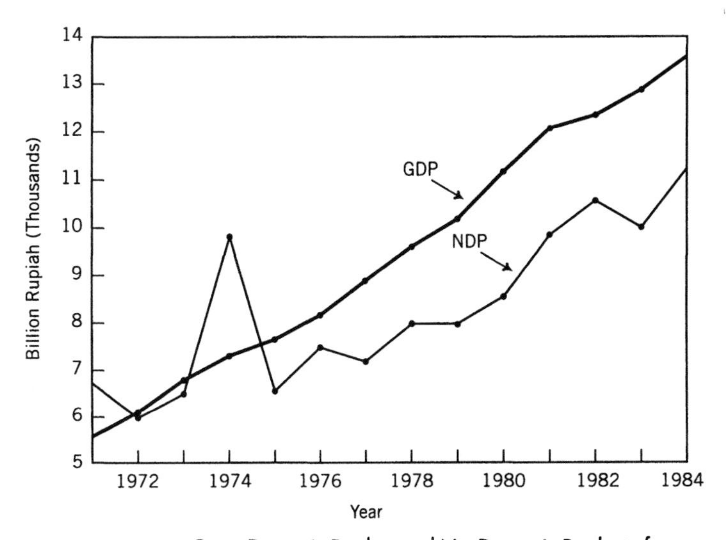

FIGURE 4.2 Gross Domestic Product and Net Domestic Product of Indonesia in constant 1973 rupiah. *Source:* Repetto et al. 1989, p. 7.

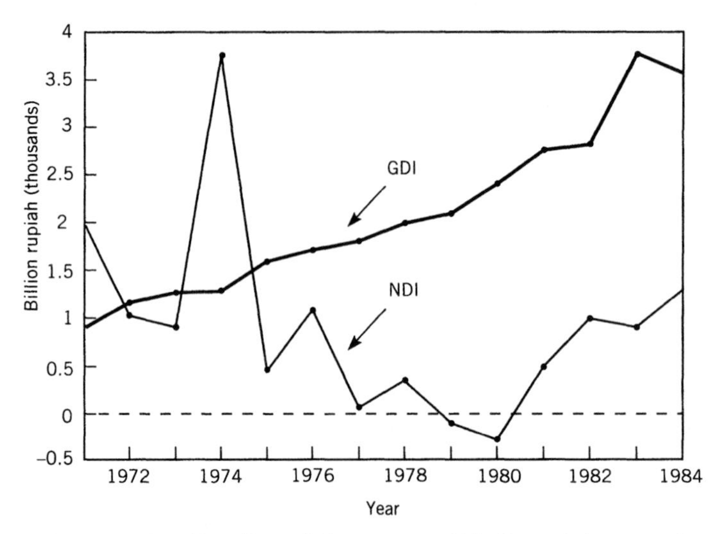

FIGURE 4.3 Gross Domestic Investment and Net Domestic Investment in Indonesia in constant 1973 rupiah. *Source:* Repetto et al. 1989, p. 9.

Several approaches have been suggested to modify standard accounting systems so that the values of resources, the regeneration costs of resources, and the services of nature will be accounted for.

SSNNP

A sustainable social net national product (SSNNP) has been defined as the net national product (NNP) minus defensive expenditures (DE) minus depletion of natural capital (DNC) (Daly 1989)

$$\text{SSNNP} = \text{NNP} - \text{DE} - \text{DNC}$$

The defensive expenditures (DE) term is included because of activities necessary to maintain the economic system that do not improve the quality of life, but are carried out to keep it from getting worse as a result of increasing economic activity.

Defensive expenditures include

Expenditures induced by the overexploitation of environmental resources in the general course of economic growth, such as the costs of all environmental protection activities and expenditures for environmental damage compensation.

Expenditures induced by the increased risks generated by the maturation of the industrial system, such as increased protection against crime, accident, sabotage, and technical failure.

Expenditures induced by the negative side effects of car transport, such as traffic accidents with associated repair and medical expenses.

Expenditures arising from unhealthy consumption and behavioral patterns and from poor working and living conditions, such as costs generated by drug addiction, smoking, and alcohol.

ISEW

When resource depletion and degradation are factored into economic trends, what emerges is a radically different picture from that depicted by conventional methods. For example, Daly and Cobb (1989) have produced an Index of Sustainable Economic Welfare (ISEW), which shows that, although GNP in the United States rose over the 1956–86 interval, the true economic welfare of Americans has remained relatively unchanged since about 1970. When factors such as loss of farms and wetlands, costs of mitigating acid rain effects, and health costs caused by increased pollution are factored in, the U.S. economy has not improved at all. The important message is that if we focus only on GNP, we may drive the economy down while we think we are building it up. Daly and Cobb acknowledge that many arbitrary judgments go into their ISEW, but they claim nevertheless that it is less arbitrary than GNP as a measure of welfare.

Energy

Energy is sometimes used as a universal currency in models that combine economic and ecological values because both natural and human-made capital have energetic values. The energetic value of a tree is the number of calories released when the tree is combusted, and it is theoretically equal to the number of calories used in the synthesis of that tree. The energetic value of a truck is the sum of all the energy, both labor and fossil fuel energy, in manufacturing that truck.

The major drawback of energy as a universal currency is that things that have an equivalent energy value may not have an equal social and economic value. For example, a family consisting of a man, wife, and two children would have approximately the same caloric value as a cow. Yet the amount of money that society would spend to heal a sick family is much more than the amount it would spend to heal a sick cow. Energy embodied in humans is much more valuable than energy embodied in cows. The "quality" of the energy is higher.

Odum (1988) has attempted to correct this problem by incorporating "transformity" into his models. The energetic value of something depends not only on its energetic content, he states, but also on the amount of energy used in its synthesis. For example, sunlight energy enters a pond and is transformed into biomass of producers (aquatic plants) that have an energy value. The plants are consumed by fish and/or decomposers (material recycle) that also have an energy value. But during each step along the way, an energy loss takes place in the form of heat which exits out of the system. As a result, the energy in plants is 10^3 more valuable than sunlight energy, and the energy in fish is 10^3 more valuable than the energy in plants.

Resource contributions multiplied by their transformities can provide a value system for human service, environmental mitigation, foreign trade equity, public policy alternatives, and economic vitality. However, transformities for such factors are difficult to determine. Use of energy for comparisons between similar systems as discussed in Chapter 6 is much more useful because any assumptions that are necessary apply equally to both systems that are compared.

Economic Criteria for Sustainability

Regardless of the approach, a minimum condition for sustainability is to *maintain the total natural capital stock at or above the current level.* An operational definition of this condition for sustainability means that

The human scale must be limited within the carrying capacity (at the standard of living desired) of the remaining natural capital.

Technological progress should be efficiency-increasing rather than throughput-increasing.

Harvesting rates of renewable natural resources should not exceed regeneration rates.

Waste emissions should not exceed the assimilative capacity of the environment.

Nonrenewable resources should be exploited but at a rate equal to the creation of renewable substitutes (Costanza and Daly 1990).

For an economy to be truly sustainable, it must be an economy in steady state. This means that the stock of human bodies and the total stock or inventory of artifacts must be held constant. Technology, information, wisdom, goodness, genetic characteristics, distribution of wealth, income, and product mix are not held constant (Daly 1993). Incentives for work would not be limited to acquisition of material goods. Rather, a pollution-free, congestion-free environment should be included as rewards for participating in the economy. Counting these benefits as part of the gross national product would reinforce the impression of their importance to a high-quality life.

ECONOMIC DEVELOPMENT

Development is a process that has an inception, a growth stage, and a maturity beyond which no further physical growth occurs. These three stages are common to virtually any individual or institution that is said to "develop."

For example, a baby is born, grows rapidly during childhood and adolescence, and then reaches physical maturity at about the age of 21. The size that any animal reaches at the age of maturity is in keeping with the proportions of its body. An elephant has much thicker legs in relation to its height than does a mouse because as a body increases in size, its mass increases exponentially while its dimensions increase linearly. If the growth hormones of a mouse were to run amok and the mouse continues to grow, it would collapse at about the time it reached the size of a small dog, because its legs would be unable to support the mass of its body.

Because a person ceases to become taller after maturity does not mean the person has ceased to develop. Although quantitative development slows or stops, qualitative development can continue. The person can continue to study, become less self-centered, grow more altruistic, and display other attributes of maturity throughout life.

Ecosystems undergo a process of development called *succession.* For example, a farmer's field, after abandonment, is invaded rapidly by annual herbs and grasses. After a year or two, the annuals are replaced by perennial grasses and shrubs. Then fast-growing, shade-intolerant trees enter the field, and are eventually replaced by slow-growing, shade-tolerant trees. Early stages of succession are characterized by rapid accumulation of biomass, but as the forest matures toward a steady-state biomass (sometimes called the "climax" stage), biomass production is approximately balanced by decomposition. An ecosystem that is mature with respect to biomass may still continue to develop qualitatively. For example, new species of songbirds may appear.

A mature ecological system may be one of the best models of a sustainable system (Costanza 1990). There is no "pollution" in mature ecosystems. All waste and byproducts are recycled and used somewhere in the system or dissipated. Closed systems that recycle pollution also are characteristic of a sustainable economic system. Economic uses must be found for pollution, so that materials are recycled instead of discarded.

Ecosystems that reach a mature state continue to be dynamic and interesting. For example, occasional disturbances such as storms are important in maintaining the high diversity of tropical rain forests and coral reefs, both being "mature" ecosystems. Analagously, economic systems that reach maturity need not be uninteresting. Change and the need to adapt will always be with us.

All types of organizations go through a series of stages during the course of their development. The U.S. Atomic Energy Commission (now the Department of Energy) provides an opportunity for first-hand observation of the growth and maturity of a research and development organization. The commission was founded in the 1940s, after the production of the first atomic bomb. In the decades that followed, the agency grew quickly and spawned almost a dozen major laboratories throughout the country, including Oak Ridge, Savannah River, Brookhaven, Argonne, and Lawrence Livermore. Great scientific excitement became palpable as atomic energy was developed for peaceful sources of energy. Then in the 1970s growth slowed as the limitations of atomic energy became evident. By the 1980s growth had ended, but other initiatives were explored, such as alternative sources of energy.

It is not only individuals, ecosystems, and social institutions that pass through a regularly recognizable series of changes: countries, empires, and entire civilizations also undergo the stages of inception, growth, maturity, and, unless innovative changes are allowed, decline. The S-shaped curve that describes this development (Modis 1992) is equally applicable to economic systems. Given this universally predictable sequence of events for all earthly individuals and natural and social organizations, it is incredible that some economists still believe that the global capitalistic economy *can* and *should* continue to expand, and give no recognition of the inevitability of the limits of growth.

Although some may deny the inevitability of limits to growth, capitalist economics is indeed in the midst of a developmental transition. Colby (1990) recognizes an evolutionary series of economic development stages designated as frontier economics; environmental protection; resource management; and ecodevelopment.

Frontier Economics (Laissez-faire Stage)

The frontier relationship of society against nature has been common to both decentralized capitalist economies and centrally planned Marxist economies. Although capitalistic and communistic societies differ in the strategies they employ for organizing development within the economy, their underlying worldviews about humans and nature are similar, usually having a vision of infinite economic growth and human progress. Technologies are developed with the purpose of increasing the power of the socioeconomic system to increase production by extracting resources from the environment, as well as to dampen the negative impacts of nature's variability on economic activities. Central to the frontier economics paradigm is the belief that environmental damage can easily be repaired where necessary and that infinite technological progress founded in human ingenuity, together with economic growth, will provide affordable ways to mitigate environmental problems.

Environmental Protection (Regulatory Stage)

The environmental protection stage begins with the suspicion that environmental problems are perhaps a little more serious than originally assumed. The principal strategy is to legalize the environment as an economic externality, in a modest variation of the frontier economics mode of development. The environmental protection perspective is defensive or remedial in practice, concerned mainly with ameliorating the effects of human activities. The approach focuses mostly on controlling damage and on repairing and setting limits to harmful human activities, but it is not concerned with finding ways to improve development and ecological resilience. Government agencies are created and are responsible for setting these limits. Environmental impact statements or assessments are institutionalized in many industrial countries as a rational means of assisting in weighing the costs and benefits of economic development before they are started. Areas of common property are set aside as state property for preservation or conservation. National parks and wilderness reserves also are created. Resource depletion and ecosystem services are generally not perceived in policy making as serious limiting factors for economic development. The interaction between human activity and nature in the environmental protection paradigm is still seen as a question of development versus environment.

Because central control was so important in previously communist countries, amelioration of environmental problems should have advanced more quickly there than in capitalistic countries. However, evidence that appeared upon the collapse of the Union of Soviet Socialist Republics indicated that the economy never advanced beyond the frontier stage.

Resource Management (Incentive Stage)

The resource management stage marks the beginning of a recognition that nature is critical in supplying not only goods, but also critical services to the economy, and that cooperation with nature is essential. The basic idea is to incorporate all types of capital and resources such as forests and fisheries as well as traditional human, infrastructural, and monetary resources into calculations of national accounts, productivity, and policies for development and investment planning. The objective is to take greater account of the interdependence and multiple values of various resources, and management of global commons resources. Not only stocks of physical resources but also ecosystem processes need to be considered as resources and capital that should be maintained, as well as used more effectively, by the use of new technology. The stabilization of population levels and reductions in per capita consumption through increased efficiency in the industrial nations are viewed as essential to achieving sustainability. It is understood that the scale of human activity is now so large that it affects the life-supporting environment, and that these impacts have a feedback effect on the quantity and quality of human life. The resource management approach is the basic theme of reports such as the Brundtland Report, Our Common Future, and the World Resources Institute's annual World Resources reports. The perspective in these reports is anthropocentric, and the concern for the life-supporting environment is based on the insight that hurting nature is also hurting humans. Basic ecological principles are essential for management to maintain the stability of the life-supporting environment.

Ecodevelopment (Cooperation Stage)

The essential step that will prevent humans from ultimately destroying the habitability of the planet is the recognition that they are part of the planet's system of stocks and flows of material and energy, and are not some remote controller, immune from the effects of their desires and decisions. Whatever humans do ultimately comes back to affect them one way or the other.

Ecodevelopment more explicitly sets out to restructure the relationship between society and nature by reorganizing human activities so as to be synergetic with ecosystem process and functions. In this stage, humans move from economizing ecology to ecologizing the economy, and realize that great economic and social benefits can be obtained from fully integrated ecological economic approaches to environmental management.

The ecodevelopment approach recognizes the need to manage for adaptability, resilience, and uncertainty, and for coping with the occurrence of nonlinear phenomena and ecological surprises. Rather than asking how can we create and then how can we remedy, ecodevelopment attempts to provide a positive, interdependent vision for both human and ecosystem development. This approach emphasizes that planning and management ought to be embedded in the total environment of the system under consideration, including all the actors concerned, which means that global system awareness must be coupled with local responsibility for action.

The term *development,* rather than meaning only growth, management, or protection, connotes an explicit reorientation and upgrading of the level of integration of social, ecological, and economic concerns in designing for sustainability. This perspective emphasizes a shift from a system in which the polluter pays to a system in which pollution prevention pays, and also stresses the need to move from throughput-based physical growth to qualitative improvement. Such development not only implies becoming more efficient in the use of energy, resources, and ecosystem services, but it also emphasizes the need for improvement in terms of synergies gained from designing agricultural and industrial process in order to mimic and use ecosystem processes in an explicit manner.

Quality versus Quantity as an Index of Progress

Quantity is the driving factor behind the unbridled free market system. More is better; bigger is better.

It can be argued that the market system recognizes quality, for quality goods have a higher market value than inferior goods. But there is more to life than increasing the quantity, size, and even quality of goods possessed. In the context of this book, these other things would include a clean environment, a diverse environment, and the room and freedom to enjoy that environment.

How can the people of a country and the world hasten the emergence of an economy out of the frontier stage, through the environmental protection stage, and into the resource management stage with an eye eventually to reaching the ecodevelopment stage? It is an educational and a political process. Part of the education will come from the suggested readings included at the end of each chapter in this book. Some ideas for the political process are the subject of the next chapter.

SUGGESTED READINGS

For overviews of approaches to accounting for ecosystem services and natural capital:

Ahmad, Y. J., S. El Serafy, and E. Lutz, eds. 1989. Environmental Accounting for Sustainable Development. The World Bank, Washington, D.C.

Costanza, R., ed. 1991. Ecological Economics: The Science and Management of Sustainability. Columbia University Press, New York.

For an analysis and critique of current economic assumptions, and their role in environmental problems:

Daly, H. E. 1991. Steady-State Economics. 2nd ed. Island Press, Washington, D.C.

Repetto, R., W. Magrath, M. Wells, C. Beer, and F. Rossini. 1989. Wasting Assets: Natural Resources in the National Income Accounts. World Resources Institute, Washington, D.C.

For articles discussing the values of various ecosystem types and functions:

Kessler, E., ed. 1992. Environmental Economics. Ambio 20(2):52–98.

For a detailed analysis of the functions of nature and their economic value:

De Groot, R. S. 1992. Functions of Nature. Wolters-Noordhoff, The Netherlands.

CHAPTER 5 Policies for Conservation

CHAPTER CONTENTS

PRINCIPLE

Neither the state nor the market alone can meet all human needs.[1]

[1]From *Centisimus Annus*, the 1991 encyclical of the Roman Catholic Church.

CHAPTER OVERVIEW

When the free market is unable to deal with problems of common interest such as health, welfare, and the environment, we must turn to government. Regulation is one approach used by governments. However, regulatory control often is inefficient because it lacks the feedback mechanisms of the free market system. Often the most effective mechanisms of government control are incentives because they provide quick reward for desirable behavior.

International environmental policy is more difficult to implement. Success rests on the willingness of governments to agree to, and abide by treaties between nations. ❧

PROBLEMS OF THE COMMONS

The Case for Government Intervention

On the economic frontier where ecosystems are an infinite source of raw materials and an infinite sink for pollution, the free market efficiently allocates resources and distributes goods and services. As the frontier becomes populated, as the economy matures, and as sinks of used resources become full, the free market begins to fail in its service to humankind. It fails because it cannot recognize the life-supporting functions of nature that are damaged by unrestrained exploitation of resources. It assumes that nature's services, including absorption of pollution, are unlimited and are free.

When the free market fails, government interference may be necessary. Government intervention is *desirable* because it can fulfill functions such as regulation of pollution that cannot be fulfilled by the unregulated free market. However, government intervention often is *undesirable,* because it lacks the feedback mechanisms of the free market (Adam Smith's "invisible hand") through which consumers continually direct producers, thereby ensuring an efficient allocation of goods and services. After a time, government bureaucracies tend to become self-serving and unresponsive to changing world conditions.

FIGURE 5.1 *The environmental crisis: Right-wing (this page) and left-wing (next page) versions. Source:* Young (1990), pp. 142–143. Reprinted by permission of the publishers from *Sustaining the Earth* by John Young, Cambridge, Mass.: Harvard University Press, Copyright © 1990 by John Young.

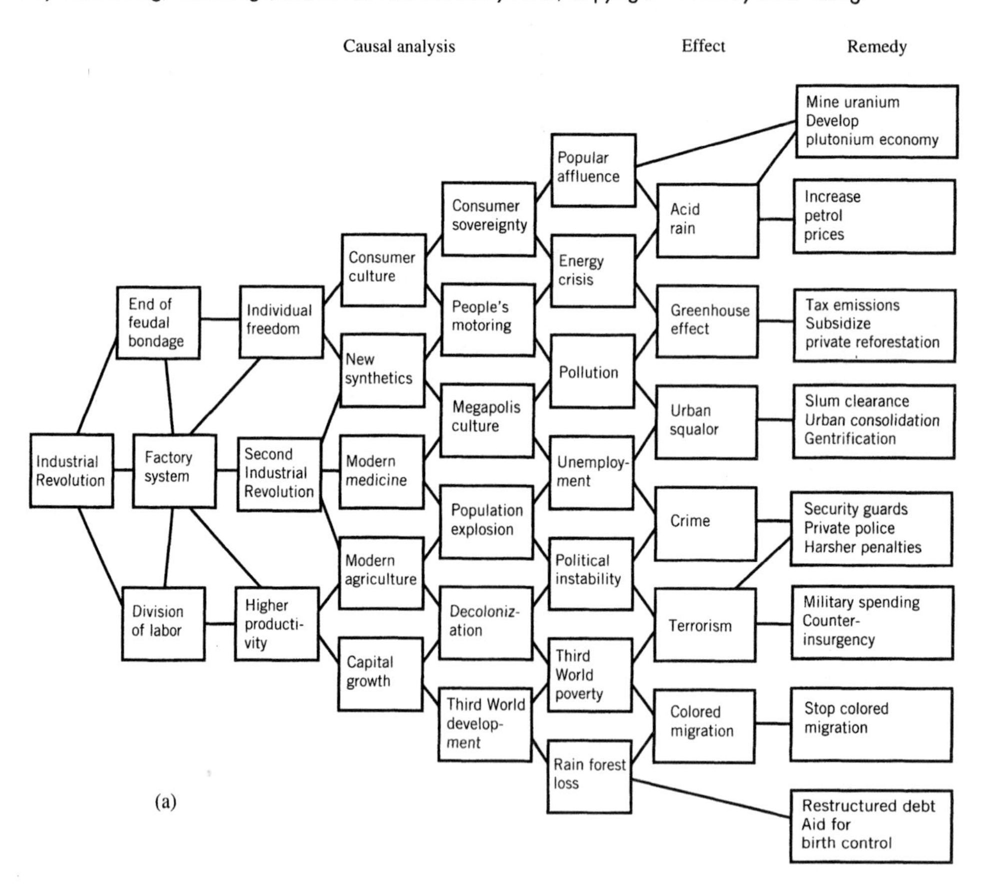

Political Philosophies

In democratic societies throughout the world, political and economic preference can be divided into two main camps, conservative and liberal. In reality, a whole spectrum of political views exists, ranging from extreme left-wing Marxism to extreme right-wing fascism. For the sake of discussion, however, it is convenient to characterize the camps simply as conservative and liberal.

In terms of values and goals, conservatives tend to place higher priority on the economic objectives of growth and efficiency than on social, distributional, or environmental objectives. In contrast, liberals tend to favor measures that, within the limits imposed by the capitalist economy, are more directly pro-poor, pro-labor, pro-equality, pro-minority, and pro-environment.

Conservatives look much more to free markets for solutions to social and environmental problems. They generally believe that the private economy left to itself does quite well. Wealth and the employment that directly results from it is created in the private sector. Government intervention aimed at bringing about better outcomes is likely instead to make things worse. Conservatives tend to minimize the seriousness of current problems and to argue that whatever social and environmental problems do exist are best dealt with by allowing market forces to produce responses from private business. They also tend to believe that lower government spending and less regulation are almost always better.

FIGURE 5.1 Continued

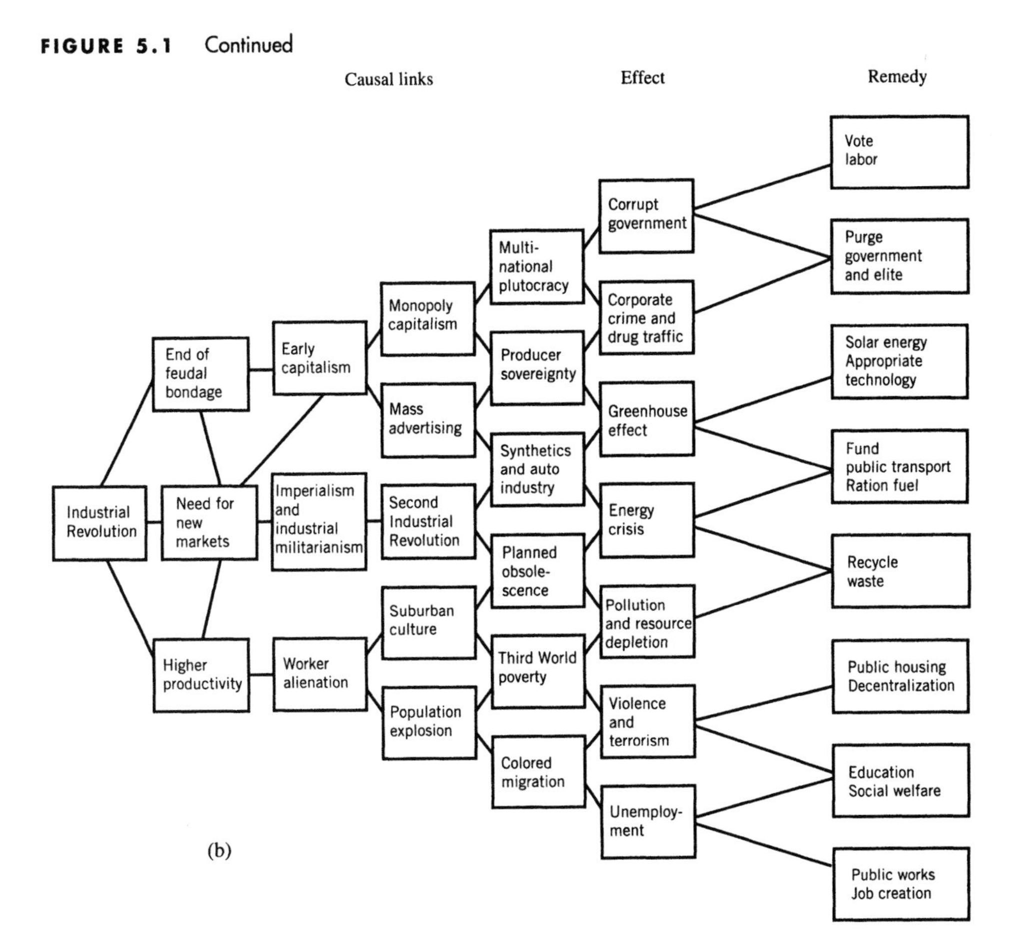

(b)

Government programs are viewed as at best unnecessary or ineffective and often as actually counterproductive because of the tax drain they cause on private enterprise. They also regard the tax revenues needed to pay for public programs as creating inequities, inefficiencies, and perverse incentives, and consider taxation beyond levels for basics such as defense and police protection as decreasing the productivity and economic well-being of all citizens, rich and poor alike. They view regulations as reducing citizen welfare.

In the other camp, liberals have a greater belief in both the need for and the possibility of effective government intervention in the private economy in order to deal with major social and environmental problems. They believe that private markets are unable to accomplish such important societal objectives as protecting the environment, producing reasonable levels of health and safety on the job, and reducing inequality to a tolerable level. They also tend to believe that well-designed and well-implemented governmental expenditure and regulatory programs can make substantial contributions to dealing with these problems (Campen 1986). Without government intervention, they state, the extremes in income between rich and poor will grow until a large proportion of the population becomes disenfranchised from their basic rights.

These basic philosophies are reflected in two contrasting versions as to the causes, effects, and remedies for the environmental crisis, illustrated in Fig. 5.1. In the right-wing version, the immediate results of the Industrial Revolution are generally favorable: the end of feudal bondage, individual freedom, higher productivity, modern medicine, modern agriculture. In the left-wing version, the negative impacts start sooner: imperialism and industrial militarism, worker alienation, monopoly, and suburban (materialistic) culture. In the left-wing version, the causal links of the environmental crises are an inherent part of the capitalistic system. In the right-wing version, the causes are external to the economic system.

Not all the remedies shown in Fig. 5.1, however, reflect the philosophies because another aspect of conservatism is to stick with what is old, while liberalism often is more open to the new. Thus, solar energy is listed as a left-wing remedy, but plutonium economy is a right-wing remedy. Yet a generation ago, it was the government that developed atomic energy: now it is often the government that must support solar energy technologies until they become economically competitive. National peculiarities also are reflected in the remedies. Aid for birth control is shown as a right-wing remedy, but in the United States, it is generally opposed by conservatives and favored by liberals.

Blending the Best of Two Philosophies

Conserving most natural resources and reducing many pollution problems are management problems of the global or local commons, and thus they often require government intervention. Government intervention is not always required, however. An example comes from mineral resources and the recycling of metals. When the cost of extracting metals from the earth becomes greater than the cost of collecting and recycling used products made of these metals, private investment will conserve the metals by pursuing an economically rewarding recycling program. Another example is that when the government is unable to provide a water supply pure enough for drinking standards (as often occurs in cities in the tropics), private suppliers of purified and bottled drinking water will offer a supply.

Although private enterprise may be preferable when it is capable of solving the problem, it is often not capable. For publicly owned resources that cannot be recycled, or for pollution control in commonly owned sinks such as air and water, private markets are inadequate. Government control is necessary.

An extreme left-wing vision for adequate control of resource management and pollution control has been described as ecodevelopment (see also Chapter 4). Ecodevelopment attempts to restructure the relationship between society and nature by reorganizing human activities so as to be synergetic with ecosystem process and functions. This emerging paradigm, ecologizing the economy, stresses the great economic and social benefits that are to be obtained from fully integrated ecological economic approaches to environmental management. Such development is unlikely to occur. Only through radical revolution could such a change occur in modern democratic societies, and a radical revolution would so profoundly destroy the economy that day-to-day survival would dominate the concerns of the populace.

Instead of searching for such an "ideal" policy strategy, it is important to reflect on the freedom and power that any government has to make radical change. No single policy measure, or even a whole series enacted together, is likely to alter significantly the underlying and fundamental structure of institutions. All governments work within the prevailing political and economic system. The basic legal, governmental, social, and economic institutions remain largely unaltered. Change is gradual, and attempts to produce change have to operate within the constraints set by the system. At best, planners can nudge the process toward desired outcomes (Rees 1988).

In the following pages, we discuss various interventions through which policymakers and planners can lead a democratic society toward more effective conservation of resources and prevention of pollution. A gradual transition is desirable, for it gives society an opportunity to try a policy and see if it works. If it does not, then the policy can be abandoned, or another one can be tried. This is a feedback mechanism that keeps policy in the direction that is most beneficial to society. The feedback is not as rapid as that of the free market system. Nevertheless, it is much more responsive than a revolutionary bureaucracy which, once established, is usually impervious to reform even in the face of evidence of failure.

Basically, a successful approach is one that combines the best of the two political philosophies: a feedback mechanism to keep control functions efficient and responsive; and governmental intervention to ensure that the interest of the commons is represented. Such a compromise is more likely to serve society better as a whole than would an approach that demands complete overthrow of the free market system, or an attitude that all government control is anathema and must be resisted at any cost.

POLICY OPTIONS

A number of policy options are available to governments; they are as follows.

Moral Persuasion/Information

The increasing exposure of conservation issues in public forums, the press, and especially television has fostered a new attitude in our society today that has led to a mentality of global awareness. A particularly important example is the book *Earth in the Balance,* by Al Gore, vice president of the United States, in which he makes an emotional plea for conservation. The concluding words are: "For civilization as a whole, the faith that is so essential to restore the balance now missing in our relationship to the earth is the faith that we do have a future. We can believe in that future and work to achieve it and preserve it, or we can whirl blindly on, behaving as if one day there will be no children to inherit our legacy. The choice is ours: the earth is in the balance" (p. 368).

The book is notable too, for not only describing the problem in forceful terms but also proposing a solution. Gore introduces his solution by describing the Marshall Plan, the strategy that helped the nations of Western Europe rebuild after World War II and fend off the spread of communism. The aid concentrated on the structural causes of Europe's inability to lift itself out of its economic, political, and social distress. It was designed not only to rebuild the ruined cities and factories, but also to change economic barriers such as trade laws that impeded the growth of free markets. Through these types of aid, it facilitated the emergence of a healthy economic pattern.

Gore has called for "A Global Marshall Plan" to guide the effort to save the environment. His plan has five strategic goals:

1. *Stabilization of the world population* through policies designed to create in every nation the conditions necessary for making the demographic transition from a dynamic equilibrium of high birthrates and death rates to a stable equilibrium of low birthrates and death rates.
2. *Rapid creation and development of environmentally appropriate technologies,* especially in the fields of energy, transportation, agriculture, building construction, and manufacturing. These technologies should be capable of accommodating sustainable economic progress without the concurrent degradation of the environment.
3. *A comprehensive and ubiquitous change in the economic "rules of the road" by which we measure the impact of our decisions on the environment.* We must establish, by global agreement, a system of economic accounting that assigns appropriate values to the ecological consequences of choices in the marketplace by individuals and companies, and macroeconomic choices by nations.
4. *The negotiation and approval of a new generation of international agreements* that will embody the regulatory frameworks, specific prohibitions, enforcement mechanisms, cooperative planning, sharing arrangements, incentives, penalties, and mutual obligations necessary to make the overall plan a success.
5. *The establishment of a cooperative plan for educating the world's citizens about our global environment,* first through the establishment of a comprehensive program for researching and monitoring the changes now under way, and second, through a massive effort to disseminate information about local, regional, and strategic threats to the environment. The ultimate goal of this effort is to foster new patterns of thinking about the relationship of civilization to the global environment.

Each of these goals is closely related to all the others, and all should be pursued simultaneously within the larger framework of the Global Marshall Plan. Finally, the plan should have as its more general, integrating goal, *the establishment, especially in the developing world, of the social and political conditions most conducive to the emergence of sustainable societies* (Gore 1992).

Moral persuasion and information are especially effective for children. Many schools have instituted programs to educate children about the importance of conservation measures such as recycling. The "Rain Forest Game" is an example. In this game, each child takes a slip of paper that assigns the child to a role in the rain forest food web. The assignment can be either a species in the rain forest or a cycling component such as an atom of calcium that moves from species to species. One child draws the role of "bulldozer." Once the rain forest is set up and the components are interacting, the bulldozer randomly eliminates one component at a time. The object of the game is to see at which point the system becomes disconnected so that it can no longer function.

Campaigns that focus on the national interest, that describe the rewards of being good to the environment, and that produce guilt feelings in those who are not, are often effective. Campaigns to use less water during the summer, to recycle newspapers, to save trees, or simply to "clean the beaches" have had significant success in many nations. Specially designated days, such as Arbor Day and Earth Day, are also effective in promoting awareness of conservation issues.

The United Nations Conference on Environment and Development (UNCED) held in Rio de Janeiro from June 3 to 14, 1992, though billed as a conference to begin solving environmental problems, in reality was a global forum on which problems and grievances were aired. Although declarations and agendas were made and signed, the major accomplishment of the conference was to generate information that through moral persuasion would lead the world to begin concrete steps toward environmental reform.

UNCED Secretary-General Maurice Strong, in his opening address to UNCED, stated:

> *The Earth Summit is not an end in itself, but a new beginning. The measures you agree on here will be but first steps on a new pathway to our common future. Thus, the results of this conference will ultimately depend on the credibility and effectiveness of its follow-up. . . .*
>
> *The road beyond Rio will be a long and difficult one; but it will also be a journey of renewed hope, of excitement, challenge and opportunity, leading as we move into the 21st century to the dawning of a new world in which the hopes and aspirations of all the world's children for a more secure and hospitable future can be fulfilled. (Haas et al 1992, p. 7)*

The agreements signed in Rio created several new international institutions. Agenda 21 created a new United Nations body, the Sustainable Development Commission. The conventions on climate change and biodiversity led to new bodies for scientific and technical advice relating to the treaties and their implementation. Other new organizations whose creation was motivated by the UNCED process, though not formally created by governments at UNCED, include the Planet Earth Council and the Business Council for Sustainable Development. Simultaneous to UNCED, a large gathering of nongovernmental organizations (NGOs) was held in downtown Rio, under the umbrella title of the Global Forum. It was a mixture of extensive NGO networking, street fair, trade show, political demonstration, and general events involving more that 18,000 participants and 200,000 visitors.

Although moral persuasion, information dissemination and publicity are having a positive impact for conservation worldwide, they are not enough. They must lead to legislation that establishes regulations, incentives, and other legal means to implement conservation.

Regulations

Direct regulations are a customary way to control environmental problems. Government can establish limits on timber exports or on taking of endangered wildlife or exotic plants. It can prohibit the use of some pesticides or set limits on the application or residues of others. Examples are the Clean Air Acts of 1965, 1970, and 1977, the Federal Water Pollution Control Act of 1972, the Wilderness Act of 1964, and the Endangered Species Act of 1973. The National Environmental Policy Act of 1969, which requires that any project to be undertaken by a government agency be evaluated in order to project its environmental impact before it is started, also is regulatory in that the evaluation *must* be made.

In 1972 the Environmental Protection Agency (EPA) was created and empowered to set environmental standards for major air and water pollutants, as well as to enforce many of the federal environmental laws. With conventional command-and-control regulatory mechanisms to control pollution, the EPA either specifies the technology that must be used for this purpose (a technology-based standard) or sets an emission rate cap that all sources must meet (a uniform performance standard). In the first case, government specifies the equipment that must be used to control pollution. An electrical utility, for example, may be required to install flue-gas scrubbers to control sulfur dioxide emissions or electrostatic precipitators to control particulate matter. Greater flexibility is provided by performance standards, which allow firms to decide how they will meet the specified goal such as a maximum allowable level of pollutant emitted per unit of product output (Wirth and Heinz 1991).

These conventional policy approaches can be effective in achieving environmental goals, but they tend to impose relatively high costs on society because some unnecessarily expensive means of controlling pollution will be used. In addition, unfairness may be introduced because the costs of controlling emissions vary greatly from one source to another.

Another frequent problem with regulations is that, although they may temporarily solve one problem, the solution is not permanent. The Clean Air Act is an example. One provision required the use of catalytic converters on all new automobiles. This was a great step forward, but the act alone did not solve the problem of air pollution. Although on the one hand we now have cars emitting fewer pollutants, we also have had an increase in the number of autos and miles driven, with the result that in some areas, air pollution continues to exceed limits.

Incentives

A program of incentives can sometimes be more effective in inducing desirable conservation behavior than a program of controls and regulations, just as it is sometimes more effective to offer a mule a carrot than to beat it with a stick to induce desirable behavior.

Promotion of long-term land stewardship through property rights is an incentive needed to improve land stewardship. Sharecropping and shifting cultivation are examples of agriculture that is environmentally degrading owing to lack of land tenure. Sharecropping in the southern United States in the late 1800s and early 1900s led to severe erosion of the lands because the sharecroppers had no interest in, or at least no capability of, providing for long-term stewardship. Sharecroppers were poor, landless farmers who were allowed to farm a piece of land belonging to someone else and in return, had to give part of the crop they harvested to the landowner as rent.

Shifting cultivators in developing tropical countries contribute to modern environmental degradation. Because they do not own the land they cultivate, they always live in fear that the government or adjacent ranchers might chase them away or reclaim the land. Thus, peasant farmers have no incentive to carry out land conservation techniques such as terracing, which would lead to sustainability of production.

Land tenure in itself does not guarantee improved land management. In some countries, land reform programs have led to the distribution of large plantations or ranches among local peasants. Often, however, the individual peasant farmer, even if he owns the land, does not have the capability or the resources to manage land in a sustainable way. Sustainable agricultural techniques often take years to develop, and the poor farmer does not have years.

One solution to the problem has been for land reform programs to title land to a *community* of farmers. The *ejidos* of southeast Mexico are an example. Following the Mexican Revolution, the division of extensive haciendas into ejidos was carried out following the dictum of Emiliano Zapata, "The land belongs to the people who work it." The term *ejido* is used for the land commonly owned by a group of peasants. Although the land cannot be sold, natural resources from the land can be. Each ejido has a General Assembly, and the rights of ejido members are inherited. An analysis of ejidos developing forest products concludes that, although several are below their economic potential, it appears that they are moving toward sustainability (Richards 1991).

Incentives can be direct or indirect. A direct incentive is one in which those who are depleting the resource are paid directly to protect these resources. For instance, some governments are hiring local shifting cultivators to act as tourist guides or to manage a forest reserve. The National Biodiversity Institute of Costa Rica (INBIO) hires local people as taxonomists to assist in scientific projects. Subsidizing public transportation such as subways and buses to relieve traffic congestion is another kind of incentive.

The key to a well-designed and well-implemented incentive program is responsiveness to changing conditions. A problem with subsidies is that they tend to become institutionalized and to outlive their intended purpose. When a program is not responsive, the negative aspects of bureaucracies increase. A great majority become harmful to the environment in ways that were never foreseen. The U.S. Forest Service has been a case in point. One of its early missions was to manage national forests to ensure that there would be enough timber to satisfy the needs of a developing society. It was believed that if the Forest Service were to ensure that standing timber were available and accessible, timber companies would have an incentive to cut the timber, thereby supplying the needs of a growing society. At the time this mission was formulated, a future scarcity of timber seemed likely. The probability seemed small that enough private landowners would invest in timber plantations or reserves to supply the country's needs.

Times have changed, however. There are many privately owned forests and timber plantations, and their number is increasing as the price of timber increases sharply, as is predicted by classical supply and demand curves. Meanwhile, with regard to the national forests, there is an increasing outcry for the Forest Service to reduce or eliminate the timber-supply function. There is an increasing need to satisfy the desire of many citizens to escape an increasingly urbanized society, at least temporarily, by placing more emphasis on the scenic and recreational values of the forests. In addition, there has been a need to restore one of the original purposes of the national forests—to protect the natural watersheds in mountainous areas in order to reduce the severity of floods and to provide clean water (Shands and Healy 1977).

The Forest Service, however, has remained generally unresponsive (Cortner and Schweitzer 1993; O'Toole 1988). As a result, the scenic and recreational values of the forests continue to be diminished. At the same time, private landowners who desire to get into the timber-cropping business are being undercut by the Forest Service's "Below Cost Timber Sales," where the cost to the taxpayer for managing the forest and opening roads to permit harvest is greater than the amount received from logging companies for rights to log the forest. In November 1993, President Clinton replaced the chief of the Forest Service, and it is hoped that this change will result in a shift of Forest Service policy on the national forests.

Another example of incentives that eventually turned out to be counterproductive were the benefits offered by the Brazilian government to ranchers who would establish

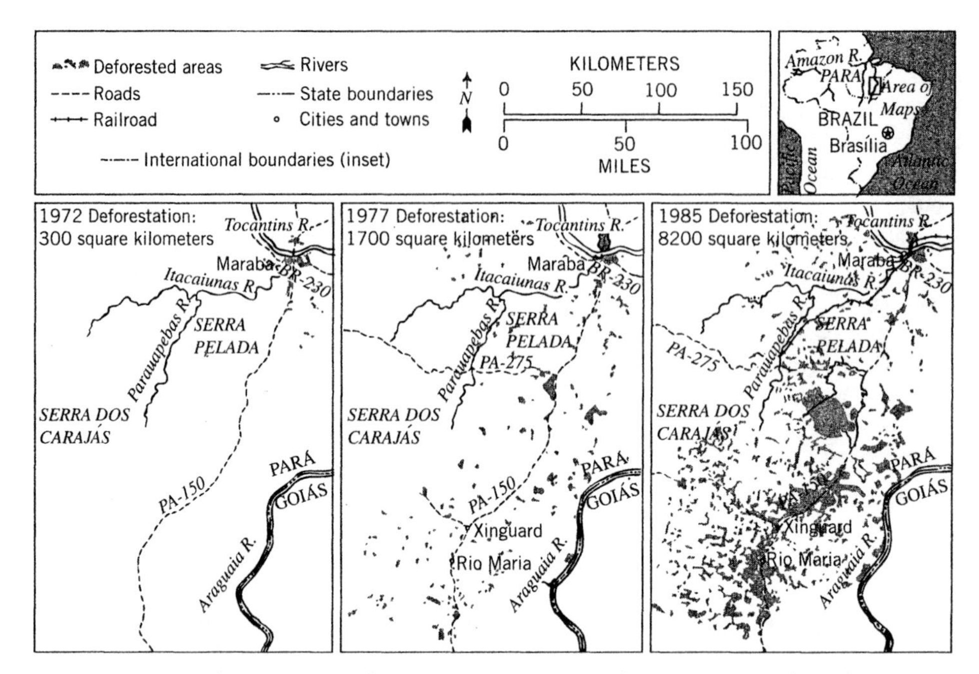

FIGURE 5.2 Deforestation in southeast Pará, 1972, 1977, and 1985. *Source:* Redrawn from Landsat.

pastures in the Amazon region. The purpose of the Brazilian measure was to encourage development of a region that the government perceived as being underutilized. The typical livestock project's profitability depended critically on the extent to which the rancher could avoid or defer commitment of his own resources by taking advantage of government tax credits and loans, and on the extent to which he could write off costs against other tax liabilities (Browder 1988).

To establish their pastures, ranchers hired crews to cut and burn the rain forest. The pace of deforestation was rapid (Fig. 5.2). The Brazilian government turned a deaf ear to the protests of environmentalists who decried losing tropical forest, which had such high value for global sustainability, merely to enrich a few speculators with political influence. Only when the outcry over Amazonian destruction reached a critical level and was voiced by Brazil's secretary of the environment did the Brazilian government begin to rescind the perverse tax incentives (Cavalcanti 1991).

These cases illustrate that even though incentives have outlived their usefulness or have been shown to be misdirected, they can be difficult to eliminate. An incentive creates a clientele, and once the clientele is established, they will fight all efforts to eliminate the subsidy, which in the mind of the beneficiaries often becomes a "right." Farm subsidies are a particularly notorious example of incentives that have proven exceedingly difficult to eliminate, despite the fact that national and international economies and trade are almost always negatively affected by such subsidies.

Incentives can be indirect as well as direct. An indirect incentive is a policy change that induces conservation practice without cash transfer. A good example is the case of a fishing community on the coast of Mexico that was overfishing its resource. Once control of the fishing rights was put into the hands of the community, the overfishing problem ended (National Research Council 1992).

U.S. policy on inheritance taxes has been modified to encourage conservation through a device called *conservation easements*—restrictions on land use. They are legal agreements that a property owner makes to restrict the type and amount of development that may take place on his or her property. Conservation easements have been established specifically to preserve agricultural, historic, scenic, open space, wilderness, and other types of sites. Conservation easements have evolved because high taxes on land with conservation value can force the owner to develop the land or to sell it to a developer. With a conservation easement, the owner gives up specific rights, such as the right to develop the land, in return for a reduction in taxes. The owner and the prospective easement holder identify the rights and restrictions on use that are necessary to protect the property. The owner then conveys the right to enforce those restrictions to a qualified conservation recipient, such as a public agency, a land trust, or a historic preservation organization.

A typical situation is that a valuable piece of farmland in an estate can trigger an estate tax so large that the land itself will have to be sold to pay the estate tax. The landowner can donate the property to a land trust and thereby avoid taxes on the land. In return, the owner and the heirs may retain certain rights, such as living in the original farmhouse, but they may not be able to subdivide or further develop the property. There are many kinds of easements, each tailored to the specific requirements of the donor and receiver. Examples, including the following, are given by Elfring (1989).

> *Hathaway Ranch, Monterey County, California: The Big Sur coastline is a jewel: dramatic cliffs, giant redwoods, white sand beaches. During 50 years, the Hathaway family assembled one of the largest ranch holdings on the Big Sur coast. Estate home builders were showing great interest. However, the family wanted the area protected. Through a trust, the family has conveyed 1179 acres to the U.S. Forest Service. (p. 73)*

Reductions in land taxes can be obtained in some states if the owner agrees to restrict its use. For example, in Georgia, a landowner who agrees to keep land in timber production for at least 10 years, or who agrees to protect environmentally sensitive land, can qualify for property tax reductions.

Removing Perverse Incentives

Many traditional investment incentives, tax provisions, credit and land concessions, and agricultural pricing policies create fiscal burdens for taxpayers while sacrificing long-term economic welfare and wasting resources (Repetto 1987). Many conservationists believe that, by simply removing these perverse incentives, markets would act to reduce the rate of resource depletion.

A number of tax benefits and subsidies are actually perverse incentives. To remedy the problem, Repetto et al. (1992) suggest the following:

1. *Reduce or eliminate depletion allowances for energy and other minerals.* Current law allows independent oil and gas producers, mining companies, and other enterprises that extract nonrenewable resources to deduct certain percentages of gross revenues from taxable income as depletion allowances. The more a resource is exploited and the faster it is exploited, the more that can be deducted. This policy counteracts the tendency of free markets to increase prices as supplies decrease.
2. *Reduce or eliminate depletion allowances for groundwater extraction.* Just as depletion allowances counteract the tendency of free markets to conserve increasingly scarce energy supplies, they also encourage nonefficient use of groundwater.

3. *Charge market royalties for hardrock mining on public lands.* Hardrock mining companies have been able to exploit mineral rights on public lands without payment of royalties. This has been a giveaway of publicly owned resources that also results in environmental damage.
4. *Eliminate below-cost timber sales on public lands.* Only when the full cost of resources is paid can we be sure they will be regenerated.
5. *Charge market rates for grazing rights on public lands.* The producers of beef should compete fairly with producers of pork and chicken for a share of the consumer's dollar.
6. *Charge market rates for state and federal irrigation water.* Taxpayer-subsidized irrigation results in inefficient use of water resources. Marginal land that is inefficiently farmed at taxpayer expense should be restored to its natural habitat.
7. *Charge market rates for federal power.* Just as with any subsidy, power subsidies favor a privileged few at the expense of the general public.

In developing countries, governmental land policies often represent a perverse incentive contributing to deforestation. For example, in many regions, settlers on government lands are required to "improve" the land in order to gain title. "Improvement" consists of removing the trees and establishing an agricultural crop or pastureland. This policy was understandable in times when government was attempting to establish sovereignty over wilderness areas. However, now that conditions have changed, and forests have become an endangered resource instead of a barrier to be overcome, this policy should be changed.

In some regions of Brazil, for instance, the policy of "improvement" has had a particularly perverse effect. In order to gain title to land for speculative purposes, ranchers hire crews to clear and burn large areas, purportedly for pasture. Often the pastures will be stocked with few cattle, and little care will be given to them because beef production is not really the goal. Control of land is the ranchers' real objective.

Policy on trees in developing countries also is frequently a perverse incentive. To protect trees, some national governments claim ownership of all trees for the state. This creates a commons which individuals often surreptitiously exploit. The policy denies individuals property rights to trees that they might otherwise take care of or establish. Some countries have begun to remove these obstacles. Nepal, which had nationalized all forests in 1956, later handed forests back to local communities as a policy designed to improve their management. Survival rates in China's tree planting program are reported to be much improved following the country's policy to allow peasants to own the trees they plant (Laarman and Sedjo 1992).

Disincentives

Disincentives are the opposite of incentives and have the advantage that they often are more easily eliminated, once conditions change. Fines for individuals or companies that illegally trade in endangered species, or penalties for logging on protected areas are examples of traditional disincentives. Tolls for driving cars in metropolitan areas or for commuting without ride sharing is another disincentive that can reduce pollution and congestion.

Taxes are usually disincentives. Taxes on alcohol, cigarettes, and other commodities that are "bad" for people are often made higher than taxes on necessities such as vegetables that are "good" for people. Taxing "bads" more than "goods" is one way for the government to influence behavior. To redirect people away from environmentally destructive

Table 5.1 *Types of Environmental Charges*

I. Effluent or emissions charges
  1. on water effluents
  2. on toxic releases
  3. on vehicular emissions
  4. solid waste collection and disposal charges

II. Charges on environmentally damaging activities
  1. recreational user fees on public lands
  2. highway congestion tolls
  3. noise charges on airport landings
  4. impact fees on installation of septic systems, underground storage tanks, construction projects with environmental impacts.

III. Product charges
  1. taxes based on the carbon content of fossil fuels
  2. gasoline taxes
  3. excise taxes on ozone-depleting substances
  4. taxes on agricultural chemicals
  5. taxes on virgin materials

IV. Deposit-return charges
  1. on vehicles
  2. on lead-acid and nickel-cadmium batteries
  3. on vehicle tires
  4. on beverage containers
  5. on lubricating oil

Source: *R. Repetto et al.,* Green Fees: How a Tax Shift Can Work, *World Resources Institute, Washington, D.C., 1992, Table 20, p. 73. Reprinted by permission of World Resources Institute.*

activities and toward socially desirable activities, taxes on resource exploitation and pollution could be increased, and to compensate, taxes on income and profits could be reduced (Repetto et al. 1992).

Taxes on "bads" that would promote conservation are illustrated in Table 5.1. Environmental charges can be classified as fees or taxes on toxic effluents, environmentally damaging activities, products whose use entails environmental costs, and fees (deposit-return charges) that are refundable when an environmentally damaging product is disposed of in an approved way. The idea that environmental charges should be levied against those who damage the environment is known as the polluter pays principle.

Effluent or Emissions Charges

Effluent and emissions charges include discharges of polluting wastes into air, water, and soil. Charges imposed on the basis of the volume and toxicity of such emissions are called *effluent charges* or *emissions taxes.* Several countries, including France and Sweden, impose taxes on airborne emissions of sulfur dioxide; Germany and the Netherlands levy charges on effluents discharged into surface water.

The beauty of effluent or emissions charges is that they contain a feedback mechanism that encourages desirable behavior. The less an industry pollutes the environment, the less the taxes on that industry will be, the greater the profit of that industry, the more money will be available for pollution control, and the less the industry will pollute the environ-

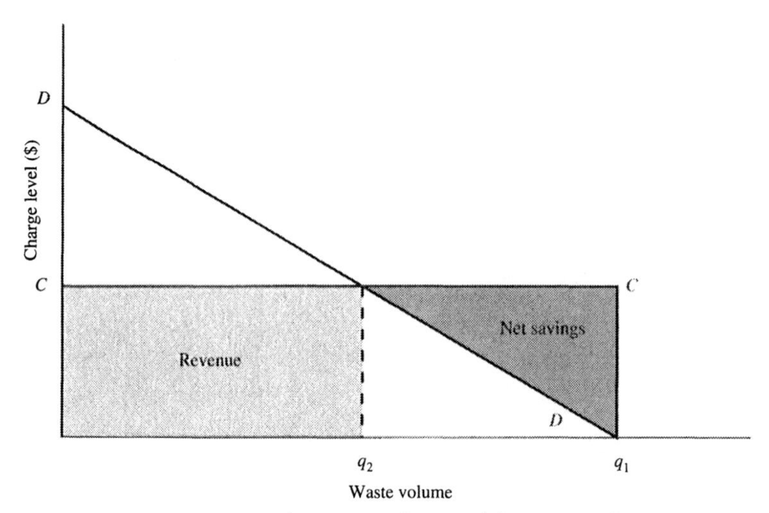

FIGURE 5.3 Revenue and net savings from a solid waste unit charge. *Source:* R. Repetto et al., 1992 *Green Fees: How a Tax Shift Can Work,* World Resources, Institute, Washington, D.C., p. 24. Reprinted by permission of World Resources Institute.

ment. This type of reinforcement is comparable to the reinforcement received by a company for offering a desirable product on the free market. The more they do it, the more money they make. In some cases, the fees could eventually lead to a complete elimination of the problem. For this reason, both conservatives and liberals favor emission charges.

Solid waste collection and disposal is rapidly becoming an important problem, as landfill areas are saturated and new areas are not available. Figure 5.3 illustrates the relationship between the appropriate charge level and incremental costs, revenues, and net economic savings. The sloping demand curve, DD, is the relationship between the charge level and the volume of waste. When services are financed through a fixed fee, the marginal charge is zero and the volume of waste is q_1. Line CC represents the incremental cost of waste disposal. It is assumed that costs per ton remain constant as the volume of wastes change. If charges are set to equal incremental costs, the volume of waste is reduced to q_2. The revenues generated are equal to the charge level per unit volume times the volume of waste collected. The net savings are the costs of waste disposal that are avoided by the reduction in volume. As the volume of garbage decreases, the net savings to the householder increase, thus compensating for the extra time and effort involved in reducing waste volume.

Charges on Environmentally Damaging Activities

Charges on environmentally damaging activities have the advantage that the costs associated with the activities fall directly on those who cause the damage rather than on the general public. This aspect should please conservatives and liberals alike.

Whenever a charge is instituted or increased for things like admission to public parks or to highways through tolls, almost always a flurry of outraged complaints is heard. These complaints, however, always emanate from the small group of users, and not from the much larger segment of the public that originally paid for the park or highway. Often, politicians or administrators will respond to such an outcry by rescinding a proposal for increased user fees. If the politicians would instead make the case to the general public

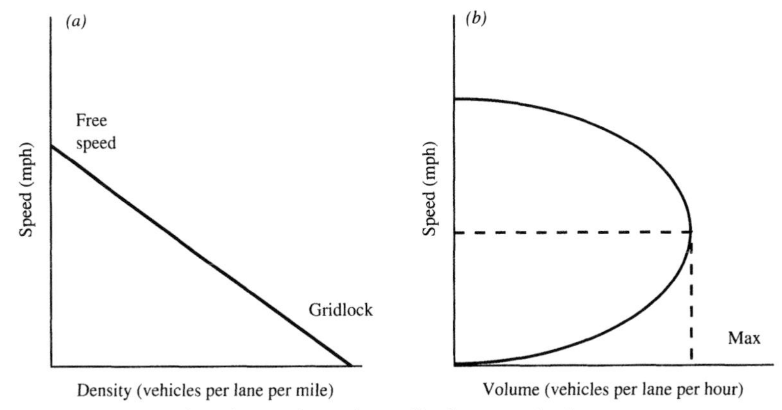

FIGURE 5.4 The relation of speed to traffic density and volume. *Source:* Repetto et al. 1992, *Green Fees: How a Tax Shift Can Work,* World Resources Institute, Washington, DC, p. 43. Reprinted by permission of World Resources Institute.

that such fees are saving them money through reduced taxes for upkeep and regulation, environmental conservation would become more successful.

Fees such as landing charges based on the amount of noise generated by airplanes, as is done in France, Switzerland, and Britain, have the same type of built-in feedback loop as effluent charges. The airlines would be rewarded for reducing the amount of noise, and these savings could be plowed back into increased noise-control measures.

Impact fees on construction projects also shift the cost of environmental damage from the general public to those who cause the damage and those who benefit from it. Housing developers complain when they are required to carry out environmental protection and amelioration associated with the construction of sewers in new urbanizations because such measures increase the cost of the houses and make them more difficult to sell. Yet it is only fair that such expenses be shifted to the contractor and eventually the home-buyer, and away from the general public who pays the taxes that otherwise would be necessary to repair the damage caused by construction.

Fees can be beneficial in other ways. Figure 5.4 illustrates how fees for rush-hour traffic can actually increase the amount of traffic handled by highways around major urban centers. Figure 5.4*a* shows the relationship between the number of vehicles on each mile of highway and the speed that is possible by those vehicles. The higher the density, the lower the speed that is possible, until gridlock ultimately stops all traffic. What density of vehicles is most desirable? Presumably that which results in the highest volume of traffic moving past a given point per unit time. The relationship between speed and volume is shown in Fig. 5.4*b*. Clearly, in areas where there is traffic congestion, it would be desirable to increase tolls so that traffic density could be reduced to the point where the volume of vehicles per hour would be maximized. Raising tolls would result in larger numbers of people taking busses and the metro, ride-sharing, bicycling, and walking. If tolls were charged only during rush hours, some drivers would be encouraged to stagger their commuting hours.

Product Charges

Product charges are levied not directly on the environmentally harmful activity itself but on a particular product used in that activity. Often there is a close link between the two. However, the product is considerably easier to monitor and tax than is the activity. An example of an environmentally harmful activity is use of gasoline because the carbon dioxide released contributes to the greenhouse effect. The product to be taxed is carbon. For any fuel, average carbon content is constant, so the weight of carbon in the fuel and that of the carbon dioxide discharged are proportional. Imposing taxes on fossil fuels is far easier than monitoring and taxing CO_2 emissions directly.

Because a carbon tax makes fossil fuels more expensive, it will alter the use of capital, labor, energy, and other economic resources. In response, businesses and households will try to lower their tax payments by reducing their use of fossil fuels and increasing their use of capital, labor, and nonfossil energy. Consumers might respond to higher electric prices by buying more efficient appliances or using those that require less electricity. Utilities might increasingly produce electricity through energy sources that emit little or no carbon, such as wind or solar power.

Taxes on virgin materials are another type of product charge. They are fees imposed on products at the point of manufacture or importation based on the quantity of virgin materials built into the product. For example, paper mills would be charged a materials fee on the basis of the percentage of their paper made from freshly cut trees as opposed to recycled pulp and paper. This would encourage manufacturers and processors to purchase equipment necessary for recycling and for the efficient collection and use of recycled materials (Gore 1992).

Deposit-Return Charges

Similar to product charges are deposit-return charges. These are product charges that are refunded when the product is returned to a designated collection point. Deposit-return charges are particularly appropriate when the policy objective is not only to discourage use of the product but also to encourage its proper disposal, including delivery to recycling facilities. In many parts of the United States, deposit-return charges are applied to beverage containers. In other countries, similar charges are levied on tires, motor oil, lead-acid batteries, and vehicles.

Environmental bonding, like deposit-return charges, encourages environmentally desirable behavior. A company or institution desiring to exploit a natural resource must post a bond. An assurance bond equal to the best estimate of the largest potential future environmental damages would be levied and kept in an interest-bearing escrow account. The bond would be returned only if the firm could prove that damages had not occurred or would not occur. If damages did occur, the bond would be used to rehabilitate or repair the environment and to compensate injured parties. The size of the bond has to be sufficient to cover any damages done to the area in case an accident occurs (Costanza 1991).

Environmental bonds could assure security and performance, or both. They could be used to ensure that a resource such as a national forest was not degraded by logging activities and that manufacturing activities conformed to certain standards of pollution emissions. Environmental bonds could be used for deep-sea oil drilling, for users of the Antarctic, and for industrial companies that pollute the air.

There are also ways to deal with the problem of hazardous waste disposal. Rather than focus on burdensome regulations requiring a cadre of inspectors to implement, a simple deposit-refund system could reward companies for proper disposal or recycling of waste. Companies would pay a large deposit when they purchase containerized hazardous chemicals. The deposit would then be returned when the used chemicals were delivered to a licensed disposal center. Those companies able to recycle toxics would pay the deposit only once as they continued to reuse the same solvents.

Tradeable Permits

Unlike a charge system, a system of tradeable permits allows the government to specify an overall level of pollution that will be tolerated. This total quantity is allotted in the form of permits among polluters. Firms that keep their emission levels below the allotted level may sell or lease their surplus allotments to other firms, or use them to offset excess emissions in other parts of their own facilities. Such a system will tend to minimize the total societal cost of achieving a given level of pollution control.

A disadvantage of tradeable permit systems is that the total cost of control is not known in advance, and the administrative costs of these systems can be very high. Tradeable permit mechanisms have been applied in the United States under the Environmental Protection Agency's Emissions Trading Program, the phasedown program of lead in automotive fuel, and chlorofluorocarbon reduction. Congress also has enacted a tradeable program for acid-rain control (Wirth and Heinz 1991).

Policy and Feedback

Feedback is what keeps a system efficient. Feedback works only when the manufacturer or resource harvester is free to respond to the market signals received and is not bound by inflexible government regulations. So to the greatest extent possible, decisions as to how to control pollution costs should be left to the individual companies or agencies. In many cases, economic incentive approaches will allow a given level of environmental protection to be achieved at lower total cost than would be possible with conventional policy. It has been estimated, for example, that the market approach to acid-rain reduction could save $1 billion per year over a dictated technological solution (Wirth and Heinz 1991).

Rather than set rigid technology-based standards, incentive-based systems allow individual firms to decide how they will achieve the required level of environmental protection. In a competitive market economy, market forces will then tend to drive these decisions toward least-cost solutions.

Policy Costs

To solve environmental problems, a whole range of policy options are available, from strict prohibitions through those with a prominent role for the market system. Table 5.2 illustrates examples for aquatic ecosystems, but the mechanisms are applicable to all types of ecosystems threatened with pollution and overexploitation. Regardless of the mechanism, each policy has a cost. Initially, the cost is borne by the manufacturer or resource extractor, but eventually it will be passed along to the consumer through higher prices.

Higher prices result in decreased consumption, and this lowers conventional indices of economic activity. A diminishing gross national product is seen to reflect badly on incumbent politicians; therefore, many legislators will vote against policies to protect the environment. If, however, as suggested in the previous chapter on economics, a clean and healthy environment and a sustainable resource base were to be given economic status and included as part of economic indices, this could sway many more politicians to support the environment. For example, if increasing purity of the air were to be counted as an increase in the gross national product equal to the amount of money spent on health problems related to air pollution, then legislators who voted for pollution control could boast about increasing gross national product instead of being guilty of decreasing it.

Table 5.2 *Institutional or Policy Mechanisms for Managing Aquatic Ecosystems and for Allocating the Use of Fish and Their Habitats.*[a]

Mechanism	Instrument of Control	Purpose or Observed Consequence
Prohibition	Exclusion of sport fish from commercial harvests	Improve recreational opportunities for anglers
	Specification of zero discharge of some toxics or contaminants	Reduce exposure of biota and humans to poisons
Regulation	Specification of low phosphorus concentrations in sewage effluents	Control eutrophication which, if intense, degrades the aquatic ecosystem
	Specification of gear and area, in fishing	Reduce fishing intensity to prevent overfishing
Direct government intervention	Control nonnative sea lamprey by lampricide, dams, etc.	Foster recovery of lake trout and other preferred species to benefit fishermen
	Development of islands and headlands with fill and dredge spoils	Provide recreational facilities and spawning areas to benefit anglers, boaters, etc.
Grants and tax incentives	Subsidy to industry for antipollution equipment	Lower pollution levels and distribute costs more widely
	Subsidy to commercial fishermen to harvest relatively undesirable species	Reduce competitive undesirable species to benefit preferred species, and their users
Buy-back programs	Government purchase and retirement of excess harvesting capacity	Reduce excess fishing capacity and compensate owners of the excess capacity
Civil law	Losers enabled to sue despoilers in civil court	Preserve ecosystem amenities for broader public, recompense losers
Insurance	Compulsory third-party insurance for claims of damage	Reduce pollution loadings because insurance premiums are scaled to loading levels
Effluent charges	Charge for waste disposal, either direct cost of treatment or indirect cost of impacts on ecosystem	Reduce pollution and/or allocate resources to high-value and/or profitable uses
License fees	Tax or charge on harvesters, scaled to level of use	Foster efficient use of resource by discouraging overcapitalization, recovering fair return for the owners (populace) of the resource
Demand management	Rates involve marginal cost pricing and/or peak responsibility pricing	Improve overall efficiency of use and foster conservation
Transferable development rights in land-use planning	Limited rights to develop one area exchanged for broader rights to develop another	Direct the development to areas preferred by government
Specific property rights, as with transferable individual quotas	Purchase of pollution loading rights to predetermined loading levels	Limit pollution and foster efficient use of resources
	Harvest rights to explicit quantities to be purchased	Limit effective fishing effort and allocate resources to high-value and/or profitable uses

[a] *Items at the top are largely administrative while those at the bottom have a prominent role for the market system. Examples relate to the Great Lakes of North America.*

Source: *Rieger and Baskerville, 1986, p. 98.*

INTERNATIONAL POLICY FOR MANAGEMENT OF THE GLOBAL COMMONS

The primary goal of international environmental policy is to protect the global commons. This section considers four avenues to approaching this goal: treaties; financial agreements; international business investment; and international trade agreements.

Treaties

Four global commons are regulated by international treaties: Antarctica; the oceans, the atmosphere; and tropical forests. Tropical forests, because they exist within the boundaries of nations, are not really a global commons. Rather, it is the biodiversity that they contain which transcends national boundaries and that is the global legacy. However, biodiversity is an abstraction, whereas tropical forests are the most significant manifestation of that abstraction.

Antarctica

Antarctica has benefited from several innovative approaches for dealing with environmental problems in the global commons. For instance, Antarctica was the first nuclear-free, military-free zone, the first territory dedicated to peace and science, the first one managed by international cooperation, and the first where no country can make claims to ownership. All these treaties have been reached by consensual agreement by all the nations who had previously claimed ownership, namely, the United States, the former Soviet Union, Great Britain, Argentina, Chile, New Zealand, Australia, France, and Norway.

In 1957 during the International Geophysical Year (IGY), the original claimant nations and several others cooperated to establish 50 research stations in Antarctica. The success of this IGY was enormous and in 1961 resulted in the creation of the Antarctica Treaty. This treaty designated Antarctica as a global commons dedicated to the peaceful pursuit of knowledge and the free exchange of scientific information. Any member country of the United Nations may join this treaty, now called the Antarctica Treaty System (ATS).

ATS members have fortified the original treaty by including agreements on conservation, including protection of wildlife, waste disposal, and mineral development. The agreement on minerals is perhaps the most controversial and is still being revised. It attempts to create institutions and rules to govern prospecting, exploration, and development of mineral deposits in Antarctica. Some conservationists oppose this agreement because the Antarctica environment is especially sensitive to pollution. They would rather have a complete ban on mineral prospecting and development, and want the region turned into a permanent park dedicated to peace and science as stated in the original 1961 treaty (World Commission on Environment and Development 1987).

There have been shortcomings. Waste from scientific camps has caused pollution problems (Mervis 1993). Despite the shortcomings, the ATS measures are more complete and more successful than any others applying to a large region of the world. They demonstrate how it would be possible to manage the global commons with multinational treaties (Douglis 1990).

The Oceans

Law of the Sea The oceans are a commons with regard to the fish that they contain, the minerals on the ocean bottom, and as a disposal site for society's waste. In recognition of common action needed to manage these resources and sinks, in 1974 over 160 countries began working on a treaty that was to be called the United Nations Convention on the Law of the Sea (LOS). By 1987 the convention had been signed by 159 nations and 32 countries had ratified it (World Commission on Environment and Development 1987).

The agreement gives any coastal country legal rights over all marine fishery resources and ocean mineral resources in an Exclusive Economic Zone, which extends 200 nautical miles from its shores. Foreign fishing and mining operations for oil, natural gas, and other minerals are allowed in these zones only with permission and licensing from the government of each coastal country. It also provides for the regulation of ocean pollution and the conservation of living resources, including increased protection of marine mammals.

Resources in the high seas are recognized as common property to be shared by the entire world and managed by the United Nations under international law. The treaty allows any country to remove minerals from the deep seabeds outside of the Exclusive Economic Zones. However, the resources or the profits from them should be shared by all countries. Several countries including the United States, Germany, and the United Kingdom oppose this provision because it forces private enterprises to share profits with others who have not invested in the infrastructure necessary to exploit the resources.

Whaling The International Whaling Commission (IWC) was established in 1946 to regulate the whaling industry. The IWC has set annual quotas to prevent overharvesting and extinction, but because it has no powers of enforcement, it has been unable to stop the decline of many whale species. Whaling countries have often ignored quotas on the basis that the population data are inadequate. Although the data may be inadequate, in such instances it is better to apply the precautionary principle: that is, assume the worst possible case because the consequences of ignoring the data could be extinction of species.

In 1970 the United States stopped all commercial whaling and banned all imports of whale products. Since then, conservation groups and many governments have called for a ban on all commercial whaling. In 1982 the IWC established a five-year halt on commercial whaling, starting in 1986. Although Japan, the Soviet Union and Norway initially planned to ignore the ban, international criticism eventually led them to reconsider. However, Japan announced it would take 875 whales each year for "scientific" purposes, and Norway and Iceland announced similar plans.

The Atmosphere

The increasing use of refrigerators, air conditioners, and similar products has resulted in the increasing amounts of chlorofluorocarbons (CFCs), chemicals that have commonly been used in these appliances and similar products. As use of refrigerants grew, an increasing amount escaped into the atmosphere and made their way to the stratosphere. There they appear to act as a catalyst that breaks down ozone in the high atmosphere (Firor 1990). Stratospheric ozone plays an important role with respect to life on earth. Ozone's structure allows it to absorb ultraviolet sunlight that would otherwise reach the surface of the earth and affect living material. Growing fear that CFCs could break down ozone in the stratosphere to a dangerous level led the international community to take unprecedented steps to control and ultimately ban the productions of CFCs and other ozone-depleting substances such as halons and carbon tetrachloride by the year 2000. The Vienna Convention and the subsequent Montreal Protocol on Substances That Deplete the Ozone Layer, adopted in 1987 and strengthened in 1990, set strict timetables for phasing out these substances and established rules governing their international trade. In 1991 new scientific evidence that the ozone was thinning more quickly than predicted prompted calls to speed up the phaseout. About 70 nations had signed the convention by July 1991.

The costs of converting to an ozone-friendly technology have caused skepticism in many developing nations. In response to this concern, the treaty was amended to create a $160 million fund, financed by the largest CFC users, to compensate developing countries for the added costs they incur in following a CFC-free path (World Resources Institute 1992).

Tropical Forests

The boom and bust export pattern in the tropical timber trade is often blamed on demand by developed countries, high import barriers, and low international wood prices. More important reasons may be rooted in the tropical countries' own policies relating to timber concessions and wood-processing industries. These policies suppress timber scarcity signals and must be revised if the trade is to promote sustained economic growth (Vincent 1992).

The major international mechanisms that address tropical forest conservation are the Tropical Forestry Action Plan (TFAP) and the International Tropical Timber Organization (ITTO), both created in the mid-1980s.

The Committee on Forest Development in the Tropics is a statutory body of the United Nations Food and Agriculture Organization (FAO). It was initially composed of 45 member countries "to study and report on the international programs in tropical forestry and on the concerted action which could be undertaken by governments and international organizations to ensure the development and rational utilization of tropical forests and related resources" (FAO 1985). The TFAP is a result of the work of this body.

Financing for TFAP projects has relied heavily on donor countries and on their perceptions of what suitable forest management should be for the recipient countries. Little attention has been given to the broader social and conservation needs of the recipient country, and how they would be affected by industrial forestry. For example, most industrial forestry projects are monoculture plantations, which do not provide for local community needs as do natural forests, nor do they provide ecosystem services such as wildlife habitat.

The ITTO is a commodity trade organization, with a secretariat based in Yokohama, Japan. Much of the organization's work is tied to projects that are funded through the aid agencies of consumer nations (Oldfield 1989). Conservation of natural forests is given attention because native forests are still the primary resource on which the tropical timber trade is based. Tropical forest plantations still contribute only a very small proportion.

Neither TFAP nor ITTO is as effective as they could be in promoting conservation and sustainable use of tropical forests. Buschbacher (1992) has suggested three institutional reforms that would help address the needs that are not currently being met.

1. A Global Forest Convention such as that considered but not pursued by the United Nations Conference on Environment and Development could improve management practices in temperate and boreal forests as well as tropical forests.
2. Consumer information mechanisms ("labels") should be instituted that would rigorously identify the social and environmental impact of forest products. Labeling should provide recognizable symbols, so that consumers can identify sustainably produced lumber and other manufactured wood products such as furniture. Labels would give

the consumer a choice. If the label also signified that the wood met quality standards, then sustainably produced lumber would command a premium price.

3. Industrialized countries must increase their contribution of resources to goals that promote conservation and social welfare, and the Southern countries should agree to conditionality and monitoring. Examples of such arrangements are given in the section on Financial Agreements.

UNCED

The United Nations Conference on Environment and Development (UNCED) held in Rio de Janeiro, June 3–14, 1992, was the most significant environmental meeting ever to occur on a global scale. Some of the most important documents, as reported by the *New York Times,* June 15, 1992, were:

Biodiversity Convention. This is a legally binding treaty that requires inventories of plants and wildlife, and plans to protect endangered species. It also requires signers to share research profits and technology with nations whose genetic resources they use. This could provide incentives and capital for the protection of habitat in developing countries. The United States did not sign in 1992, but after his election in 1993, President Clinton signed the document.

Global Warming Convention. This is a legally binding treaty that recommends curbing emissions of carbon dioxide, methane, and other "greenhouse" gases thought to warm the climate by trapping the sun's heat close to the earth. The rich countries are asked to aim at returning to 1990 levels, although they are not required to do so. They do have to report periodically on their progress. The United States signed but insisted that the document not set targets for reducing carbon dioxide emissions.

Declaration on Environment and Development. This is the nonbinding statement of 27 principles (Table 5.3) for guiding environmental policy that emphasizes protecting the environment as part of economic development, safeguarding the ecological systems of other nations, and giving priority to the needs of developing countries that are the "most environmentally vulnerable." It was adopted by consensus.

Agenda 21. This is a nonbinding blueprint for what governments and agencies should do to protect the environment while letting Third World industries grow in the twenty-first century. It was adopted by consensus, after developing countries abandoned their demand for specific commitments of aid from industrialized nations to pay for the plan.

Statement of Forest Principles This nonbinding document recommends that countries assess the impact of economic development on their forests and take steps, both individually and with other countries, to minimize the damage. It was adopted by consensus. Negotiations for a legally binding treaty collapsed when developing countries complained that it focused on stopping them from burning tropical forests but did nothing to control logging in the United States, Europe, and Canada.

Financial Agreements

International treaties and conventions such as those resulting from UNCED are one type of mechanism for dealing with environmental problems that transcend national boundaries. Financial agreements are also a tool for dealing with conserving the global commons.

Table 5.3 *United Nations Conference on Environment and Development: Rio Declaration on Environment and Development.**
[Adopted at Rio de Janeiro, June 14, 1992] †Cite as 31 I.L.M. 874 (1992) †

I.L.M. Content Summary

TEXT OF DECLARATION–I.L.M.PAGE 876

Table 5.3 *Continued*

ADOPTION OF AGREEMENTS ON ENVIRONMENT AND DEVELOPMENT

The Rio Declaration on Environment and Development Preamble

The United Nations Conference on Environment and Development,
Having met at Rio de Janeiro from 3 to 14 June 1992,
Reaffirming the Declaration of the United Nations conference on the Human Environment, adopted at Stockholm on 16 June 1972, and seeking to build upon it.
With the goal of establishing a new and equitable global partnership through the creation of new levels of cooperation among States, key sectors of societies and people,
Working towards international agreements which respect the interests of all and protect the integrity of the global environmental and developmental system,
Recognizing the integral and interdependent nature of the Earth, our home,
Proclaims that:

Principle 1

Human beings are at the centre of concerns for sustainable development. They are entitled to a healthy and productive life in harmony with nature.

Principle 2

States have, in accordance with the Charter of the United Nations and the principles of international law, the sovereign right to exploit their own resources pursuant to their own environmental and developmental policies, and the responsibility to ensure that activities within their jurisdiction or control do not cause damage to the environment of other States or areas beyond the limits of national jurisdiction.

Principle 3

The right to development must be fulfilled so as to equitably meet developmental and environmental needs of present and future generations.

Principle 4

In order to achieve sustainable development, environmental protection shall constitute an integral part of the development process and cannot be considered in isolation from it.

Principle 5

All States and all people shall cooperate in the essential task of eradicating poverty as an indispensable requirement for sustainable development, in order to decrease the disparities in standards of living and better meet the needs of the majority of the people of the world.

Principle 6

The special situation and needs of developing countries, particularly the least developed and those most environmentally vulnerable, shall be given special priority. International actions in the field of environment and development should also address the interests and needs of all countries.

Principle 7

States shall cooperate in a spirit of global partnership to conserve, protect and restore the health and integrity of the Earth's ecosystem. In view of the different contributions to global environmental degradation, States have common but differentiated responsibilities. The developed countries acknowledge the responsibility that they bear in the international pursuit of sustainable development in view of the pressures their societies place on the global environment and of the technologies and financial resources they command.

Principle 8

To achieve sustainable development and a higher quality of life for all people, States should reduce and eliminate unsustainable patterns of production and consumption and promote appropriate demographic policies.

Principle 9

States should cooperate to strengthen endogenous capacity-building for sustainable development by improving scientific understanding through exchanges of scientific and technological knowledge, and by enhancing the development, adaptation, diffusion and transfer of technologies, including new and innovative technologies.

Table 5.3 Continued

Principle 10

Environmental issues are best handled with the participation of all concerned citizens, at the relevant level. At the national level, each individual shall have appropriate access to information concerning the environment that is held by public authorities, including information on hazardous materials and activities in their communities, and the opportunity to participate in decision-making processes. States shall facilitate and encourage public awareness and participation by making information widely available. Effective access to judicial and administrative proceedings, including redress and remedy, shall be provided.

Principle 11

States shall enact effective environmental legislation. Environmental standards, management objectives and priorities should reflect the environmental and developmental context to which they apply. Standards applied by some countries may be inappropriate and of unwarranted economic and social cost to other countries, in particular developing countries.

Principle 12

States should cooperate to promote a supportive and open international economic system that would lead to economic growth and sustainable development in all countries, to better address the problems of environmental degradation. Trade policy measures for environmental purposes should not constitute a means of arbitrary or unjustifiable discrimination or a disguised restriction on international trade. Unilateral actions to deal with environmental challenges outside the jurisdiction of the importing country should be avoided. Environmental measures addressing transboundary or global environmental problems should, as far as possible, be based on an international consensus.

Principle 13

States shall develop national law regarding liability and compensation for the victims of pollution and other environmental damage. States shall also cooperate in an expeditious and more determined manner to develop further international law regarding liability and compensation for adverse effects of environmental damage caused by activities within their jurisdiction or control to areas beyond their jurisdiction.

Principle 14

States should effectively cooperate to discourage or prevent the relocation and transfer to other States of any activities and substances that cause severe environmental degradation or are found to be harmful to human health.

Principle 15

In order to protect the environment, the precautionary approach shall be widely applied by States according to their capabilities. Where there are threats of serious or irreversible damage, lack of full scientific certainty shall not be used as a reason for postponing cost-effective measures to prevent environmental degradation.

Principle 16

National authorities should endeavour to promote the internalization of environmental costs and the use of economic instruments, taking into account the approach that the polluter should, in principle, bear the cost of pollution, with due regard to the public interest and without distorting international trade and investment.

Principle 17

Environmental impact assessment, as a national instrument, shall be undertaken for proposed activities that are likely to have a significant adverse impact on the environment and are subject to a decision of a competent national authority.

Principle 18

States shall immediately notify other States of any natural disasters or other emergencies that are likely to produce sudden harmful effects on the environment of those States. Every effort shall be made by the international community to help States so afflicted.

Table 5.3 *Continued*

Principle 19

States shall provide prior and timely notification and relevant information to potentially affected States on activities that may have a significant adverse transboundary environmental effect and shall consult with those States at an early stage and in good faith.

Principle 20

Women have a vital role in environmental management and development. Their full participation is therefore essential to achieve sustainable development.

Principle 21

The creativity, ideals and courage of the youth of the world should be mobilized to forge a global partnership in order to achieve sustainable development and ensure a better future for all.

Principle 22

Indigenous people and their communities, and other local communities, have a vital role in environmental management and development because of their knowledge and traditional practices. States should recognize and duly support their identity, culture and interests and enable their effective participation in the achievement of sustainable development.

Principle 23

The environment and natural resources of people under oppression, domination and occupation shall be protected.

Principle 24

Warfare is inherently destructive of sustainable development. States shall therefore respect international law providing protection for the environment in times of armed conflict and cooperate in its further development, as necessary.

Principle 25

Peace, development and environmental protection are interdependent and indivisible.

Principle 26

States shall resolve all their environmental disputes peacefully and by appropriate means in accordance with the Charter of the United Nations.

Principle 27

States and people shall cooperate in good faith and in a spirit of partnership in the fulfilment of the principles embodied in this Declaration and in the further development of International law in the field of sustainable development.

* *[Reproduced from UNCED document A/CONF.151/5/Rev. 1, June 13, 1992. UNCED recommended that the U.N. General Assembly endorse the Declaration at its 47th Session.*

[The Introductory Note for UNCED at 31 I.L.M. 814 (1992) discusses the Rio Declaration. Three other UNCED documents are reproduced in I.L.M.: the Framework Convention on Climate Change, 31 I.L.M. 848 (1992); the Convention on Biological Diversity, 31 I.L.M. 818 (1992); and the Statement of Principles for a Global Consensus of the Management, Conservation and Sustainable Development of All Types of Forests, 31 I.L.M. 881 (1992).]

Buying Reserves

Governments of developing countries do not have enough capital to adequately manage their nation's biological resources such as rain forests and coral reefs. One suggestion on how to finance the cost of conserving rain forests is to create parks and reserves in less developed tropical countries, financed by contributions from richer nations. The money would come from a progressive tax based on each nation's GNP (Myers 1992).

Trust Funds

Another approach to financing the management of the global commons is to create a Global Commons Trust Fund (Stone 1992). International taxes and fines on pollution beyond national borders could serve to reduce effluents and form a base for financing the

Trust Fund. There would be problems, however. Any plan to apply mechanisms for effective international regulation of the global commons needs to be politically acceptable. It also has to be organizationally, economically, and socially feasible to implement. Last but not least, it has to recognize and permit diversity in attitudes and circumstances.

The organizational infrastructure needed to do those jobs is already in place. It includes, along with the United Nations Environment Program (UNEP), the Global Environmental Facility (GEF), the World Meteorological Organization (WMO), the World Wide Fund for Nature (WWF), the U.N. Food and Agriculture Organization (FAO), and the International Union for Conservation of Nature (IUCN). However, the heavy international regulation needed to implement a Global Commons Trust Fund makes the plan very difficult to implement. The need for regulation, taxation, and subsidies at the international level raises the question of who will police the world. The United Nations is perhaps the most appropriate institution, but the bureaucratic problems would be immense.

Debt-for-Nature Swaps

The developing nations of the world owe more than a trillion dollars to public and private lending institutions as well as governments and multilateral development banks. Repayment of the debt decreases the funds available to debtor nations for investing in infrastructure and for remedying social and environmental problems. Nonrepayment of foreign loans has produced a decline in foreign investment in many of the Third World nations.

Debt-for-nature swaps are a mechanism through which debtor nations can stimulate conservation, while at the same time reducing their debt. Here is how they work: A United States or European bank holds commercial loans from a Third World country that is unable to pay them. The bank therefore puts these loans up for sale on the secondary market at a very attractive discount. For the lending bank, this may be the only way of getting at least some of the money back. International conservation institutions buy these discounted loans and then negotiate them with the government of the debtor nation, usually by forgiving the purchased debt in exchange for commitments of equivalent sums in local currency to a conservation fund. These funds can be used to purchase land for nature reserves (Hultkrans 1988).

Debt-for-nature swaps have both advantages and disadvantages.

ADVANTAGES

Partial foreign debts are forgiven, thus decreasing the nation's total debt.

Debtor governments can use local currency, rather than having to sell more natural resources to repay with dollars.

Swaps increase the ability of conservation agencies to preserve land with limited funds.

In the United States, the lending banks are permitted a charitable deduction on their income tax, equivalent to the discounted price of the debt.

To take advantage of the swaps, governments must act immediately to address conservation problems.

The important issue of sovereignty is not a real problem because the owner country retains ownership.

DISADVANTAGES

Debt-for-nature swaps demand that the local governments allocate resources to be applied to conservation projects that are defined with little or no popular participation.

Input from the local populace is necessary to gain their enthusiasm and support for the project.

Debt-for-nature swaps have not been used widely enough to relieve the nation's foreign debt significantly. However, in some cases they have encouraged the donation of new money from private sources.

The swaps may have inflationary effects in the debtor country. To fund the internal conservation obligations of a debt-for-nature swap, the government of a debtor country may simply print more money.

In order to qualify for a tax deduction, the U.S. bank or agency must make the donation to a U.S.-based organization, which retains primary control over the funds and the funded activities in the debtor nation. This situation appears to create the problem of foreign control of domestic resources, and it becomes a great disincentive. However, the funds can be funneled through an organization such as the Nature Conservancy which serves merely as a conduit.

Many regard debt-for-nature swaps as a strategy for private appropriation of resources regarded as a common heritage.

Some observers claim that the nature swaps divert attention from the main conflict: the extraction and transference of a significant part of the labor, natural resources, and wealth of Third World countries to the capitalist economy of exploitative countries. Some countries have declared that the debts should already be declared to be repaid. The argument is that the prices of resources and labor exported to the capitalist economy were far lower than their replacement costs. The "debt" merely compensates for the difference between the price paid and the true value of the resources, including the replacement costs of those resources, and the services of nature that were lost when the resources were harvested (Schmidheiny 1992).

The Beni Biosphere Reserve in Bolivia is an example of a debt-for-nature swap. In 1987 Conservation International (CI) purchased $650,000 of Bolivian commercial bank debt for $100,000 from an affiliate of Citibank. In return for CI's agreement to extinguish the debt obligation, the Bolivian government committed itself to establishing a currency fund equaling $250,000 to pay for the operating costs of managing an existing biosphere reserve. The government also agreed to create a buffer zone around the reserve equaling approximately 3.2 million acres (Kelly 1989).

In many cases, debt-for-nature swaps have been a powerful tool for conservation, but success can lead to their undoing since they can increase the market value of the debt on which they bid.

Global Environmental Facility

International investments must turn from expenditures that enrich the politically elite while despoiling the environment, toward rehabilitating the ability of nature to provide for all. Investments must shift from human-made capital accumulation toward natural capital restoration. In addition, technology should be aimed at increasing the productivity of natural capital more than human-made capital. "If these two things do not happen, we will be behaving uneconomically, in the most orthodox sense of the word" (Daly 1991c. p. 34).

The role of multilateral development banks in the new era would be increasingly to make investments that replenish the stock and that increase the productivity of natural capital. Instead of investing mainly in sawmills, fishing boats, and refineries, development banks should now invest more in reforestation, restocking of fish, and sustainable agriculture. Because natural capacity to absorb wastes is also a vital resource, investments that preserve that capacity through pollution reduction also should become a higher priority.

As a start, the World Bank, the United Nations Environmental Program, and the United Nations Development Program, have an exploratory program of biosphere investment known as the Global Environmental Facility which will provide programs for investing in the preservation or enhancement of four classes of biospheric infrastructure or nonmarketed natural capital. These are: (1) protection of the ozone layer; (2) reduction of greenhouse gas emissions; (3) protection of international water resources; and (4) protection of biodiversity.

International Aid

The international aid that the United States supplied to Western Europe and parts of Asia after World War II has been praised as successful and evaluated as essential to the flourishing of democracies in these regions (Gore 1992). International aid from the United States, Europe, and Japan to tropical countries in recent years has been much less successful. Although there has been some success in temporarily slowing short-term disasters such as starvation in Somalia, long-term successes are difficult to identify. Part of the problem is due to market distortions such as dumping of agricultural surpluses by donor countries, which discourages local agricultural entrepreneurs. "Tied aid," assistance given on the condition of purchases from the donor nation, often forces poor countries to buy materials of dubious value or to pay more for useful goods than they would have in open competitive bidding. In 1990 assistance to developing countries from the Organization for Economic Cooperation and Development (OECD) totaled $62.6 billion, yet developing countries were estimated to lose at least $150 billion in revenues owing to barriers to trade (Schmidheiny 1992). Many poor countries would be better off if industrial nations discontinued such aid and instead opened more markets for products from the poor countries.

Reform is difficult, because large international aid organizations may be bound by counterproductive rules. Aid often is intended to help the disadvantaged of a society, but large aid organizations often must deal with the national government where interests and concerns are different from those at the local level. Smaller nongovernmental organizations are sometimes more effective in understanding and reaching the people who really need the help.

International Business Investment

International business investment is a potentially powerful tool for dealing with environmental problems that transcend national boundaries.

Areas for Action

In addition to internalizing environmental and social costs, sending appropriate market signals, allowing more liberal trade, and increasing investment in natural resources, there are at least six additional areas where action could be taken (Schmidheiny 1992).

Population Growth To control population growth, it may be necessary first to lower infant mortality. The idea that to produce fewer people more children must be kept alive may seem to contradict common sense. However, families will have fewer children only if a mother knows that her existing children will have a good chance of surviving into a relatively healthy and productive adulthood.

Infant mortality rates have fallen, and contraceptive use has increased where governments have educated both women and men, enlarging their options for participation in public and business life. However, it is important that all people have access to basic needs such as food, land, housing, and health care. Business can help governments grow in a socially and ecologically appropriate way by creating jobs and providing education. Policy reforms that allocate more resources for social and environmental needs are implemented more easily in a flourishing economy.

Poverty, Migration, and the Environment Business has a vested interest in land use that is economically and environmentally sound. Business investment in sustainable agriculture and forestry, together with their related industries, can relieve poverty and aid the environment in rural areas, while at the same time slowing down migration to cities that results in urban congestion.

Indebtedness International debt, coupled with rising interest rates, have been a major impediment to economic growth, to reduction of poverty, and to improvement in environmental conditions. Many countries such as Mexico have lessened their problems of debt by privatizing government monopolies in industry and banking, thereby cutting public deficits, attracting foreign and local investment, and encouraging internal saving rates.

Ineffective Rules Why is there less investment than needed in most developing countries? The attractiveness and reliability of the "rules of the game" of a given country determine its business climate and either encourage or discourage local and foreign investors. The main elements of an attractive investment climate are macroeconomic stability; free, open markets; clear property rights; and political stability.

Small Enterprises From a business perspective, one of the key ways of alleviating poverty in the developing world and spreading entrepreneurial talent is through encouraging the growth of small and medium-sized enterprises. Besides their finer-tuning to local social conditions and environmental restraints, small enterprises are much less susceptible to the large-scale corruption that so often plagues huge development schemes. However, a decentralized network needs the diffusion of useful information and credit to those who need it. It is here that business can play a role, through local or regional nodes of production and management expertise.

Education and Training Training entrepreneurs in foreign countries is a relatively new idea. Many observers believe that entrepreneurs are born, not made. Nevertheless, education and training in modern professional management methods and practices are essential for preparing future successful businesspeople. Governments and business together should develop special programs of education for ecologically and socially compatible technology.

Ecotourism as an Example

The idea behind ecotourism is to offer expeditions to unspoiled, inaccessible parts of the world that would interest natural history enthusiasts, or tourists looking for learning through adventure. As ecotourism developed, it became apparent that these expeditions resulted in economic benefits for local communities. Travel agencies needed local guides who knew remote areas well. They also needed people to prepare food, drive jeeps, carry packs, and offer overnight lodging. At the most popular ecotourist sites, they needed more and better accommodations.

Ecotourism has boomed in some parts of the world. In 1980 around 60,000 U.S. citizens visited Costa Rica. By 1988 the number had grown to 102,000 (Miller and Tangley 1991). Costa Rica's stable democratic tradition made visitors feel secure, and within just a few days, an energetic ecotourist could visit lowland rain forests, high-elevation cloud forests, dry forests, active volcanoes, river rapids, and remote, palm-lined beaches.

Ecotourism is also experiencing growing pains. In some cases, much of the tourist dollar goes into foreign rather than local pockets. When local residents are hired only for menial, low-paying jobs while foreign travel agencies make bountiful profits, tourism generates resentment rather than support for national parks and forests. High levels of tourist traffic also can have negative effects on parks. A high need for firewood to cook meals for tourists can contribute to deforestation. Pollution from beverage cans, film containers, and other human activities can destroy scenic values. Crowds ruin the experience for many visitors.

On balance, ecotourism has a very high potential for conservation. To achieve a pattern of success, however, conservationists will need to act to prevent potential problems.

International Trade Agreements

International trade agreements are a fourth type of policy with long-term potential for global environmental conservation. Without adequate provisions, however, they could have a short-term damaging effect on the environment. Consequently, conservationists are divided about whether to support them.

Free Trade Versus Environmental Protection

International agreements and treaties are generally based on the philosophy that world economic development can be improved by eliminating barriers to free trade. A country imposes trade barriers in order to protect a particular segment of its economy. For example, at one time the United States had trade barriers against the importation of certain classes of cotton clothing. The purpose was to protect the U.S. textile industry against competition from Asian countries, where the goods could be manufactured at a much lower price owing to low wages. Although the barriers did protect the textile industry, its overall result was to increase the cost of living for the rest of Americans and to deprive some Asians of legitimate work. Elimination of the barriers put many U.S. textile mills out of business, but that loss was considered to be outweighed by the lowered cost of goods to the entire American public and a greater world stability resulting from higher employment in Asia.

Some environmentalists believe that free trade is a dire threat to the environment. Goldsmith (1990) fears that if the present course is continued, "then the entire world will effectively be transformed into a vast Free Trade Zone, within which human, social and environmental imperatives will be ruthlessly and systematically subordinated to the purely selfish, short-term financial interests of a few transnational corporations" (p. 204). He believes that the solution "is not to increase the freedom of commercial concerns but, on the contrary, to bring those concerns back under control—to limit the size of markets rather than expand them; to give local people control of their resources." Goldsmith supports the view that development should be restricted. If development continues in the Third World, he believes that much of the planet will be "rendered unfit for human habitation" (p. 204).

An alternative view holds that Third World countries now too poor to afford environmental protection will begin reducing pollution once incomes rise to a level sufficient to meet basic needs. Grossman and Kruger (1991) studied air pollution in the urban areas of 42 countries and found "that concentrations increase with per capita Gross Domestic Product at low levels of national income, but decrease with GDP growth at higher levels of income. The turning point comes at about $5000. Since per capita GDP in Mexico is now at this level, Grossman and Krueger conclude that "we might expect that further growth in Mexico, as may result from a free trade agreement with the U.S. and Canada, will lead the country to intensify its efforts to alleviate its environmental problems" (pp. 19, 20).

Rather than trying to stop growth, a majority of environmental and conservation organizations have directed their influence to ensuring the sustainability of the development created by free trade. In testimony before Congress, the Environmental Defense Fund advocated "strategies that combine trade liberalization with measures that protect the environment." Still, the Sierra Club and Greenpeace fear that free trade agreements will cause more environmental harm than good.

Pollution Havens Some observers are concerned that free trade will result in the creation of "pollution havens." For example, as a result of the North American Free Trade Agreement (NAFTA), it is feared that businesses may move to Mexico to escape environmental regulation, and that U.S. and Canadian producers will face competition from cheap Mexican goods manufactured without the costs of environmental protection.

The 100-km wide area on either side of the border is cited as an example of what could go wrong. This partial free trade zone established in 1965 is known as the "border area." Assembly plants called *maquiladoras* have been encouraged to locate there so that they can take advantage of low labor rates. The worst problems center around water. Sewage treatment is inadequate or nonexistent. Juarez, for example, has no sewage treatment at all. As wells go dry from overuse, people increasingly draw water from the Rio Grande River, which also doubles as a disposal for sewage and agricultural runoff contaminated with pesticides. Pathogens for polio, dystentery, cholera, typhoid, and hepatitis found in the water could lead to a public health disaster at any time, especially during periods of low water. Another major concern is hazardous waste. U.S. companies are required to transport this waste back over the border for disposal but don't comply owing to lax enforcement. Beyond the threat to the human population, these toxic substances along with water shortages and clearing of land are threatening the area's once rich wildlife.

The border area has overdeveloped at the expense of underdevelopment in the rest of Mexico. NAFTA hopefully would reduce environmental problems on the border by taking away the artificial incentive for companies to locate there. Language contained within NAFTA dealing with investments should discourage future "pollution havens."

Environmental Standards Conservationists are also concerned about the preemption of environmental standards. Perhaps the best known example of this concern is the recent tuna–dolphin dispute.

Acting on the requirements of its 1988 Marine Mammal Protection Act, the United States prohibited the import of yellowfin tuna from Mexico because the incidental taking rate of dolphin (actually porpoise) exceeded that of domestic vessels by more than 1.25 times. Mexico reacted by filing a complaint with the GATT court and won. To satisfy the court, the United States would have to weaken its law designed to prevent the unnecessary drowning of dolphins in fishing nets. Despite the ruling in its favor, Mexico decided to join the United States and eight other nations in negotiations through the Inter-American Tropical Tuna Commission. In April 1992, an agreement was reached to reduce the killing of dolphins by 80 percent during this decade.

Public Scrutiny A third concern about free trade centers on lack of public scrutiny. This is the fear that because trade negotiations generally occur behind closed doors, public input will be blocked on issues that may be environmentally sensitive. To address this issue in the case of NAFTA, an Environmental Commission was established. Among the purposes of the commission are promoting cooperative environmental problem solving, providing information, and ensuring effective public participation. The initial reception of the idea was enthusiastic, but continued input from environmental groups will be needed to make it function effectively.

Other potential problems that could arise from global free trade, unless specific agreements are reached, are as follows:

Trade in endangered species. Unrestricted trade in endangered species will lead to their extinction.

Trade in dangerous substances such as hazardous waste. Unrestricted allowances will result in dangerous temptations for countries desperate for cash.

Food safety and related health problems. Under unrestricted free trade agreements, countries could not forbid food imports despite dangerous loads of pesticides.

Regulation and protection of common resources. Unrestricted trade agreements could encourage the mining of the global commons such as ocean bottoms.

Trade involving production processes that have a deleterious effect on ecosystems or health. Unrestricted trade agreements could not prohibit the importation of goods, even though their manufacture caused sickness among workers or caused environmental devastation owing to the disruption of ecosystem processes.

Trade resulting from subsidies that distort the economy. A ban on beef from cattle raised in pastures made from tropical forests would be impossible under unrestricted free trade rules.

Many of these issues have been dealt with under free trade agreements. It is important that they be considered in future agreements to lower international barriers against trade.

GATT

In 1944 the United States hosted an international monetary conference at Bretton Woods, New Hampshire, which saw the birth of the United Nations, the World Bank, and the International Monetary Fund. At the conference, the U.S. and British governments maintained that trade conflict was one of the root causes of World War II. To encourage and regulate world trade, they proposed an International Trade Organization with binding rules. For reasons of sovereignty, the U.S. Congress rejected the treaty, known as the Havana Charter.

In 1947, 25 countries met in Geneva to negotiate a General Agreement on Tariffs and Trade (GATT). Because this pact was a nonbinding agreement, U.S. Congressional approval was not necessary. The United States embraced GATT as a step that promoted the benefits of liberalized trade and the free enterprise system. As co-signatories of the agreement, the United States and Canada agreed to reduce tariffs through multilateral negotiations. Following GATT, both countries enjoyed a prolonged period of prosperity in which they continued to participate in periodic rounds of negotiations that lowered tariffs further. During this period, increasing concern about conserving natural resources led to agreements with trade provisions covering the protection of whales, quarantine of plants, and other topics.

Although the United States and Canada mutually signed these side agreements, inclusion of provisions for environmental protection within GATT itself was opposed because such provisions were seen as restrictions to free trade. Such problems were among the factors that led to the stalling of the talks begun in Uruguay in 1986 (Spero 1990).

How should GATT rules be changed so that the agreement can cope with the conflict between concerns for free trade and concerns for the environment? The following principles should be introduced into GATT law to address environmental issues (Schmidheiny 1992).

- *Transparency.* Notification requirements need to be introduced so that all environmental regulations with potential trade impacts become internationally known, and their provisions made unambiguous. No international forum currently exists to which environmental regulations may be reported and where trade implications can be assessed. Such a forum could be created under GATT.
- *Legitimacy.* Environmental measures that restrict trade should be legitimate, and thus backed by strong scientific evidence. International panels of scientific experts should be established to test the legitimacy of such measures. Where environmental threats are particularly serious or irreversible, GATT should adopt precautions, erring on the side of prudence.
- *Proportionality.* Trade-restrictive measures should not go beyond what is absolutely necessary to produce a desired environmental result. Internationally accepted advisory panels and criteria would be required.
- *Subsidiarity.* Every time an environmental goal can be achieved without a measure affecting trade, trade-related measures should be avoided.

NAFTA

To increase economic growth and stimulate progress on world trade talks in the mid-1980s, Canadian Prime Minister Brian Mulroney and U.S. President Ronald Reagan initiated bilateral free trade talks between the countries. Following successful negotiations, the U.S.-Canada Free Trade Agreement went into effect in January 1989. The agreement eliminates all bilateral tariffs by 1998, and it reduces some nontariff barriers as well. The countries agreed to submit trade disputes to a binational panel of trade experts.

Shortly after the Free Trade Agreement went into effect, the Canadians asked for a reduction in the emissions of coal-burning plants in the United States because of the pollutants that were "exported" across the border by prevailing winds and were falling in Canada as acid rain. Using the good-will developed during the trade negotiations, Prime Minister Mulroney persuaded President Bush to begin talks that resulted in a separate agreement to limit air pollution originating in one country and adversely affecting the other.

Meanwhile, south of the United States, Mexico was undergoing a period of economic turmoil because of its inability to repay its huge foreign debt. Thus, environmental laws that had been on the books were widely ignored. Conservation was an idea that Mexico could ill afford at this juncture. The concept of free trade would also have faced harsh criticism as "bargaining with the imperial power to the North." However, half a century of bad experience with nationalization, especially in banks and the oil industry, left Mexico ready to reexamine its stance on free trade.

Mexico renegotiated payments on its debt and, in an effort to stimulate exports, joined GATT in 1986. In 1988 Mexico passed the General Law for Ecological Equilibrium and Environmental Protection. In 1990 President Bush announced the Enterprise for the Americas Initiative, calling for closer U.S.–Latin American ties. The initiative was shaped by the U.S. view that what was needed most was not monetary foreign aid but thorough market reform.

A short time later, Canada joined the negotiations. Following "fast-track" procedures that prevent amendments as filibusters, the North American Free Trade Agreement (NAFTA) was ceremoniously initialed in October 1992 by George Bush, Mexican President Carlos Salinas de Gotari, and Canadian Prime Minister Mulroney. For countries that had worked so long to keep out the influence of the United States, Canada and Mexico had undergone revolutionary change.

Canada and Mexico are the United States' first and third largest trading partners. (Japan is second.) Under NAFTA, the three countries will form the world's largest common market, exceeding the European Community by $1.3 trillion.

Impact of NAFTA Low wages in Mexico are expected to attract industry, resulting in a loss of labor-intensive manufacturing jobs from the United States and Canada. For Mexico, the increase of relatively low-skill, low-wage jobs will translate into an improved standard of living. To remain competitive, the United States and Canada will need to improve efficiency, quality, and education. During a campaign speech in 1992, candidate Bill Clinton stated that NAFTA will "provide more jobs through exports. It will challenge us to become more competitive. It will certainly help Mexico. . . . A wealthier Mexico will buy more American products. As incomes rise there, that will reduce pressure for immigration across the border into the United States."

Many people in Caribbean regions fear that NAFTA will hurt local economies. Countries like the Dominican Republic are anxious that industrial growth will be diverted away from them and toward Mexico. Even Puerto Rico is concerned. Ironically, Puerto Rico pulled itself out of a period of economic stagnation several decades ago with the initiation of Operation Bootstrap, which attracted industry to Puerto Rico with low taxes and low wages. In the meantime, taxes and wages have increased to the point where it would make economic sense for some companies to locate or relocate in Mexico rather than Puerto Rico.

Other segments of the United States besides labor are also showing resistance. Sugar cane growers, for example, are pressing for an exemption on the grounds that cheaper Mexican sugar would drive them out of business. If exemptions were to be given to this special interest group, exemptions should be given to all special interest groups, and the purpose of the whole agreement would be defeated. There are additional reasons for phasing out sugar production in some regions of the United States such as Florida. There, sugar plantations are one of the chief causes of pollution and diversion that damage the Florida Everglades, and this in turn is destroying the spawning grounds of coastal fisheries, the coral reefs, and Miami's water supply.

Accomplishments NAFTA has brought together free trade advocates and environmentalists from very different backgrounds. From the perspective of ecology, NAFTA provides a way for the United States, Mexico, and Canada to cooperate in improving their environment. Indeed, NAFTA represents a step forward in the integration of the international trade system and environmental protection measures. Through cooperation on free trade, North Americans may also achieve the common goals of economic prosperity and conserving their resources.

Successful implementation of NAFTA may have a positive influence on including environmental provisions in GATT and may help move the stalled negotiations forward. Rather than taking a confrontational approach to dealing with armed strife in Central America or narcotics trade in South America, the Enterprise for the Americas Initiative proposes providing less environmentally harmful economic alternatives created by western hemispheric free trade.

CITES

The Convention on International Trade in Endangered Species (CITES) was adopted in 1973 by a diplomatic conference in Washington, D.C., because endangered species, like tropical forests, are part of the global commons. Their price does not reflect their value. Consequently, there must be a mechanism, other than the forces of supply and demand, to regulate usage. CITES was established to carry out this mission.

CITES is the principal government-level wildlife conservation authority in the world today. The secretariat is housed in the United Nations offices in Switzerland. A total of 122 countries belong to CITES. Members are required to abide by majority rule. Thus, if the member nations agree to list a species under Appendix One of CITES rules, which declares it officially endangered, then no member may permit imports or exports of that particular species. In cases where a ban is less than total, the secretariat has discretionary authority.

Unfortunately, bureaucratic problems inherent in international trade agreements are illustrated by the ivory trade which CITES is regulating in an effort to protect the African elephant. The secretary of CITES strenuously opposed listing the African elephant under Appendix One for years. The alternative was a system of "trade controls" and quotas imposed in 1986. This system allowed any member nation to export as much ivory as it wanted, as long as it informed the secretariat in advance as to how much it planned to export so that it could be held accountable. The system also exempted carved ivory, on the grounds that it was too difficult to monitor. The result was the creation of carving factories in Dubai, Singapore, Taiwan, Macau, and Zaire, where workmen performed rudimentary work on tusks that were then shipped legally to Hong Kong, where the 4000 or 5000 master craftsmen working there finished them. In addition, governments were allowed to legalize confiscated ivory. Because of these abuses, and because of pressure from international conservation organizations, the member nations of CITES finally voted to list the African elephant under Appendix One, thereby giving it world status as an endangered species and prohibiting all ivory trade (Reisner 1991). However, many African countries such as Botswana, Zimbabwe, and South Africa strongly opposed the move. They felt that conservation is more effective when a limited harvest is allowed, because this provides an economic incentive for local peoples to manage the herds sustainably. Because of the listing, the elephants lost their market value, and local people lost economic incentive to conserve them (Douglas–Hamilton and Douglas Hamilton, 1992).

National Policy as an Obstacle to International Policy

International policy can achieve only limited objectives when national policy is the primary obstacle to conservation. The forest policy of many developing countries is a case in point.

Policies that affect forest cutting prescribe how private buyers, leaseholders, and concessionaires contract rights to extract timber and other goods and services from public lands. There are also policies that are part of a national industrialization program, which affect trade in logs and wood products. Deficient forest policies and weak public forest administration interact to result in loss of forest resources. Weak forest administration results in inadequately conceived and enforced forest policies. Poor policies, in turn, lead to unmanageable pressures on the forests and to meager generation of public revenues from timber royalties and other forest charges. Both results hurt the image and morale of public agencies responsible for forest administration. Their ineffectiveness keeps them low in national funding and political priorities. Simultaneously, the wealthy and politically powerful timber concessionaires ensure that government forestry agencies listen carefully to their point of view. This is made easier in many developing countries by the low salaries paid to the civil service.

Policies outside the forestry sector also can have a strong negative influence on conservation. This happens when government agencies concerned with agrarian reform and land development regard forested regions primarily as "empty lands" to accommodate expanded infrastructure projects, plantation crops, annual crops, pastures, and colonization schemes. Pressures for colonization are increased by policies that protect or increase large agricultural landholdings. Colonization in marginal areas is facilitated by vague or unrecognized tenure claims (Laarman and Sedjo 1992).

Why do such policies remain in place if they do not promote rational forest protection and management?

First, there may be a failure to understand and appreciate the economic, social, and humanitarian impacts of policies.

Second, there may be understanding, but there is inadequacy of institutional structure. An example is the structure of fees and other charges on tropical timbers. Many professional foresters in forestry ministries know that more revenue should be collected, that each tract of timber should be appraised and charges set accordingly, and that policies should encourage greater utilization per hectare. However, they lack personnel and budgets to accomplish these tasks.

Third, there is the corruptive influence whereby wealth and power are able to bend forest policies for personal gain. In the above example, the forest ministries know how to improve their system of timber charges, but the private concessionaires prevent charge increases by making the correct donations to influential politicians or bureaucrats.

Two immense and frequently criticized projects planned by national governments but financed in part internationally are Brazil's Trans-Amazon highway, which has opened the Amazon region for settlement, and Indonesia's Transmigration Program, which has relocated millions of land-hungry Javanese to the Outer Islands (Ascher and Healy 1990). Because of agricultural methods inappropriate for the nutrient-poor and easily eroded soils, the projects have cost much more in terms of resources destroyed and wasted than they made in terms of new agricultural production.

Reform for misguided aid such as that which abetted these projects is often stymied by lack of guidance for improvement. However, there is at least one policy guideline that international agencies could use to prevent the wasteful destruction of tropical forests: Ensure that any new projects minimize the construction of roads through old-growth forest. It is the roads bulldozed out of the mature forest that have allowed access to the shifting cultivators, the loggers, the ranchers, the miners, and all the others who contribute to forest destruction. The tropical forest is not a hospitable place. It is not easy to penetrate without an army of bulldozers, graders, and trucks, and such machinery often becomes available to national governments only as a result of international loans or aid.

If the aim of loans or aid is development, let it be spent in areas that have already been cleared and where the roads are already built. For example, one can drive for hour after hour through scrubby secondary forest in the eastern part of Pará State in the Amazon region of Brazil. This wasteland is mostly abandoned or scarcely used pasture that once was primary forest. Yet in most regions, there already are towns and roads. These are the places where aid should be channeled for development of a more sustainable agriculture described in the next chapter.

CONCLUSION

An important stimulus to sustainable development and conservation is a system of open, competitive markets in which prices are made to reflect the costs of environmental as well as other resources. Open markets are great motivators because

Competition encourages producers to use as few resources as possible, if resources are priced properly.

Producers are encouraged to minimize pollution because it represents wasted resources, particularly if pollution is given a price and producers are made to pay the full cost for its control and the damage it causes.

Competition is the primary driver for the creation of new technology, which is needed to make production processes more efficient and further reduce pollution.

Capitalism, in its ideal form, is the most efficient method of allocating resources, when the resources are privately owned and when they can be bought and sold on the free market. However, when resources (and sinks) are a commons, actions for the common good are needed to ensure that resources are not wasted, polluted, or destroyed.

Actions for the common good can be achieved through three basic types of policy.

Command and control. These are basically government regulations, including performance standards for technologies and products, and effluent and emission standards, both for industry and for natural resource management, including agriculture. However, regulations are inflexible and often expensive. They are most useful when there is a serious threat to health or safety, including threats from pollution.

Self-regulation. These are initiatives by corporations or sectors of industry to regulate themselves through standards, monitoring, and pollution targets. Self-regulation may be cheaper to society in general than regulations. However, it can be undermined by companies that increase their competitive advantage by cheating or by ignoring industrywide agreements on self-regulation.

Economic incentives. These are efforts to alter the prices of resources and of goods and services in the marketplace via some form of government action that will affect the cost of production and consumption, or that will induce desired action. Where particular problems can be addressed by economic measures, these incentives are effective. An example would be conservation easements in which private land set aside for conservation is taxed at lower rates.

Policies that reward conservation can be more effective than those that punish wasters and polluters. But it is important to make any policy that is finally imposed flexible. Policies that promote conservation today may be less than desirable tomorrow.

It is easier to implement policies that reward conservation on a national than on an international level. International treaties often are more difficult because of vested national interests. Destructive policies by developing countries toward their own resources are the most difficult to influence. Eliminating international aid that abets national policy of resource destruction would be an important step toward implementing conservation in developing countries.

SUGGESTED READINGS

For a popular account of the earth's environmental problems, and a proposed solution from the viewpoint of the vice president of the United States:

Gore, A. 1992. Earth in the Balance: Ecology and the Human Spirit. Houghton Mifflin Co., Boston.

For a proposal to learn how long-term, large-scale interactions between environment and development can be better managed to increase sustainability:

Clarke, W. C., and R. E. Munn. 1986. Sustainable Development of the Biosphere. Cambridge University Press, Cambridge.

For an overview of UNCED, the Earth Summit held in Rio de Janeiro in June 1992:

Haas, P., M. A. Levy, and E. A. Parson. 1992. Appraising the Earth Summit: How Should We Judge UNCED's Success. Environment 34(8):7 and ff.

For a perspective on development and the environment from the business point of view:

Schmidheiny, S., with the Business Council for Sustainable Development. Changing Course: A Global Business Perspective on Development and the Environment. MIT Press, Cambridge, Mass.

For a review of the economic and political aspects of deforestation, and recommendations for policy improvements:

Laarman, J. G., and R. A. Sedjo. 1992. Global Forests: Issues for Six Billion People. McGraw-Hill, New York.

For an account of how and why a government agency strayed from its role in protecting the commons:

O'Toole, R. 1988. Reforming the Forest Service. Island Press, Washington, D.C.

For a picture of the role of nongovernmental organizations in conservation activities throughout the world:

Snow, D. 1992. Inside the Environmental Movement. Island Press, Washington, D.C.

For an analysis of green fees:

Repetto, R., R. C. Dower, R. Jenkins, and J. Geoghegan. 1992. Green Fees: How a Tax Shift Can Work for the Environment and the Economy. World Resources Institute, Washington, D.C.

For an analysis of market mechanisms of pollution control:

Pezzey, J. 1988. Market Mechanisms of Pollution Control: "Polluter Pays," Economic and Practical Aspects. Pp. 190–242 *in* R. K. Turner, ed., *Sustainable Environmental Management: Principles and Practice.* Westview Press, Boulder, Colo.

Publications that guide the landowner to an understanding of the mechanisms of conservation easements and land trusts:

Diehl, J., and T. S. Barrett. 1988. The Conservation Easement Handbook. Trust for Public Land, San Francisco.

Small, S. J. 1988. Preserving Family Lands: A Landowner's Introduction to Tax Issues and Other Considerations. Powers & Hall Professional Corp., Boston.

For contrasting viewpoints on the question, "Does free trade harm the environment?"

Bhagwati, J. 1993. The Case for Free Trade. Scientific American 269(5): 42–49.

Daly, H. E. 1993. The Perils of Free Trade. Scientific American 269(5): 50–57.

CHAPTER

MANAGEMENT OF NATURAL RESOURCES

CHAPTER CONTENTS

CHAPTER OVERVIEW

Conservation management is accomplished by minimizing management subsidies and maximizing the services of nature. Management subsidies are all the things that are done to an ecosystem to increase its commercial production, from spraying with chemicals to plowing and road building. The services of nature are all those functions performed by naturally occurring ecosystems such as nutrient cycling and disease control. ❧

PRINCIPLE

Resource production systems (farms, forests, etc.), which resemble the natural ecosystems of a region,[1] require fewer subsidies than systems that are quite different.[2] As a result, they are more stable and more resilient, both ecologically and economically.

[1]An example is a pecan orchard in Georgia which resembles the native oak forests in structure and function.

[2]An example is a pasture created by destroying rain forest.

DEFINITIONS (PIMM 1991)

Stability: A system is stable if all the variables return to previous conditions after displacement from them. Because steady-state conditions exist in nature only for short periods of time, stability must have a reference time frame.

Resilience: Resilience is the length of time for a variable that has been displaced to return to its prior state. Long return times mean low resilience, and vice versa.

SUSTAINABILITY

Definitions

The objective of resource conservation (including conservation of clean air and water) is to manage resources sustainably.

There are many definitions of sustainability, some of which have come from agronomy. For example, "Sustainable agriculture involves the successful management of resources for agriculture to satisfy changing human needs while maintaining or enhancing the quality of the environment and conserving natural resources" (CIMMYT 1989). Definitions often incorporate the following characteristics: long-term maintenance of natural resources and agricultural productivity; minimal adverse environmental impacts; adequate economic returns to farmers; optimal crop production with minimized chemical inputs; satisfaction of human needs for food and income; and provision for the social need of farm families and communities (National Research Council 1989). Management for sustainability of other resources, including forests, fisheries, and wildlife, would have similar definitions.

The World Commission on Environment and Development (1987) has offered a definition of sustainability that encompassed *all* resources, including a habitable environment: "Sustainable development is development that meets the needs of the present without compromising the ability of future generations to meet their own needs."

These definitions of sustainability are broad and general enough to be embraced by everyone. However, they are not specific enough to guide effective action. For example, although everyone would agree that resources should meet the needs of the present, defining exactly what these needs might be is difficult. All would agree that three meals a day and a roof over one's head are essential, but how much more is really needed? Does everyone in China need a bicycle? Does everyone in the United States need a car? If everyone in the United States needs a car, who is to say that every Chinese does not need a car? Who decides who needs what? It is not merely a semantic problem. As long as capitalism is the world's dominant economic system, needs will never be satisfied because capitalism rests on the necessity to continually create new and unfulfilled needs.

Some authors define sustainability as a system characteristic or quality. For example, Conway (1987) defined sustainability as the ability of the system to maintain productivity in spite of a major disturbance such as that caused by intensive stress or a large perturbation. Sustainable systems are not necessarily static, but they are usually dynamic, changing or adapting across time and space.

Other definitions depend on particular perspectives.

Ecological Sustainability

In an absolute sense, the word "sustainable" has no time horizon; it means going on forever. Because nothing goes on forever, however, nothing is absolutely sustainable.

Naturally occurring ecosystems are the nearest thing to absolutely sustainable systems. They are "ecologically sustainable" in the sense that there are no artificial energy, nutrient, and water subsidies. A forest with no human inhabitants is ecologically sustainable because it can maintain itself with only the sun and rain as inputs. (Strictly speaking, nutrients from the atmosphere and the bedrock should also be included.) There are no inputs such as fertilizers from humans.

As soon as humans begin to extract resources from a system, for example, game or timber, output of the ecosystem is greater than the input, and the system is no longer sustainable. In some cases, the human impact may be negligible, and the resource system can still be considered ecologically sustainable. An example of an ecologically sustainable resource system might be the Columbia River salmon fisheries for the pre-Columbian Indians. The Brazil nut groves from which nuts are gathered provide another example of an ecologically sustainable extractive system. As in pristine forests, the only inputs into a Brazil nut forest are solar energy, rainfall, and nutrients from the soil.

No countries in the world can survive based on ecologically sustainable production systems alone. To maintain modern civilizations and present-day populations, production systems must be modified and subsidized. Ecological sustainability is impossible for advanced societies.

Economic Sustainability

A resource system that is not ecologically sustainable can still be economically sustainable. Any production system that yields a profit is a potentially sustainable system, at least economically. For example, all farmers who made a profit last year and plant again this year are engaged in economically sustainable agriculture, at least in the short term, and within their present economic system.

Table 6.1 *Agricultural Production Resources Derived from Internal and External Sources*

FACTOR	INTERNAL RESOURCES	EXTERNAL RESOURCES
Light	Sun	Artificial lights
Water	Rain or small irrigation	Large dams, centralized distribution
Nitrogen	Fixed from air, recycled in soil organic matter	Applied synthetic fertilizer
Other nutrients	From soil reserves recycled in cropping system	Mined, processed, and imported
Weed and pest control	Biological, cultural, and mechanical	Chemical herbicides and insecticides
Seed	Varieties produced on farm	Hybrids or certified varieties purchased annually
Machinery	Built and maintained on farm or community	Purchased and replaced frequently
Labor	Family living on farm	Contract labor
Capital	Family and community reinvested locally	External indebtedness, benefits leave community
Management	Information from farmers and local community	Information from input suppliers, crop consultants

Source: *Adapted from Francis and King 1988, p. 68. By permission of Elsevier Science Publishers, Copyright © 1988.*

The problem with an economic definition of sustainability is that it does not specify how far into the future such sustainability will continue. If the profitability is maintained by government subsidy, it will cease to be sustainable when the subsidy is withdrawn. If the profitability is dependent on cheap oil prices, the sustainability will cease when the price of oil rises. If the profitability is dependent on a fad crop such as cocaine, sustainability depends on social and political forces.

When people talk about sustainable systems, they usually are thinking about economic sustainability over the long term. Some people might consider 10 years as being "long," whereas others would consider "long" as being a minimum of 100 years. The length of time for which a system is sustainable depends on many factors, including consumer preference and political stability of the government. An important factor in determining the economic sustainability of a system is the degree to which it uses internal resources, as opposed to external resources. Internal resources vis-à-vis external resources for a farm system are contrasted in Table 6.1. The more a production system uses internal resources, the more sustainable it will be. *Low input* and *resource efficient* are the key concepts that distinguish sustainable agriculture (Neher 1992).

Technological Sustainability

Technologically sustainable systems are an artifact of academic research institutions, but they should be mentioned because of the great deal of publicity they often receive in the media. A technologically sustainable system is a system that can remain productive as long as unlimited funds are available to support the production technology. Agricultural experiment stations such as that at Yurimaguas, Peru (Sanchez et al. 1982) are an example. Continuous production depends on a continuing hi-tech analysis of the production system, and a continuing supply of the fertilizers, pesticides, and so on, called for by the analysis. Such conditions frequently do not occur beyond the bounds of the experiment station. Technological sustainability when it requires large inputs of resources runs counter to the goals of conservation.

Holistic Management

The term *holistic management* has often been used as a goal toward which resource managers should strive. What does holistic management mean?

The concept of holistic management is a reaction to the trend in recent years to maximize just one thing—often productivity or gross economic income. In agricultural systems, the trend toward maximizing gross but not necessarily net economic income has led to an increased number of large farms at the expense of small farms, increased use of large, capital-intensive machinery, and more intensive use of fertilizers and pesticides. The result has been soil loss and environmental contamination. In some countries, it has caused a deterioration of social and cultural conditions. For example, the conversion of individual farms in southern Brazil into huge corporate soybean production systems resulted in the displacement of small farmers. This in turn forced the farmers either to move into *favelas* in the city or to migrate into environmentally sensitive areas such as the Amazon forest.

The management of publicly owned forest systems for the almost exclusive purpose of timber harvest also has increased the cry for a more "holistic" philosophy. The dominance of wood production as a goal has seriously decreased the value of forests for wildlife, stream quality, recreation, and conservation of biodiversity.

In holistic management, the wider range of environmental and social goals is considered as well as conventional economics. Rural life-styles, community standards, and environmental quality are given importance. However, holistic management is difficult to implement because most environmental and social needs are not given a value in conventional accounting systems.

Sustainable Development

It is not possible to achieve "sustainable agriculture," "sustainable fisheries," and the like, within an economic development system that is itself not sustainable. If the world is to avoid becoming an environmental disaster, it must concentrate on management of the economy in a way that is sustainable. But much of the world, including some areas of the United States, are considered underdeveloped, and in need of economic development. Therefore, much thought and effort in the early 1990s has been about working toward "sustainable development."

"Sustainable development" is an oxymoron, however, for something that is sustainable is something that goes on forever. In contrast, development is a process that has a beginning, an adolescence, a maturity, and a senescence. Just as a human being goes through a series of stages that constitute development, so does an ecosystem and so does human society. Development of an ecosystem is called *succession,* whereas development of society is called *progress.*

Many similarities exist between succession and progress. In both, the first colonists, the first arrivals, are generalists. They perform adequately all the functions necessary for survival. The early colonizers in a successional sequence capture energy and nutrients well enough to form propagules that can disperse to other disturbed areas, reproduce, and maintain the species. The early human pioneer on the frontier must know how to raise the crops, tend the cows, fix the machinery, and deliver the wife's baby.

Early colonizers are usually not efficient, however. They can survive when the resources on the frontier are abundant or when there is little competition for the resources. Indeed, the farmer can lack efficiency and still survive. The early colonizer in ecological succession can succeed because there is little competition for resources such as sunlight and nutrients. More important than efficiency to an early colonizer, both human and ecological, is the ability to use resources quickly and to grow quickly.

As development proceeds and succession progresses, more organisms appear and competition for resources increases. It becomes more difficult to be a generalist when the frontier is taken over by a bunch of specialists, all of whom are cooperating with one another. The single pioneer of the western frontier cannot compete with a community consisting of blacksmiths, plumbers, electricians, doctors, scientists, and so on. They are all very efficient in their specialty, and their ability to cooperate makes the extracting of resources much more efficient. In a like manner, natural systems, as they evolve, tend to be taken over by specialists. It is not specialism per se that ensures success, but rather specialization within the context of the overall functioning system. "Evolution of a niche" was an older way of looking at it. But in the modern context, evolution of a niche doesn't mean simply exploiting a certain set of resources previously unexplored. Rather, it means developing an efficient, functional relationship with other organisms. That function will replace whatever or whoever performed that function earlier in evolution, succession, or frontier development.

Succession may occur through autogenic change such as facilitation, tolerance, or inhibition (Connell and Slatyer 1977), where modification of initial environmental conditions by pioneer species enables other species characteristic of a later successional stage to become established. Similarly, human progress depends on what the last generation did to the environment. They may facilitate, or they may inhibit future development, or they may just keep things about the same.

Although an ecosystem "climax" is never really achieved, in the sense that the system is in steady-state equilibrium with regard to material flows and to populations, measures of ecosystem characteristics such as net primary productivity and nutrient recycling become asymptotic toward a value, or exhibit slight variations around a value, as the system passes from immaturity to maturity.

In a similar manner, as a society passes from immaturity to maturity, conventional indicators of development must approach a steady state value. However, other indices such as quality of life can continue to improve.

Improving Sustainability

Despite the lack of agreement on the definition of *sustainability,* or *sustainable development,* nearly everyone can agree on what sustainable development is *not.* It is not management that depletes the reproductive stock of a forest or a wildlife population; that causes soil erosion so severe that the subsoil is exposed; that causes pollution of the air and water; that causes migration of displaced farmers to the cities, resulting in increased urban congestion and the accompanying social ills.

Lack of consensus on what sustainability means does not prevent us from taking action to remedy these problems. In the following section, we examine resource systems to understand what can be done to this end.

ENERGY IN ECOSYSTEMS

Energy as a Currency

To analyze and understand a system, the system must be expressed in terms of some universal unit. Economic systems are modeled and understood in terms of dollars, whereas demographic systems are modeled and understood in terms of numbers of individuals.

Understanding of resource systems has been hampered by a lack of a common "currency" with which to quantify the dynamics. Money is one of the worst "currencies" because it fails to put any value on the vital, life-sustaining functions of ecosystems. *Productivity,* that is, the net yield of biomass, has often been used. Flows of carbon through the ecosystem have sometimes been substituted because it is the reduction of carbon during photosynthesis and the oxidation of carbon during respiration that control biomass accumulation. Nevertheless, biomass and carbon accumulation have weaknesses because the value of carbon varies depending on how it is stored. For example, a patch of grass might contain the same amount of carbon as a human brain, but a model that treated equally the carbon in both has limited usefulness.

Energy has been used as a currency with which to analyze ecosystems, but it, too, has drawbacks. A kilo of wood in a tree has a certain caloric content. A kilo of fine furniture has the same caloric value, but it is worth more because additional energy has gone into the kilo of furniture, over and above the solar energy used to create the wood. Odum (1988) has compensated for this weakness by using a currency called *emergy,* or embodied energy. This not only considers the energy value of a system component, but also takes into consideration the amount of energy required to synthesize that component. The biomass of organisms high on the food chain has a greater concentration of emergy than biomass lower on the food chain. The emergy value of nonorganic components of societal systems also can be measured by calculating the amount of energy necessary to manufacture the components.

Despite the conversion factors, emergy remains controversial because of the many assumptions needed to calculate the energy embodied in highly abstract artifacts of civilization such as a computer network or a university library.

In spite of its weaknesses for analyzing complex systems of society, energy is still a useful currency when comparing similar aspects of similar systems. For comparing systems of natural resource management, energy is very useful. Even if it is necessary to make calculations based on sometimes arbitrary assumptions such as the amount of energy embodied in a tractor used to plow fields, if the same assumption is made for all the systems being compared, then the comparisons become useful in a relative sense.

Output/Input of Production Systems

Grain, wood, fish, or any other output of resource systems such as farms, forests, or fisheries are the product of four sources of energy: (1) the sun through photosynthesis; (2) fossil fuel energy (the sun's stored energy) through fuel for tractors, pesticides, fertilizers, irrigation, etc.); (3) human energy; and (4) hydroelectric and atomic energy. Naturally occurring ecosystems derive all their energy from 1 and have little or no output of harvestable crops. Managed systems derive part of their energy from 1 and part from 2, 3, and sometimes 4, which are often called *energy subsidies*. Energy subsidies can compensate in part for energy lost through harvest, or for ecosystem functions destroyed by management.

In managed systems that resemble natural systems, a large proportion of energy comes directly from 1, and less energy subsidies are required. A system with a high proportion of 1 will be efficient economically and relatively sustainable ecologically. A system that is different from natural systems will require more of 2 and 3 because it works against nature instead of taking advantage of functions that occur in naturally occurring ecosystems.

According to the resource management principle, we should manage our resource systems so that their structure and function resemble, as much as possible, the naturally occurring ecosystems of the region. By so doing, we will be working *with* nature rather than working *against* it. In other words, we should do things the easy way.

"Easy" does not necessarily mean "easy" from the human perspective. Rather, it means "easy" from the viewpoint of the total amount of energy expended producing the resource. When we "work with nature," that is, when we take advantage of nature's free services, the amount of energy needed is relatively low, and the disturbance to the system is minimized. An example is raising bison on a short-grass prairie. When we "work against nature," that is, when we oppose the naturally occurring functions of the ecosystem, the disturbance and the consequent intrusiveness on the system is maximized. An example is cattle ranching in the Amazon.

Many of our modern resource management systems are "easy" on humankind but are "hard" energetically and "hard" on the environment. Pimentel et al. (1973) have given an example of how recent agriculture has become increasingly easy for human beings, while becoming hard on the environment. Between 1945 and 1970, average corn production in the United States rose from 34 to 81 bushels per acre (Fig. 6.1), while labor input per acre

Table 6.2 *Changes in Energy Inputs, Yield, and Return/Input Ratio for Corn Production in the United States Between 1945 and 1970 (in kilocalories per acre for each year)*

INPUT	1945	1950	1954	1959	1964	1970
Labor	12,500	9,800	9,300	7,600	6,000	4,900
Machinery	180,000	250,000	300,000	350,000	420,000	420,000
Gasoline	543,500	615,800	688,300	724,500	760,700	797,000
Nitrogen	58,800	126,000	226,800	344,400	487,200	940,800
Phosphorus	10,600	15,200	18,200	24,300	27,400	47,100
Potassium	5,200	10,500	50,400	60,400	68,000	68,000
Seeds	34,000	40,400	18,900	36,500	30,400	63,000
Irrigation	19,000	23,000	27,000	31,000	34,000	34,000
Insecticides	0	1,100	3,300	7,700	11,000	11,000
Herbicides	0	600	1,100	2,800	4,200	11,000
Drying	10,000	30,000	60,000	100,000	120,000	120,000
Electricity	32,000	54,000	100,000	140,000	203,000	310,000
Transportation	20,000	30,000	45,000	60,000	70,000	70,000
Total inputs	925,000	1,206,400	1,548,300	1,889,200	2,241,900	2,896,800
Corn output	3,427,200	3,830,400	4,132,800	5,443,200	6,854,400	8,164,800
Output/Input	3.70	3.18	2.67	2.88	3.06	2.82

Conversion factors (values in kilocalories): labor = 21,770/40 hr wk; 1 gal. gasoline = 36,225; nitrogen, 1 pound = 8400; phosphorus, 1 lb = 1520; seed, 1 lb = 3600; insecticides and herbicides, 1 lb = 11,000; corn, 1 lb = 1800. 1 bushel of corn = 56 lbs.

Source: *Adapted from David Pimentel et al, "Food Production and Energy Crisis,"* Science 182 (1973): 445. Copyright © 1973 by the AAAs.

decreased by more than 60 percent (Table 6.2). Thus, from the human viewpoint, agriculture has become much easier.

When we take the viewpoint of impact on the system, another pattern emerges. Table 6.2 shows that the ratio of energy expended to the energy produced between 1945 and 1970 changed from 3.70 to 2.82. (The table does not include solar energy, which presumably remained constant.) We are using more energy to produce less corn. The causes of the increased energy use are included in Table 6.2. Figure 6.2 shows that for these energy inputs, we have passed the point of diminishing returns. Thus, although in 1970 we continued to increase the total amount of corn produced, we did it wastefully.

A comparison of the energy efficiency of a wide range of resource production systems is given in Fig. 6.3 (Steinhart and Steinhart 1974). Output/input ratios range from 1/20 for feedlot beef, where each calorie of beef is very expensive energetically, to 20/1 for shifting cultivation, where each calorie of produce is very cheap.

The difference in ratios is due to: (1) the position of the product on the food chain; and (2) the proportion of input derived from "the services of nature."

Position on the Food Chain

Cattle, fish, pigs, chickens, and other animals are higher on the food chain than grains, grass, and other plant products. As energy passes through the food chain from sunlight

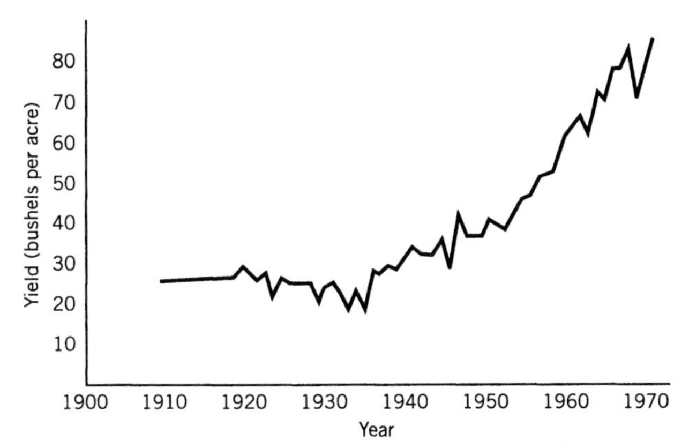

FIGURE 6.1 Corn production (bushels per acre) in the United States, 1909–71. *Source:* David Pimentel et al., "Ford Production and Energy Crises," *Science* 182 (1973): 445. © 1973 by the AAAS.

through autotrophs to heterotrophs, a proportion of the energy is lost at each level. It takes more solar energy to produce a thousand calories of beef than it does to produce a thousand calories of grass because cattle do not eat 100 percent of the grass in the pasture in which they graze, and of the grass that they do eat, they do not assimilate 100 percent.

Proportion of Input from the Services of Nature

Some resource systems depend almost entirely on the services of nature. The Brazil nut production system in the Amazon region of South America is an example (hunting and

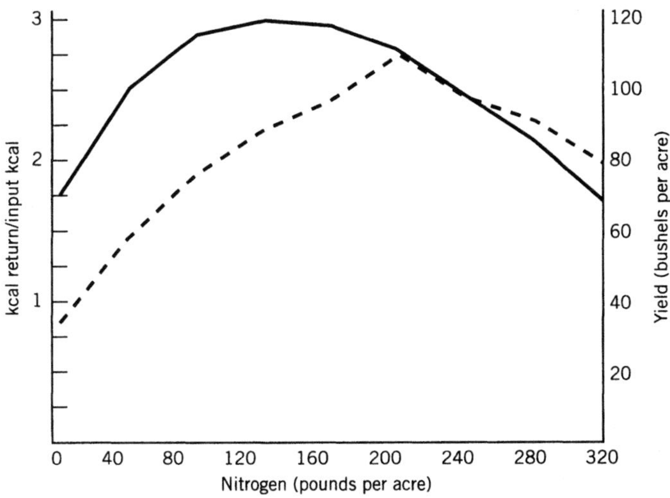

FIGURE 6.2 Corn yields .(bushels per acre, dashed line) with varying amounts of nitrogen (phosphorus = .34 lb. per acre) applied per acre *Source:* David Pimentel et al., "Ford Production and Energy Crises," *Science* 182 (1973): 446. © 1973 by the AAAS.

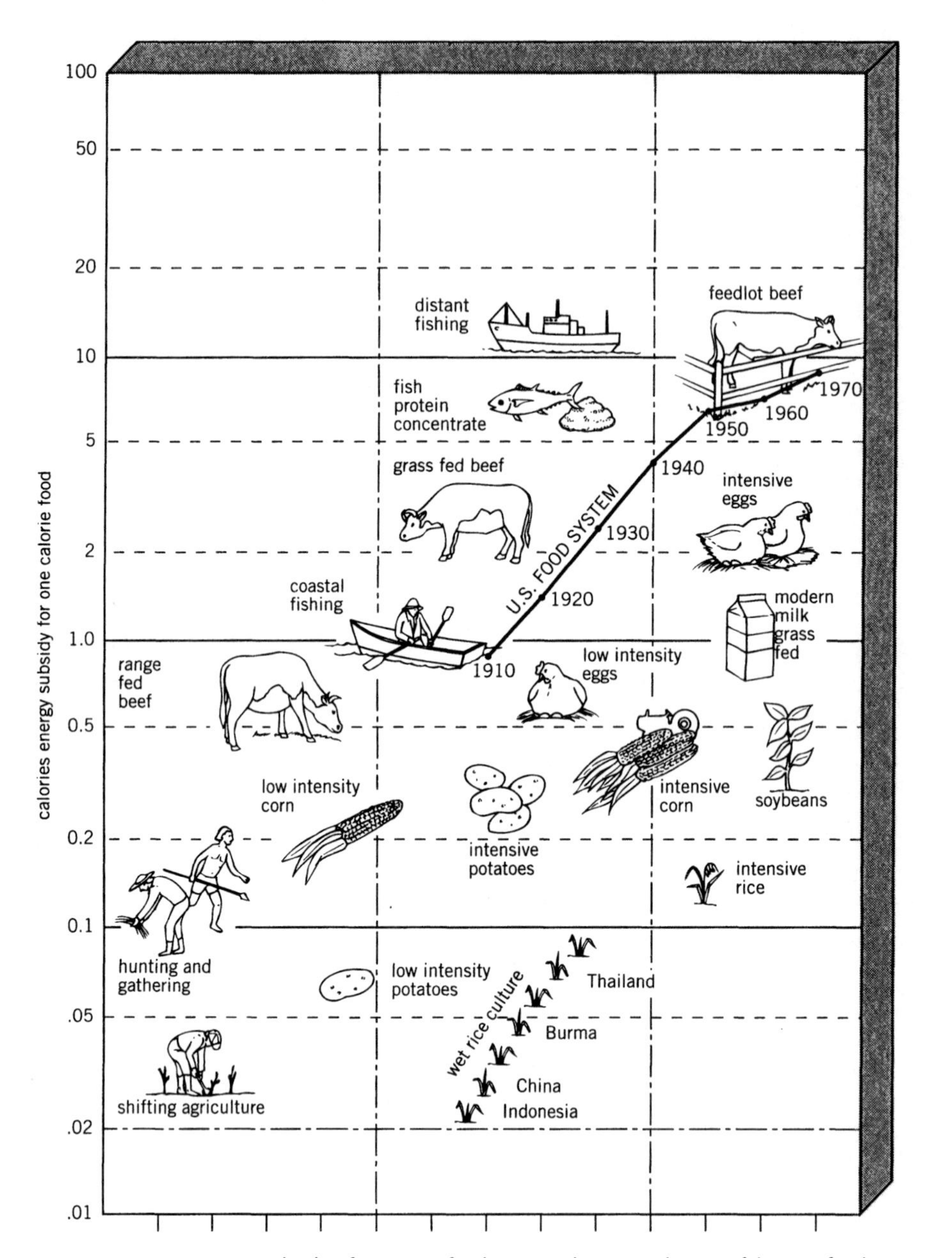

FIGURE 6.3 Energy subsidies for various food, crops. (The energy history of the U.S. food system is shown for comparison.) *Source:* Steinhart and Steinhart (1974), p. 84. Reprinted by permission of Wadsworth Publishing Co., Belmont, Calif.

gathering, Fig. 6.3). There is virtually no management of the system. Brazil nut collectors traverse trails and take nuts from trees for which they have collection rights. There is no planting of seedlings, no fertilization is done, and no weed control is required. The amount of energy and nutrients lost to the system as a result of harvest is only a small proportion of the total standing stock of energy and nutrients in the system. Nevertheless, the absolute quantity of nuts harvested per hectare often can be comparable to the yield of corn per hectare, but in contrast to corn, Brazil nut production is much more sustainable.

The Brazil nut forest is natural capital. Its huge biomass is like a large bank account. Although it may yield only 5 percent of its "principal" (standing stock), that amount is considerable because the principal is large. The natural capital performs the services of nature. When the forest is cut and replaced by a corn field, the natural capital is destroyed. Corn production can be sustained only through energy subsidies. In the conversion from Brazil nuts to corn, natural capital is replaced by imported capital in the form of petroleum or its derivatives, such as fertilizer and pesticide.

Resource-harvesting systems heavily dependent on energy subsidies are energetically inefficient. "Distant fishing" (Fig. 6.3) is very inefficient energetically because of the energy costs of the boat, fuel, labor, and so on, required to catch the fish. The direct sunlight energy that goes into the production of such fish, through photosynthesis of the algae eaten by the fish, comprises a relatively small proportion of the total energy costs of the production system.

Nevertheless, products that are *energetically* inefficient such as fish from distant fishing and feedlot beef are harvested because it is *economically* profitable to do so. People enjoy eating fish and beef, and are willing to pay relatively high prices to get them. Distant fish and feedlot beef are harvested because the energy required to get them (petroleum energy) is relatively cheap. If the environmental costs of highly subsidized resource systems, such as feedlot beef, and the depreciation of natural capital, such as ocean fisheries, were taken into account, beef and fish would be luxury items.

Services of Nature

The high output/input ratio of products or systems such as paddy rice, potatoes, and shifting cultivation is due to the fact that advantage is taken of the services of nature. Such services are not accounted for in the output/input ratios of Pimentel and Steinhart. This lack of accounting may be misleading regarding the desirability of a production system. Shifting cultivation is an example.

Shifting cultivation is based on an increase in soil fertility resulting from the nutrients in the biomass of the trees that are cut down. Gathering these nutrients from the atmosphere and from the soil and storing them in the biomass are services of nature. These services occur as the tree photosynthesizes, grows, and produces biomass. When the shifting cultivator cuts and burns the trees, he takes advantage of these nutrients and gets a lot of food calories for a very few "work" calories expended. He is taking advantage of the services of nature (nutrient scavenging).

The output/input ratio of shifting cultivation is high—20/1 in Fig. 6.3. However, when calculations for shifting cultivation include the energetic value of the trees that are cut down to fertilize the soil, the ratio shifts to about .005/1 (Jordan 1987a). Steinhart's calculations (Fig. 6.3) make shifting cultivation look so efficient because they do not include the energetic value of the trees lost through cutting and burning.

Low-intensity potatoes also appear cheap energetically in Fig. 6.3. Again, this is because the calculation does not include the services of nature. For example, in the uplands of Mindanao, the Philippines, deforestation is progressing rapidly on the upper slopes of mountains above 1500 m. (See photo-essay, The Farmer First Approach to Agricultural Development in the Uplands of Mindanao, The Philippines in Chapter 8.) When a forest is cut and burned, potatoes grow very well for about a year in the nutrient-rich, parasite-free "virgin" soil. Providing "virgin" soil is a service of nature which is not accounted for in the production of the potatoes. After about a year, potato cultivation is abandoned because of wilt. New land must be cleared for further potato production. Potatoes have been called the scourge of montane forests in Southeast Asia because of the deforestation caused by their cultivation.

Taking advantage of the services of nature through cutting and burning the forest makes a lot of sense to indigenous peoples when population density is very low and the existence of the forest is not threatened. Today in the tropics, however, many forests are threatened with destruction. The use of a large amount of energy in the form of tree biomass to produce a small amount of energy in the form of subsistence crops or to grow cheap potatoes for the market makes little sense when such production systems are of short duration (one year for potatoes) and when they destroy species that have potential values for all of humankind.

In contrast to shifting cultivation and potato production, traditional paddy rice is a resource production system that is heavily dependent on the services of nature and yet does not waste or destroy these services. The service of nature in rice paddies results from the fluctuating water level in the paddies. Water flow is driven by gravity and regulated by a system of dams and terraces. When the water is high, an azolla-algae association in the water often is important in nitrogen fixation. When fish are included in the paddies, they solubilize nutrients bound in algae. When the water level subsides and the soils dry, the azolla dies and decomposes and its nitrogen is released for use by the rice. In contrast to shifting cultivation which destroys the structure and function of the natural capital, the paddy rice system maintains structure and function that are well adapted to local environmental conditions. The paddy system takes advantage of the services of nature to recycle nutrients.

Lack of Equilibrium

All biological systems are open; that is, they exchange materials and energy across their boundaries. A forest, even one that is never touched by humans, has inputs and outputs. It receives energy (input) from the sun. It loses energy (output) as heat during decomposition of plants and animals by bacteria and fungi. Nutrients as well as energy move in, through, and out of ecosystems. Nutrients enter an ecosystem from rainwater, from decomposing bedrock, or from gases in the atmosphere. Nutrients leave an ecosystem through leaching, erosion, or volatilization.

A popular phrase among environmentalists is that certain types of development such as clear-cut logging "disturb the balance of nature." This phrase seems to imply that ecosystems are, or at least should be, in some sort of equilibrium or steady state, where input equals output and the number of individuals of each species is more or less constant. The fact is that ecosystems and populations are not in steady state, never have been, and never will be. Ecological systems are far from equilibrium.

The assumption that ecosystems are in equilibrium resulted from lack of perspective. In years past, biologists and ecologists often studied a forest or a lake or a population of plants or animals within an ecosystem for a summer or two and then published the observations in a journal. Studies were not of long enough duration for long-term dynamics to emerge from the data. Because changes were not apparent, ecologists assumed that ecosystems and the populations within them were in equilibrium, or that they were in some orderly progression (succession) toward equilibrium.

The assumption has outlived its usefulness (Botkin 1990). Ecological systems are continually in flux, albeit often within limits, on all scales from the microscale in a soil aggregate, through the fungal community on the forest floor, to the landscape itself, which is always changing under the influence of climatic trends, geologic activity, and human actions.

It is the exception rather than the rule that inputs into a natural ecosystem during a given year are quantitatively equal to the outputs. Sometimes global events occur, such as the occasional change in ocean currents called El Niño, which periodically affects climatic events throughout the world. Continual changes in populations such as the four-year cycle of the snowshore hare and lynx in Canada (Ricklefs 1993) may be related to imbalances in ecosystem dynamics or to controls within the biological community itself. The irregular fruiting of many dipterocarp (family *Dipterocarpaceae*) species in Southeast Asia has been hypothesized to be an evolutionary adaptation to predation pressure (Janzen 1974). Instead of producing the same amount of fruit each year and thereby allowing seed predators to adjust to a steady level of production, these trees produce a large crop very irregularly. The seed predators adjust to the low levels of production in off-years. Consequently, during "mast fruiting" years, the predators are saturated, and plenty of seeds are left over to ensure reproduction.

The irregularity of natural cycles poses problems for resource managers. For economic systems to work efficiently, there must be a reliable and steady source of basic resources. Thus, the farmer, the forester, and the fishery manager all strive to achieve a steady, or a steadily growing, crop. Because nature's cycles such as those of the weather and of insect pests are irregular, the manager frequently must fight against the natural tendencies of ecosystem dynamics. The farmer's continual fight to maintain a fixed production output is a struggle against the patterns of nature, which tend to be cyclic or oscillatory.

Energy Subsidies

In naturally occurring ecosystems, energy from the sun captured through photosynthesis is the almost exclusive source of energy. A naturally occurring ecosystem relies entirely on the sun's energy for sustainability. There are no energy subsidies. All the functions of the ecosystem—nutrient cycling, trophic flow of energy, reproduction, and maintenance—are carried out with energy fixed by photosynthesis in the leaves of plants living in the ecosystem. The flow of energy and the cycle of nutrients through the system are maintained through the metabolism and functioning of the plants, animals, and decomposers. The energy is sufficient to maintain the system, provided there is no export from the system, or just a very small export as would occur through the gathering of nuts by indigenous tribes. The amount of energy flow may not be the same from year to year, but there is enough energy to keep the system intact.

Naturally occurring species in native ecosystems are able to survive without energy supplements because they use a large proportion of photosynthate for scavenging nutrients, repelling predators, and competing with neighbors. As a consequence, their yield of products commercially useful to humans is often low. Natural systems have evolved to maximize survival, not commercial productivity.

As civilization developed, natural ecosystems were at first exploited; that is, products were removed with no replacement or repair of the system that produced them. At low levels of exploitation, ecosystems suffered no lasting damage. As human populations grew, exports increased, and systems ran down and ceased to produce. Humans learned that to keep an agricultural field productive of crops, fertilizers had to be added, and fields had to be kept clear of weeds and pests. To do this required effort or energy—human energy, draft animals, tractors, or the energy of factories where herbicides and pesticides are manufactured.

To maintain an energy flow as in crop production and harvest, an expenditure of energy is necessary. When the energy of the sun is insufficient to compensate for the products removed from the system, the system must be subsidized. Resource systems will produce a large amount of timber, grain, cattle, or fish only if there is an input of energy. For example, agricultural systems are maintained in the face of competition from natural populations of weeds and pests through the use of energy in the form of pesticides and herbicides. The agricultural crops "capture" more energy thanks to human help, and thereby survive the competition with wild species. The agricultural crops are the better competitors only because they have energy supplements from humans.

A wide range of strategies is available for resource management. Some options require a large proportion of input from energy subsidies, whereas others require a small proportion. Management systems that keep intact, as much as possible, the natural functions of the ecosystem require relatively low energy subsidies. The growing of annual crops in flooded riverbanks (varzeas) where yearly floods set succession back and also deposit nutrient-rich sediments is an example of such a system. Another is selective logging, where the loggers leave intact the basic structure and function of the forest.

The more a resource system resembles the naturally occurring systems of an area, the less subsidies will be needed to maintain productivity. That is why resource production systems (farms, forests, etc.) that resemble the natural ecosystems of a region require fewer energy subsidies than systems that are quite different, and in this sense are ecologically and economically superior. Management strategies in which a large proportion of energy input is based on the sun and a small proportion is based on subsidies are generally less harmful to the environment and are more sustainable. Such management systems can be said to be taking advantage of the services of nature, or working with nature instead of against it.

Management systems that drastically alter the structure of an ecosystem require a relatively large proportion of energy subsidies. If an agricultural system is quite different from the type of ecosystem that naturally occurs in a region, the amount of supplementary energy necessary to win the competition is relatively high. For example, cultivating wheat in former marshlands is energetically expensive because of the necessity of drainage. Without drainage, the wheat could not compete against sedges and reeds. Cultivating vegetables in the desert is energetically expensive because of the necessity for irrigation. Without irrigation, vegetables could not grow, or compete against desert species. Conservation is practiced when resource management systems rely to the greatest extent possible on the energy derived directly from the sun and to the least extent possible on energy derived from subsidies.

In many cases, naturally occurring ecosystems are more stable and more resilient than human-modified systems because the biomass, that is, the energy reserves, are greater in the natural systems. Buffering capacity against disturbance is determined in part by the structure of the system. In relatively sustainable resource systems, the structure is large in relation to inputs and outputs. The amount of energy stored in the standing stock biomass is large relative to the amount of energy input and output.

A bank account is a good analogy. If the amount of money coming into and going out of the account is small in relation to the principal, the possibility of overdrawing is small. The account has stability. In the case of an unusually large withdrawal, there is sufficient principal to cover the loss. The account can be maintained until the withdrawal is re-deposited. The system has resilience.

Many grain-producing agricultural systems have a relatively large amount of energy input and output in relation to the stock. In addition to inputs of sun and water, there are cultivation inputs that protect the crops against pests and weeds and provide fertilization.

Annual grain-producing systems have a low natural buffering capacity and, if subsidies are not available, a low resistance and a low stability.

Feedback Systems and Indirect Effects

Patten and Odum (1981) have argued that naturally occurring ecosystems have internal feedback systems that keep the functioning of the system within limits. When natural ecosystems are converted into production systems of commodities, the feedback systems are destroyed. Regulation of system functions such as dampening insect outbreaks must be carried out through means applied from outside the system.

The classic example of a feedback system with a fixed goal is the thermostat-furnace system that heats buildings. The fixed goal is the temperature that is set on the thermostat. A house whose temperature is controlled by a set thermostat is stable with regard to heat. The amount of heat moving into the house via the furnace ducts equals the amount of heat escaping from the house through cracks in the insulation.

Engelberg and Boyarsky (1979) have argued that ecosystems are not cybernetic systems because they lack a fixed set point. But the criterion of natural feedback systems is not that they have a fixed set point; rather, the criterion is that they remain stable within certain fixed limits (Patten and Odum 1981).

An example of a feedback system in a tropical rain forest is as follows. As a rain forest matures, the canopy of the largest trees forms a cover that increasingly shades the forest floor. Decomposition of leaf and wood litter on the soil surface slows down, and nutrients are accumulated in a layer of humus. Because of the accumulation, fewer nutrients are available in the soil for uptake by plants. Gradually, the low availability of nutrients affects the trees. A big, senile one dies and forms a "gap"—that is, an opening in the canopy through which light streams to the forest floor. Because of the increased temperatures, decomposition increases. More nutrients become available, and there is less competition for them because one of the big trees is gone. Smaller trees increase their growth rate. The system is maintained, though not at a fixed set point.

Indirect effects can be as important or more important than direct effects in maintaining system stability. On islands such as Isle Royal in Lake Superior, the dynamics of the browse-moose-wolf (Moffat 1993) are particularly illustrative. When the wolf pack is low, the moose herd grows and can overexploit the available browse, leading to an overall decline in health of the moose population. In the absence of wolves, the major interaction is between the browse and the moose. The interaction is direct in that when the browse is decimated, the herd begins to starve. In contrast, when wolves are present, there is an indirect feedback between the wolves and the browse: the moose herd grows; the browse is overexploited; the browse declines; the health of the moose herd declines, resulting in a larger number of easy prey (sick moose); more moose are taken by the wolves; grazing pressure on the browse declines; browse productivity increases; the moose herd grows; and so on. The populations of these components do not remain steady; rather, they fluctuate within limits tolerable to all populations concerned.

The concept that naturally occurring ecosystems are feedback systems that can regulate their own structure and function (within limits), whereas human-modified systems can do so only with energy supplements is an important idea for the resource manager. The more a system resembles a naturally occurring ecosystem of the area, the easier it will be to manage.

Carrying Capacity

The carrying capacity of an area of land is the number of consumers such as cattle that the land can support more or less indefinitely. If a pasture is well watered and fertilized, the carrying capacity is limited by the rate at which the grass can fix the sun's energy. There is a thermodynamic limit to the number of cattle that can graze, and this limit is set by the maximum photosynthetic capacity of the grass.

Failure to understand that there is a thermodynamic limit to carrying capacity sometimes occurs among uncritical advocates of biotechnology. For example, in a sarcastic editorial entitled "A Milk-Free Zone," Koshland (1994) paints a ridiculous picture of those who question the benefits of hormone-induced increases in milk production. However, he does not address the question of the source of this additional milk. If it results from the cow eating more grass, then what is the advantage of one hormone-induced cow over two nontreated cows?

The carrying capacity of farmland or rangeland sometimes is calculated under assumptions of optimal conditions: the weather is perfect, there are few pests and weeds, and those that exist are easily controlled. However, stocking rangeland, for example, under assumptions of optimal conditions, results in overstocking and consequent degradation of the range when environmental conditions are less then optimum. Suboptimal conditions could be caused by many events, including drought, floods, and invasion of exotic and nonpalatable weeds. Ecosystems should not be managed so as to produce the maximum carrying capacity. Management for sustainable yield occurs when the resource base is exploited at less than the rate that could occur under optimum conditions because when "bad" years occur, there is margin for recovery, and the resource base can better tolerate the load put on it. And "bad" years, it must be remembered, can occur because of sociopolitical events as well as natural disasters. When the price of petroleum increases dramatically during a "bad" economic year, there will be less production from systems that depend on energy subsidies.

In the case of humans, the carrying capacity depends not only on the number of consumers but also on the intensity with which each consumer uses the resource. A frequent topic of debate among committees concerned with the world's population is the "carrying capacity of the earth." "How many people can the earth support?" is the most frequently asked question. The data presented in Chapter 2 suggest that the world can support many more than at present, at the subsistence level. How many can it support at the level of consumption of wealthy American suburbs? On a sustained basis, perhaps less than at present.

Super Organisms

For decades, geneticists have been breeding high-yield grains, but the high yield has come at the expense of other survival functions such as pest and cold resistance. Humans have supplied the survival functions for their cultivated plants and animals in return for the higher yield. Such substitution has been economically worthwhile as long as the price of the yield has been higher than the cost of the supplement. But when the costs of subsidies and their environmental effects are high, substitution may be undesirable. Because of the high costs of subsidies such as irrigation, biotechnologists are trying to breed certain characteristics such as drought resistance back into crops. The problem is that to achieve this goal, one must breed back into the crop the characteristics that were bred out to achieve the high productivity.

By breeding in frost resistance, drought resistance, pest resistance, and the like, some of the high productivity will necessarily be lost. There is no free lunch. Energy cannot be created. The plants can photosynthesize only so much carbon, and the energy stored in the fixed carbon compounds can be used for growth or for survival. If more is to be used for survival, less is available for production, and vice versa. Increasing yields can come about only through decreasing ability to survive without energy supplements. Any increasing ability to survive without energy supplements can come about only through decreasing yields.

Another example of this principle of "no free lunch" is the failure of the quest for fast-growing trees. There are indeed trees that put on volume or height rapidly, but it is at the expense of wood density. Fast-growing trees have very light wood. One ton can have a volume of 4 m^3 or more (density is 0.25 g/cc). In contrast, the heavy, slow-growing mature trees of tropical forests can have a wood density of 1.0 or more. Each cubic meter weighs a metric ton. Thus, a fast-growing tree with a wood density of 0.25 will seem to grow four times as fast as one with a density of 1.0, but in reality, they are growing (producing biomass) at the same rate. This is not to say that trees that increase their volume rapidly are not valuable. They are, especially when restoring degraded areas. In addition, light wood is valuable for pulp. But it is important to realize that these trees are special only in that they have light wood. They have no magic wonder gene that allows their leaves to capture more energy than leaves of other trees.

Sometimes it is argued that fast-growing trees have large canopies that capture more than the average tree's share of light and by this means grow faster. There may be *individual* trees that are fast growing in the sense that they synthesize more biomass per day than other species, because of a large canopy. But fewer such trees can be planted per acre, and it is production per unit area, not per unit plant, that usually interests resource managers.

Limiting Factors

Productivity, that is, the amount of carbon that can be captured by an ecosystem, is limited. The limiting factors vary. In desert ecosystems, water is usually the factor that limits rates of photosynthesis (on a per hectare basis). In high-altitude or high-latitude ecosystems, degree days (average daily temperature times number of days during the growing season) are limiting. Table 6.3 lists a variety of natural ecosystems and the factors that often limit their productivity. In resource production systems that are well supplied with all other factors of production, light becomes the limiting factor.

The idea that only a certain amount of light energy can be captured per unit area per unit time is often ignored by some publicists of genetic engineering who offer a promise of new strains producing ever-increasing yields. It is necessary for the resource manager to understand that there is an upper limit to the amount of energy that can be fixed per unit time per unit area. No matter how the biotechnologists rearrange the genes, they cannot increase the amount of energy captured, unless the basic cycle of photosynthesis is changed.

Ecosystems in which growth is not limited by nutrients, water, and pests have a limitation imposed by photosynthesis. If a forest or field full of plants is well watered and fertilized, its production is limited by the interaction between light and the chlorophyll in leaves. As light passes through the first leaf in the canopy, some of it is absorbed by chlorophyll and the rest passes through to the next leaf. Depending on the thickness of the

Table 6.3 *Various Natural Ecosystems of the World and Factors That Frequently Limit Production When the Ecosystems Are Changed into Cropping Systems. (Light becomes limiting when other factors of production are supplied in excess.)*

Ecosystem	Limiting factor
Forest ecosystems	
Tropical	Phosphorus
High latitude	Nitrogen
Monsoon	Moisture
Pine forests	Nutrients and/or moisture
Cloud forests	Excess moisture
Montane	Erosion
Pacific Northwest	Moisture during summer
Mediterranean shrub	Moisture during summer
High altitude or latitude beyond tree line	Temperature (degree days), wind, thin soil, decomposition (slow release of nutrients).
Savannah and grasslands	Moisture, fire, nutrients, and/or aluminum toxicity
Coastal and mangrove	Salt water
Riverine	Periodic flooding
Wetland	Excess water
Desert	Moisture

leaf, the extinction of light to the point where it is no longer capable of promoting positive net photosynthesis (the light compensation point) usually occurs between 5 and 10 leaves—that is, where the leaf area index (LAI) is between 5 and 10. (LAI is the area of leaves per area of ground directly below the leaves.) The chlorophyll in a canopy may be distributed among many layers of thin leaves, or a few layers of thick leaves. But the total amount of chlorophyll is limited by the amount of light, and the supply of light is something that cannot be changed (except with artificial lighting, usually uneconomic).

Because of the limitations to photosynthetic efficiency, only so much light energy can be fixed by plants, and well watered, fertile ecosystems adapt so that virtually all of the usable light is used. There is an upper limit on the energy available to an ecosystem, and if nutrients, water, and pests are not limiting, then light is limiting. (CO_2 may be limiting in special cases, but only negligibly so.)

Certain grasses such as corn and sugar cane have a photosynthetic cycle (the C_4 cycle) that has a higher efficiency under certain conditions, owing to differences in biochemical pathways and to structural differences in the leaves. Production of biomass per unit area is greater for species with this specialized cycle. If biotechnologists are able to increase photosynthesis through modifications of the photosynthetic cycle, then both increased rates of production and increased survival ability are possible.

To take advantage of plants with increased photosynthetic ability would require increased supplies of nutrients and water. Because in most parts of the world, nutrients and water, not light, limit production, super plants would be of little benefit. It will be the economic cost of supplying nutrients and water, and keeping away pests that might limit production, not the ability of plants to capture light energy.

Natural ecosystems may not really be limited by the factors in Table 6.3 because they are adapted to the conditions. For example, the forests that grow on infertile

Oxisols and Podsol sands in the upper Rio Negro region of Venezuela have been described by Jordan (1989) as being stressed or limited because of low soil fertility. The idea has been criticized because there is no evidence that the forests in that region actually are stressed by low levels of nutrients, even though the soils are very low in nutrient content. The forests may not be stressed in the conventional sense (exhibiting small structure, extremely low productivity, dieback, etc., see Woodwell 1970), because there evolved in the roots and soil, nutrient-conserving mechanisms that resulted in an ecosystem with an above-ground structure and function that resembled a forest on much more fertile soil. When the Rio Negro forests are cleared and planted to crops, these mechanisms are destroyed, and nutrients quickly become limiting to the crops. However, because the native forest species are adapted to the stressful conditions, it may not be correct to say that these conditions are limiting *to these species.* Early explorers of the Amazon region mistakenly believed that the large size of the trees indicated fertile soil that was good for cropping. They did not realize how the species had adapted to the nutrient-poor conditions (Jordan 1982).

THE SERVICES OF NATURE

Management for sustainability should strive to decrease dependence on external subsidies such as mechanical manipulation and application of agrochemicals. To do that, farmers and foresters, stockmen and park managers, should rely more on the services of nature. How common are these services?

There has been a lot of excitement among biologists in recent years about the importance of positive interactions between species, in contrast to the previously dominant emphasis on negative interactions, primarily competition. These positive interactions, generally called mutualisms, often seem to be critical to the sustainability of naturally occurring ecosystems. In the context of resource systems such as rangelands, production forests, farms, and fisheries, mutualisms can be considered services of nature and used to advantage in resource management systems. In many cases, management of resource production systems to preserve these mutualisms will result in greater sustainability—that is, less need for energy subsidies.

Mutualism

The simultaneous evolution of two species as a result of each species forming part of the environment of the other is known as *coevolution.* A continuum of interactions have coevolved, from disadvantages to benefits (Lewis 1985). Interaction that is disadvantageous to both species is *competition;* interaction disadvantageous to one species but neutral to another is *amensalism;* disadvantageous to one species but beneficial to another, *agonism;* neutral to both affected species, *neutralism;* beneficial to one but neutral toward the other, *commensalism;* and benefits both species, *mutualism.* Of these, mutualism is the most omnipresent: all terrestrial higher plants, vertebrates, and arthropods are involved in one diffuse mutualism, and many are involved in several (Janzen 1985).

Mutualism has been more common in evolution than other interactions because both species have a selective advantage over those not involved in mutualisms and mutualisms confer an advantage to both participating species. Mutualism can be classified in five categories (Boucher 1985):

1. *Nutrition/digestion.* This includes interactions that increase the efficiency of resource harvest from the environment. Examples are nitrogen fixation through associations between *Rhizobium* and roots of leguminous plants, and other types of symbiotic nitrogen fixation; mycorrhiza, an association between plants and a fungus through which the fungus obtains carbon for energy and the plant obtains an increased ability to take up nutrients from the soil; and microbial species living in the guts of ruminant animals and other specialized species.
2. *Protection.* Ants and sometimes other insects protect the plants in which they live. When an animal begins to browse a protected plant, the ants swarm out and sting the intruder.
3. *Pollination.* Plants depend on insects for pollination and outcrossing. Insects derive energy from the nectar provided by pollen.
4. *Seed dispersal.* The spread of seeds away from the parent tree by birds, bats, and small mammals increases the probability that some seeds will survive and grow. The fleshy part of some fruits probably is an inducement for the animals to eat the fruit and carry the seed inside to a new location, where the seed is deposited during defecation.
5. *Resource management.* The development of agriculture can be viewed as a change from predation to mutualism in humanity's relation with its food organisms.

Mutualism has been studied from two contrasting aspects. First, *The Biology of Mutualism* (Boucher 1985) takes a biological perspective, examining mutualisms in relation to the population dynamics of the species comprising the association. In contrast, *Biodiversity and Ecosystem Function* (Schulze and Mooney 1993) examines mutualisms from the point of view of function in natural and human-managed ecosystems. Both volumes contain scores of examples of the types of interactions that keep natural ecosystems functioning and that should be preserved to ensure the sustainability of production systems.

From the population biology point of view, it is theoretically important for the ecosystem manager to conserve the keystone mutualists and mobile links (Gilbert 1980). Keystone mutualists are those organisms that play a critical role in the functioning of a community. Mobile links are animals that play significant roles in the persistence of keystone plant species that support separate food webs.

Despite the attractiveness of the keystone concept, it is often difficult to make a case that one species is of overriding importance. Almost every beginning scientist thinks that the particular species he or she studies is a keystone species because through intensive study, the scientist sees very clearly its functional role and how it influences other species, while for other species, that insight is lacking. Because of the difficulty of gaining a consensus regarding keystone mutualists in an ecosystem threatened by exploitation, efforts to preserve biodiversity *in general* may be a more effective strategy for conservation.

From the ecosystem point of view, it is especially important for the ecosystem manager to conserve mutualisms that play a critical role in the cycling of nutrients and flow of energy. The nutrient recycling function is an important service of mutualists that occurs primarily in the soil organic matter. Nutrients in the leaf and wood litter lying on the forest floor remain bound in insoluble forms until they are decomposed or mineralized by bacteria and fungi. Mineralization of nutrients benefits plants, because it renders nutrients available for uptake by plants. The microbes benefit from the interaction by obtaining carbon, their energy source, from the litter.

In specific examples of this type of phenomenon, the distinction between Janzen's (1985) diffuse mutualisms that permeate a community and Patten's (1991) indirect effects

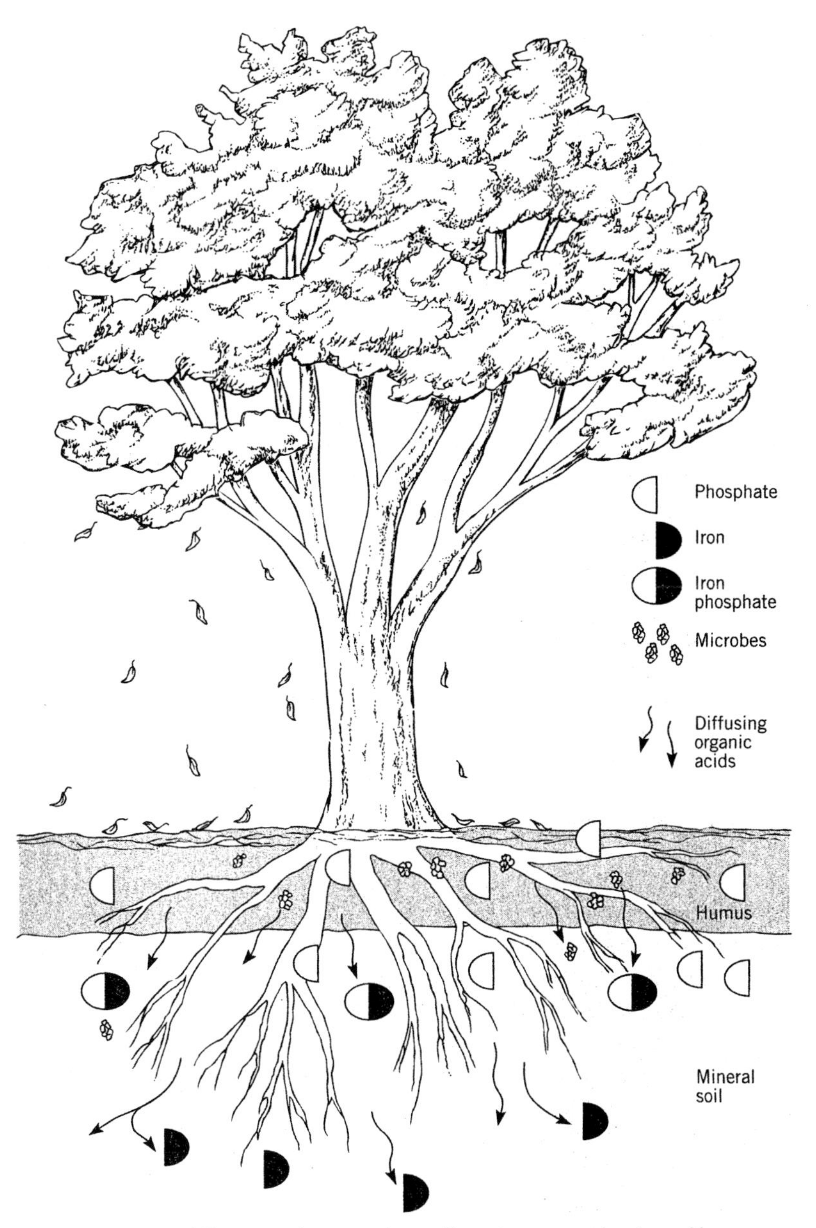

FIGURE 6.4 A diffuse mutualism or indirect effect. Organic acids released by microbes in the humus chelate iron, thereby keeping phosphate in a soluble form that can be taken up by the trees and used for further production of leaves. When there are no organic acids, iron reacts with phosphorus, immobilizing it.

in ecosystems has become blurred. Figure 6.4 shows an example of a phenomenon that can be considered a diffuse mutualism because of the reciprocal benefits both species receive. It is also an example of indirect effects.

Figure 6.4 presents a tree growing on a tropical Oxisol. These soils are high in iron and aluminum. Insoluble iron and aluminum phosphates are formed when soluble phosphorus enters the mineral soil. Phosphate in this form is unavailable to plants. In the undisturbed forest, however, P availability is not a problem. Leaf litter provides an energy source for the microbes that carry out decomposition. As a result of their metabolism, the microbes produce organic acids that chelate or react with the iron and aluminum. The organic acids use up the binding power that otherwise would be used to bind P. The organic acids may also solubilize the P in iron and aluminum phosphates. Because of this microbial activity, more P remains available for plant growth, plants grow more, and as a result, they produce more litter (Jordan 1989).

The interaction between trees and microbes is a mutualism. The plants benefit by getting more phosphorus, and the microbes benefit by getting more energy. It is basically the same interaction as the mycorrhiza-plant interaction, except the participants are not in direct contact. Boucher et al. (1982) define a mutualistic interaction in which the two species are physically unconnected as a nonsymbiotic mutualism.

The phenomenon also demonstrates indirect effects. When the microbes metabolize, they indirectly benefit the tree by mobilizing P. When the trees shed dead leaves, they indirectly help themselves, or other trees in their community, by directly helping the microbes.

In the context of conservation, it is moot to argue about whether such phenomena are mutualisms or indirect effects. The point is that we should strive to design production systems that take advantage of this function, this mutualism, this indirect effect, this service of nature. This service of nature, when part of an agroforestry system, makes that production system more sustainable. In contrast, when the phosphorus mobilization service of nature is destroyed by cutting down the forest and planting corn and rice, phosphorus becomes a severely limiting factor.

Evolution of Mutualism

As species adapt to their abiotic habitat, they influence and change that habitat. For example, plant roots produce carbon dioxide as a result of metabolism, and the CO_2 dissolves in soil water to form carbonic acid. This leaches underlying bedrock and results in a new chemical environment for the roots. Those plants best adapted to the new environment are the ones that survive and continue to change the environment.

Species also have a biotic environment: other species. If one species provides a favorable environment for another, and vice versa, a positive reinforcement can lead to further adaptation. The root-mycorrhiza interaction is an example.

Coevolution in natural systems occurs at several levels. Individual species of fauna and flora have adapted to each other and have evolved various degrees of mutualisms from casual to obligatory. Communities of floral and faunal species have adapted to the soils and climate of an area or region, and changed as new environment selected for more efficient communities.

By more efficient communities, we mean those that capture the most energy. Natural systems evolve toward maximizing energy capture, when energy capture is defined as the total net carbon captured (reduced) per hectare per year through the process of photosynthesis. Systems that capture more energy are better competitors (Odum and Pinkerton 1955). When the captured energy, stored as reduced carbon, is oxidized during respiration, the system uses that energy to extract more nutrients and water from the soil. As a result, the system is able to use a greater proportion of incoming solar radiation. Systems that have a higher amount of energy capture also have more energy to devote to mutualistic interactions. Such interactions "cost" energy, but they result in a system that can outcompete systems where such interactions are lacking.

Mutualism in Production Systems

A dominant paradigm in traditional resource management research has been competition. Competition between individuals of the same species has been the factor that determines optimal stocking. Optimal stocking has been the goal in forestry, agriculture, fisheries, wildlife management, grazing, and other resource systems.

Stocking always involves a compromise. For example, in forestry and agriculture, the farmer or forester wants to plant as many individuals as possible per hectare. However, if the plants are very close together, eventually an individual plant will become big enough to compete with its neighbor for some resource such as light, water, or nutrients. Planting individuals too closely results in competition that can stunt the growth of the whole stand and reduce overall yield below the amount that could be gained by a less dense spacing. Obtaining a maximum yield from a field or forest necessitates a planting density that is neither too close so that competition inhibits development nor too far so that there is underutilization of resources.

In management for sustainability, cooperation or mutualism rather than competition is the dominant factor. Mutualism in production systems often has been called facilitation (Vandermeer 1989). Facilitation is a process in which two individual plants or two populations of plants interact in such a way that at least one exerts a positive effect on the other. Basically, it is the same as commensalism. If both individuals or populations exert a positive effect, the process is the same as mutualism.

A classic example of facilitation occurs when corn and beans are intercropped under conditions where application of fertilizers is not economically feasible. Under such conditions, the total yield from one hectare of intercropped corn and beans can be greater than the total yield from half a hectare monoculture of corn plus a half hectare monoculture of beans. The overyielding effect may be due to nitrogen limitation on corn grown in monoculture, compensated for in the intercrop system by nitrogen fixation by root symbionts of the beans in excess of that required by the beans. The beans derive benefit from the interaction by using the corn as growing posts or stalks.

Interactions between species are not mutually exclusive. Facilitation and competition can occur simultaneously. The above intercropping example of corn and beans demonstrated facilitation. However, under those same conditions, a hectare of intercropped corn might have lower yield than a hectare monoculture of corn, indicating that competition as well as facilitation exists between corn and beans.

Not all multiple cropping systems are productively superior to monocultures. However, polycultures are sometimes practiced instead of monocultures because of economic factors, even though the polycultures may be productively inferior. For example, sun-grown coffee may have higher production than shade-grown coffee, but sun-grown coffee needs lots of fertilizers, whereas the shade-grown coffee may not need any because of the slower growth and the nutrient recycling services provided by the overstory trees. Shade-grown coffee is more sustainable because it needs fewer subsidies.

Cultural and consumer factors may also play a role. Slower grown shade coffee may command higher prices in some markets, so the higher price per unit must be weighed against the lower unit production per hectare of land.

Facilitation can have important benefits in tropical soils where almost all the phosphorus in the mineral soil is bound in insoluble iron and aluminum compounds. Guedes (1993) found that sorghum grown together with pigeon pea (*Cajanus cajan*) in an Ultisol grew significantly faster than sorghum grown alone, when the roots were in close proximity. With in vitro experiments, Ae et al. (1990) showed that piscidic acid produced in the roots of pigeon pea has the capability to solubilize phosphorus bound with iron. The increased growth of sorghum may be facilitated by pigeon pea, when the roots of both species are intertwined.

Intercropping can reduce herbivory. Vandermeer (1989) has classified herbivory-reducing interactions into three categories.

1. *The disruptive-crop interaction.* In this interaction, a second species disrupts the ability of a pest to efficiently attack its host. This occurs when the herbivore is less likely to find its host plant because of chemical or physical confusion by the second species or, having found a patch of host plants, is more likely to leave that patch because of encounters with nonhost plants that are disagreeable to the herbivore.
2. *The trap-crop interaction.* Here, a second species attracts a pest that would normally be detrimental to the principal species. For example, corn, when planted in strips in cotton fields, may attract the cotton bollworm away from the cotton (Lincoln and Isley 1947), though not always (Vandermeer 1989).
3. *The enemies interaction.* Here, the intercropping situation attracts more predators and parasites on the herbivores, thus reducing these pests through predation and parasitism. Hansen (1983) demonstrated increases in the abundances of several predator species in an intercrop system of maize and cowpea in Mexico, suggesting an explanation for the overyielding of that system.

Man as a Mutualist

The services of nature occur as a result of mutualisms, the cooperative ventures between species that make the ecological community a functional whole rather than just a random assortment of individuals. Mutualisms in a natural community are akin to barters in a rural community: the farmer pays the doctor for curing a sick child with a basket of fresh tomatoes. Both the farmer and the doctor do what they each do best, and the exchange between them results in a community that works more efficiently than if the doctor had to raise his own food, and the farmer had to practice medicine. Those social communities that employ exchange have been more "successful" than those in which everyone does everything for himself.

Just as humans are part of their social community, they are part of the system through which energy flows into the earth from the sun and back out as heat (Fig. 6.5). Although humans recognize their mutualistic duties toward their communities, they often act like parasites on the resource systems that sustain them.

Biologists have observed that in nature the ideal parasite is careful not to kill off its host. If it does, the parasite itself is doomed. An evolutionarily "smart" parasite will take something from its host but will ensure that enough is left to guarantee the host's survival. Some of humankind's actions so far have shown them to be "dumb" parasites. They are killing off the resource base that ensures their survival. However, other human actions have shown them to be "smart" parasites. Sometimes they do give back something to the earth in return for what they take. When this occurs, they are no longer parasites; they are mutualists.

Stability and Diversity

Natural systems often evolve toward diversity, whereas management of resource systems by humans is toward simplification. Diversity can lend stability to systems, and in this sense, evolution toward diversity is a service of nature.

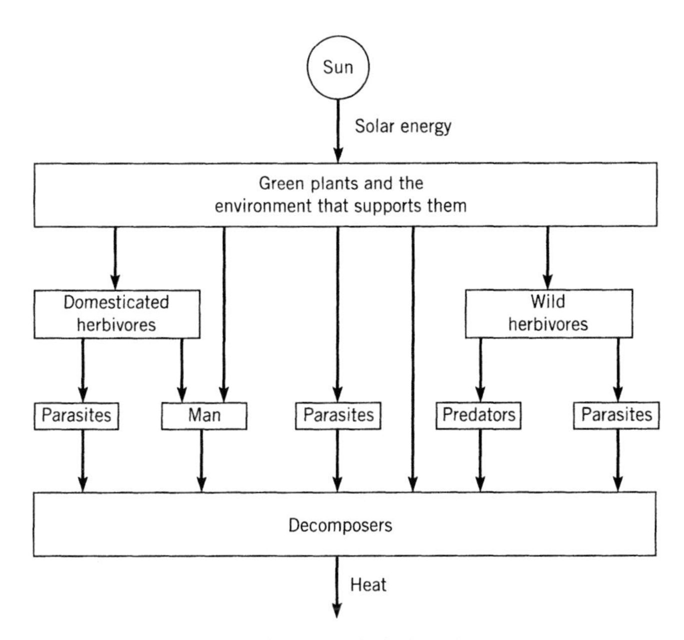

FIGURE 6.5 Energy flow through the biosphere.

Engineers know that to ensure the reliability of a system, built-in redundancy is required. For example, many modern jet airplanes have three independent hydraulic systems because aircraft control is dependent on a functioning hydraulic system. If one system fails, two backups remain. Often each hydraulic system will be different in order to ensure that a failure external to the hydraulic system will not incapacitate all the systems. Redundancy and diversity of structure and function in an aircraft system are related to the reliability of aircraft performance.

Reliability in engineering systems is comparable to stability in natural systems. Redundancy is comparable to diversity. However the relationship between diversity and stability in ecological systems is less well understood than that between reliability and redundancy in engineered systems. Some population biologists have argued that in nature there is little or no relation between complexity and stability (May 1972, 1973a,b; Pimm 1982). Complex food chains, they maintain, are as likely to be unstable, or perhaps more unstable, than simple food chains. Their studies have been used to rebut the thesis that complex agroforestry systems are more stable than simple crop monocultures and that diversity does not result in stability.

In fact, the studies do not contradict the hypothesis that complex ecosystems are more stable than simple systems. The reason is that the analyses of Pimm, May, and other population biologists focus on species or populations that may go extinct following a perturbation to the system. In contrast, the resource manager or engineer is more concerned with whether the whole system continues to function, despite a malfunction of one of its parts. From the point of view of resource management, there is more interest in the stability of the system output than of the system components per se. It is analogous to the pilot who is not particularly worried about one hydraulic system, but only about whether he has enough hydraulic systems to get him home, should one or two of the systems fail.

The reliability of hydraulic systems is not greater in airplanes with three systems than in those with one system. However, the reliability of airplanes with three hydraulic systems is greater than the reliability of airplanes with only one.

Because of redundancy, more complex and diverse ecological systems are more stable energetically than simple systems. Contributing to the relative stability of energy flow and nutrient cycling within ecosystems is the dynamic of individual populations. Populations of herbivores, predators, and omnivores can appear, grow, shrink, and become extinct. As conditions vary, one population will replace another. Ecosystems with many populations can have greater stability of energy flow and nutrient cycling than ecosystems with low population diversity. As one population declines, its function within the ecosystem switches to another species or population of the same functional group. The greater the diversity of species within the ecosystem, the more alternative pathways for energy flow and nutrient cycling, and the greater the opportunity for switching from one pathway (species or mutualism) to another.

The high diversity of populations in natural systems can damp out oscillations in ecosystem dynamics caused by fluctuations in populations. Thus, if one particular species of tree in a forest suffers from insect attack, another species may take advantage of the unused resources. For example, the chestnut blight in eastern North America virtually wiped out the native American chestnut (*Castanea dentata*), but chestnut oak (*Quercus prinus*) and red oak (*Quercus rubra*) appear to have at least partially taken over the functional niche left by chestnut (Keever 1973). Total biomass and annual primary productivity of the forests may be diminished only slightly by the demise of chestnut.

A good index to compare the stability of output between diverse systems is net primary productivity (NPP)—the total biomass produced per unit time after losses owing to respiration. The index integrates almost all the functions of the ecosystem, such as nutrient cycling, energy flow, water transport, and size of the populations within the ecosystem. When stability is measured in terms of changes in the amount of plant production, diverse communities are usually more stable than simple communities (McNaughton 1977, 1978; Tilman and Downing 1994).

Diversity and Efficiency

Every species has its own unique requirements for nutrients, light, and water for optimum growth. Some species may require high levels of nitrogen but can survive with low amounts of calcium. Other species may require lots of calcium but can survive with low levels of phosphorus. Some species do well in the shady understory, whereas others demand full sunlight. Some species demand much water, but others get by on very little.

No species, when growing in an even-aged monoculture, uses all the available resources; some resources will be "wasted." If the crop does not require high levels of potassium, some may be leached into the groundwater. If the plant does not require a lot of water, some of the soil moisture may be evaporated instead of transpired. If the plant demands full sunlight for its leaves, a certain amount of diffused light will reach the soil surface unutilized. It can be argued that one reason why multilayered forests (diverse forests with leaves at the herb level, in the understory, subcanopy, canopy, and emergent level) are more common in nature than single-species or low-diversity forests with leaves concentrated at one level is that there are many "wasted" resources in the latter type of forests. In nature, species have evolved to take advantage of unexploited resources or to "fill vacant niches." A structurally diverse community can utilize a greater proportion of incoming solar radiation owing to its complex spacing of leaves, thereby producing higher yields (May and Misangu 1982). Denser canopies also lessen the impact of high-intensity rains on the soil.

Because of the more complete utilization of all the "niches," structurally complex and diverse communities may have advantages over simple monocultures (Hart 1980; National Research Council 1982). A diverse plant community is likely to have a well-developed, complex root system that extends well below the soil surface, impedes erosion and nutrient loss, and exploits more efficiently the stocks of nutrients and water distributed throughout the entire soil horizon. Complex communities such as agroforestry systems may benefit from the presence of large trees that enrich the topsoil by sending down deep tap roots that can exploit nutrients released by the weathering of parent rock and redeposit them as leaf litter on the soil surface.

Because diversity may inhibit the exponential growth of herbivore populations, protection is another benefit of a diverse system. A diverse habitat is also more likely to include species that produce allelochemicals effective against herbivores and weeds (Gliessman et al. 1981).

Not all these advantages occur in all structurally diverse ecosystems. For example, Brown (1982) and Ewel et al. (1982) tested this principle in a series of tropical communities ranging from a simple maize monoculture through a complex wooded garden. They found that during the study, the net primary productivity was not related to ecosystem complexity. They also found that leaf damage from insects was remarkably similar in all plots, even though there were substantial species-to-species differences. Herbivores seemed to consume a nearly constant proportion of the total amount of leaves present. They also found that a sweet potato monoculture was as effective in intercepting solar radiation and reducing the impact of rainfall as more diverse ecosystems, but a maize monoculture was less effective. The big advantage of structural diversity of their sites was that exploitation of the soil reservoir was much more complete in the complex ecosystems than in the simple ones. Fine roots and total roots were much more abundant in the complex ecosystems. The quantity of roots may be a good indicator of an ecosystem's potential for nutrient uptake, and the presence of all size classes of roots indicates its ability to resist erosion.

Because this study could not document other advantages of structurally diverse ecosystems at the study site does not mean that they never occur. Risch (1981) found that beetle populations were lower in polycultures than in similarly treated monocultures. An entire volume on intercropping (Keswani and Ndunguru 1982) describes many experiments in complex agroecosystems, some of which demonstrated the advantages of complexity and others which did not.

An important problem associated with these types of experiments is that the area of the "monoculture" is relatively small—often a quarter hectare or less. In contrast, monocultures where serious pest outbreaks have occurred often are thousands of hectares in extent, such as the gmelina plantations at Jari in Brazil (Jordan 1985a), or the low-diversity spruce-fir forests of eastern Canada (Erdle and Baskerville 1986).

The evolution of natural systems toward diversity is a service of nature in that it increases the efficiency with which the factors of production are utilized.

Succession

The ecological changes that occur during the course of succession that result in the ecosystem being a more hospitable environment result from the services of nature. "Facilitation" during succession is a service of nature.

In the southeastern United States, thousands of square miles of agricultural land were abandoned during the 1920s and 1930s because of the low productivity of cotton. Low productivity was due to the boll weevil and to the increased susceptibility of the cotton plants to attack because of the depleted fertility of the soils.

An abandoned agricultural field in the Southeast is an inhospitable environment. The soil is very low in nutrients. The cations like potassium have been mostly leached down to the subsoil. The nitrogen has been volatilized. The phosphorus has been bound by the high levels of iron and aluminum in the soils. In addition, the microclimate is severe, and the physical properties of the remaining soil result in difficult conditions for the establishment of many plants.

In the Piedmont region of the Southeastern United States, early successional forests are dominated by pine. The reason may be that pine are better able to survive on the nutrient-depleted fields abandoned during the depression years. Ectomycorrhizal fungi associated with the pine may facilitate nutrient uptake, especially phosphorus cycling, by the pine. Nitrogen-fixing legumes also are common in early successional forests. Potassium may be accumulated in the trunks of early successional trees.

Once organic matter begins to build up on and in the soil, conditions become more favorable for other species. Nutrients become readily available as the litter and humus decompose. Extremes in microclimate decrease. The carbon compounds in the humus are used by soil-dwelling organisms, and their activity improves the physical properties of the soil.

Many of the species that replace pine in the successional sere are symbiotic with endomycorrhizal fungi, which may be highly efficient in exploiting nutrients bound in the humus. Hardwoods replace pine in the successional sere, possibly because their endomycorrhizal fungi may enable them to outcompete the pines for the increasing stocks of nutrients. In addition, pines may inhibit their own seedlings through shading or through formation of duff and humus too deep for a new root to penetrate.

The settlers of the southeastern United States destroyed the originally fertile soil through clearing and many years of cotton production, but once cultivation was abandoned, the restoration began through natural processes. The amelioration of the microclimate and soil by successional species is a service of nature.

EXAMPLES OF MANAGEMENT FOR CONSERVATION

To increase the sustainability of resource production systems, they must be managed to decrease energy subsidies. We should substitute technology that is energetically efficient for that which is wasteful, and substitute that which is environmentally benign for that which is intrusive. Often this can be done by using technology that relies more on what we have called the services of nature. But exactly how is this done?

This section gives examples from resource production systems.

Agriculture

Through the history of civilization, agriculture has changed from small, mixed-species plots cultivated with simple tools to large, monocultures cultivated with heavy equipment and chemicals requiring high-energy inputs. As soil is cultivated more intensively, important changes often take place: the amount of soil organic matter decreases; soil aggregates are broken down; increased soil erosion, increased soil compaction, increased leaching, volatilization, and immobilization of nutrients, and increased surface runoff of water occur. As a result of these changes, crop growth is less vigorous, and plants are more susceptible to pests and disease.

When the farmer was well off, he could counteract these changes with fertilizers, pesticides, and contour plowing. When the farmer was not prosperous, erosion, insects, drought, and disease often damaged the crop. Even when crops were well cultivated and treated with chemicals, side effects damaged the environment beyond the boundaries of the farm (Table 6.4).

In recent years, the trend toward intensified agriculture has begun to reverse itself. Increasing amounts of research have been dedicated to agricultural techniques that are less injurious to the soil and to the environment. More and more farmers are adapting these methodologies. Technology that relies more on understanding how nature works than on conquering nature has been called alternative agriculture (National Research Council 1989).

Alternative agriculture resembles what has sometimes been called organic farming or biodynamic agriculture. These approaches lessen the need for agrochemicals and mechanical disruption of the soil (American Society of Agronomy 1990). Alternative agriculture encompasses a wide variety of techniques that have been classified in a number of ways. Francis (1986) has described multiple cropping patterns, that is, the growing of two or more crops on the same field in a year. Multiple cropping can be sequential cropping or intercropping. Sequential cropping is the growing of two or more crops in sequence on the same field per year. The succeeding crop is planted after the preceding crop has been harvested. Crop rotation is sequential cropping of two or more crops in succeeding years.

Table 6.4 *Modern Agricultural Practices That Have Contributed to the Lack of Sustainability*

PRACTICE INITIATED	REASON PRACTICE INITIATED	PROBLEMS CREATED BY PRACTICE
Mechanization	Labor inefficiency	Erosion, energy dependency, capital expenses, interest payments, larger farms fewer farmers
Inorganic nitrogenous fertilizer	Low crop yield	Groundwater contamination, farm specialization, pests, erosion, energy dependency, high input expenses, less economic resilience
Pesticides	Crop loss to pests	New pests, resistant pests, water pollution, human poisoning, energy dependency, high input expenses
Hybrids and genetically narrow varieties	Crop yield and nonuniform traits	Aggravated pest problems, loss of local adaptations chemical dependency, high input expenses

Source: *Adapted from J. D. Soule and J. K. Piper,* Farming in Nature's Image, *Island Press, Washington, D.C., 1992, p. 52. By permission of Island Press, Copyright © 1992.*

Intercropping is the growing of two or more crops simultaneously on the same field. Intercropping can be mixed, when two or more crops are grown simultaneously with no distinct row arrangement. There can also be row intercropping when two or more crops are grown simultaneously in rows. Strip intercropping is growing two or more crops simultaneously in different strips wide enough to permit independent cultivation but narrow enough for the crops to interact agronomically. Relay intercropping is growing two or more crops simultaneously during part of the life cycle of each. A second crop is planted after the first has reached its reproductive stage of growth but before it is ready for harvest.

Whereas multiple cropping focuses on the crops that are planted, *conservation tillage* is a term that refers to a variety of practices used to cultivate the soil. The practices range from no-till, in which the soil is not broken at all, to a reduction in the frequency or depth of plowing. Reduced tillage allows the eventual formation of soil strata that are similar to those of native soils. It can enhance populations of such soil organisms as earthworms, insects, fungi, and microbes. Some of the organisms favored in no-till fields include predatory insects that can significantly reduce numbers of pest larvae. A healthy, active soil community shows relatively high efficiencies of nutrient cycling. Nutrients from decomposing organic matter become available slowly and are thus less vulnerable to losses by leaching, volatilization, or immobilization. Maintenance of the soil community in a relatively undisturbed state can result in improved crop vigor, lower levels of some diseases and pests, favorable soil structure, and retention of nutrients (Soule and Piper 1992).

In no-tillage agroecosystems, crop residues typically remain on the soil surface rather than being turned under. Surface litter helps mycorrhizal associations to develop. Once an organic horizon becomes well developed, the need for inorganic fertilizer inputs diminishes greatly. In addition, crop-derived mulch provides a buffer that protects the soil surface from wind and water erosion, and increases soil water retention.

On the negative side, conservation tillage depends on no-till planters to insert seeds in the unplowed soil, and on the use of herbicides instead of plows to control weeds. Some of the energy saved by minimizing tillage is offset by these costs. Some herbicides can reduce the populations of beneficial soil organisms, and leachates can pollute groundwater. Increased emphasis is being devoted to developing herbicides that decompose quickly upon contact with the soil.

A common element in all of these techniques of alternative agriculture is management to preserve or to increase soil organic matter, the decomposing residues of plant and animal material. As a nutrient store, soil organic matter gradually releases essential elements at a rate that may approximate demand by growing plants. The efficient movement of nutrients from decomposing organic matter to plant roots is desirable because it decreases leaching, volatilization, and fixation of the nutrients. In many soils, the organic matter buffers growing plants against sudden changes in their chemical environment and preserves moisture in times of drought. The carbon in the organic matter is an energy source for soil fauna that keep the soil in a friable, easily penetrated physical condition, well-aerated and free draining. It thereby provides young seedlings with an excellent medium for growth. Other benefits of soil organic matter may include humic and fulvic acids, and other compounds that may stimulate the growth of plants (Anderson 1985). Use of animal manure complements plant residues in that it decomposes more quickly and releases nutrients faster.

Although biodynamic agriculture is often ecologically superior, it is not always immediately economically feasible. Its advantages often begin to become apparent only after several years or more. However, in some cases long-term biodynamic farming is just as financially viable as conventional farming (Reganold et al. 1993).

Integrated pest management (IPM) differs from "regular" pest management in that it is part of the overall crop production system that seeks to optimize the social, environmental, and economic aspects of farming systems, and does not focus exclusively on getting rid of pests. For example, it considers an economic threshold that indicates when a pest population is approaching the level at which control measures are necessary to prevent a decline

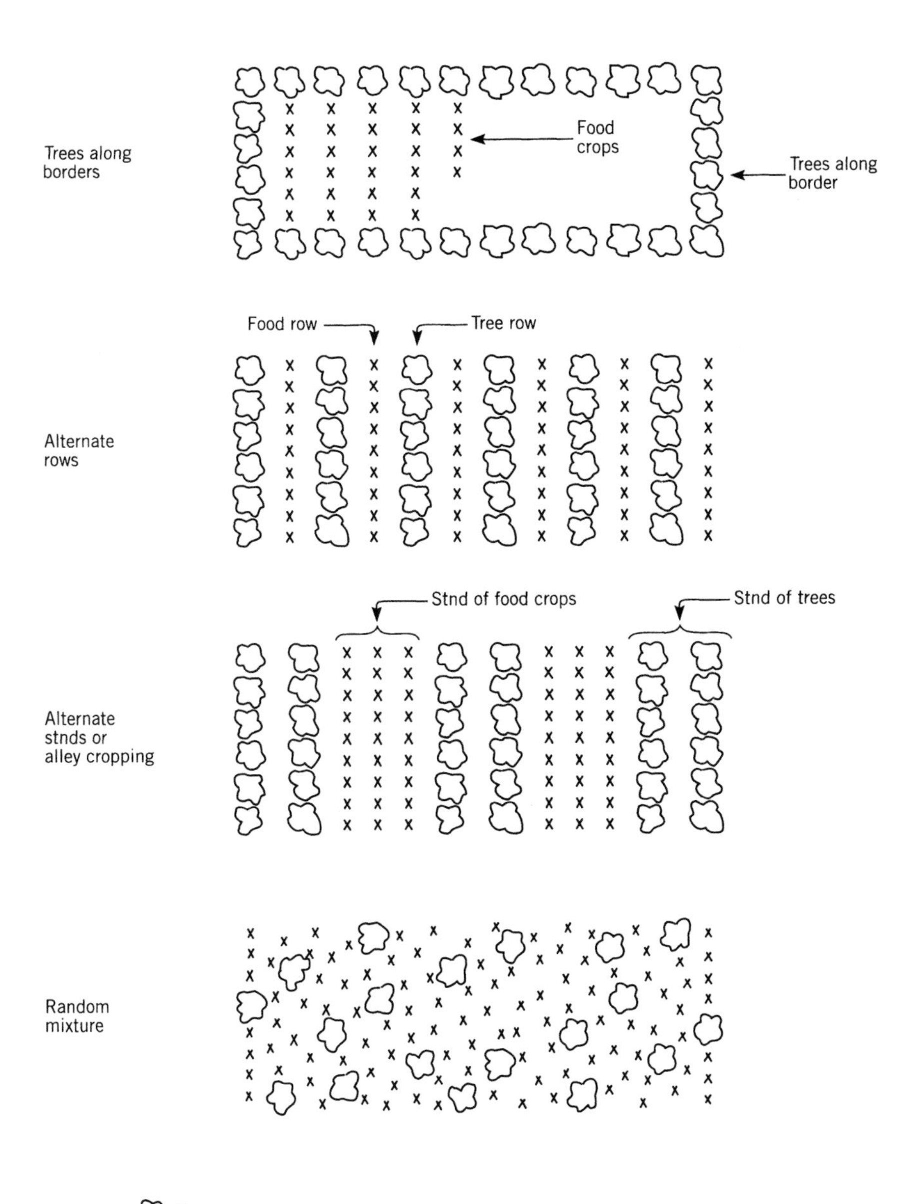

FIGURE 6.6 Spatial arrangement of crops in agroforestry. *Source:* MacDicken and Vergara, p. 40. John Wiley & Sons, Inc., New York, copyright © 1990. Reprinted with permission.

in net returns—that is, when the predicted value of the impending crop damage exceeds the cost of controlling the pest (National Research Council 1989). It is based on an understanding of predator-prey and host-parasite interactions and insect life-history patterns in nature. Integrated pest management incorporates several tactics of pest control. One is to rely on natural control factors such as pathogens, parasites, predators, and weather unfavorable for pests. Some examples are tiny wasps whose larvae attack and kill the pests; hoverflies and beetles, whose grubs eat their way through parasites; mites, which suck the fluids of pests; and molds which grow on live insects, debilitating them.

Table 6.5 *Processes by Which Trees May Maintain or Improve Soils*

Processes that augment additions to the soil:

*Maintenance or increase of soil organic matter through carbon fixation in photosynthesis and its transfer via litter and root decay.

*Nitrogen fixation by some leguminous and a few nonleguminous species.

*Nutrient uptake; the taking up of nutrients released by rock weathering in deeper layers of the soil.

*Atmospheric input; the interception of nutrients in dust and rainfall, and their gradual release to the soil upon leaf shedding.

*Exudation of growth-promoting substances in the rhizosphere.

Processes that reduce losses from the soil:

*Protection from erosion and thereby from loss of organic matter and nutrients.

*Nutrient retrieval; trapping and recycling nutrients that would otherwise be lost by leaching; including through the action of mycorrhizae associated with tree roots.

*Mobilization of some nutrients through root exudates.

*Reduction of the rate of organic matter decomposition by shading.

Processes that affect soil physical conditions:

*Maintenance or improvement of soil physical properties such as structure, porosity, moisture retention capacity, and permeability, through maintenance of organic matter and effects of roots.

*Breaking of compact or indurated layers by roots.

*Modification of extremes of soil temperature through a combination of shading by canopy and litter cover.

Processes that affect soil chemical conditions:

*Reduction of acidity, through addition of bases in tree litter.

*Reduction of salinity.

Soil biological processes and effects:

*Production of a range of different qualities of plant litter through supply of a mixture of woody and herbaceous material, including root residues.

*Timing of nutrient release; the potential to control litter decay through selection of tree species and management of pruning and thereby to synchronize nutrient release from litter decay with requirement of plants for nutrient uptake.

*Reduction of weeds, and thereby lessening the need for herbicides that may affect soil fauna.

*Direct transfer of matter between root systems by mycorrhizal bridges.

Source: *Adapted from Anthony Young,* Agroforestry for Soil Conservation, *T. J. Hardwick, Publisher, Wallingford, Oxon, United Kingdom, 1989, p. 97. Reprinted by permission of CAB International.*

The IPM practitioner evaluates whether there will be sufficient pests to justify control by the grower, whether the pests will last long enough or remain dense enough to lower yields, and whether natural controls will intervene. The actions taken may be cultural methods, biological controls, the use of toxic chemicals, or a combination of these. Cultural methods include manipulation of the density and diversity of the vegetation, cultivation, sanitation, variation in planting and harvesting dates and the varieties planted, and alteration of fertility or irrigation levels. Classical biological controls may employ predators, parasites, pathogens, and nematodes and may include foreign exploration to find natural enemies. Chemical methods employ substances that disrupt the functioning of behavioral pheromones (chemicals emitted by insects that facilitate orientation and mating) and judicious use of pesticides as a last recourse.

Integrated pest management can require a sophisticated data-gathering and monitoring system. Optimum systems and the application of their data for control are areas that present much opportunity for research.

Agroforestry

Agroforestry is a means of managing or using land that combines trees or shrubs with agricultural, forestry, horticultural, and animal husbandry practices. Ideally, agroforestry systems are more stable and sustainable than annual monocultures. Their production is distributed more evenly over a longer period of time. More even production can provide increased stability of cash flow to farmers. This is especially important in less developed countries, where farmers may have difficulty storing or marketing their produce.

The integration of trees into agricultural systems may result in more efficient use of sunlight, moisture, and plant nutrients than is generally possible by monocropping of either agricultural or forestry crops. Processes by which trees may maintain or improve soils are given in Table 6.5, and adverse effects are listed in Table 6.6. The economic and social advantages and disadvantages of agroforestry are listed in Table 6.7.

Despite the advantages, agroforestry may often not be economically viable (Lal 1991). One advantage of agroforestry—little need for manufactured fertilizers—may be almost irrelevant in areas where chemical fertilizers are cheap. Agroforestry relies heavily on labor and so is usually more economic where labor is cheap but where capital and energy are expensive.

Types of Agroforestry

A wide variety of practices can be considered to be "agroforestry" (Table 6.8). Some of the spatial arrangements in these systems are shown in Fig. 6.6. Detailed descriptions and explanations of the various systems are given by MacDicken and Vergara (1990).

Some types of agroforestry compromise between modern and alternative agriculture. Alley cropping, in which trees are planted in hedges and grain crops are planted in the alleys between the hedges, is an example. The trees in the hedges are usually fast growing legumes that have the ability to sprout. Several times during the growing season, they are lopped off at the stem or the branches are cut back. The litter falls into the alley, where the mulch serves to improve the soil and to supply nutrients. Because the grain crops are planted in rows, they can be cultivated with tractor and plow, with conservation tillage, or with draft animals.

More complicated forms of agroforestry are common in the less developed regions of the world, especially the humid tropics, where landholdings are small, labor is cheap, agricultural chemicals are expensive, and machinery is usually not available. Agro-

Table 6.6 *Adverse Ecological Effects of Trees in Agroforestry Systems*

Loss of organic matter and nutrients in tree harvest
Competition for nutrients, water, and light between trees and crops
Production of substances that inhibit germination or growth (allelopathy)
Acidification by trees that produce mor-type humus
Reduction in area that can be cultivated in crops
Habitat or alternative hosts for pests, especially birds

Source: *Adapted from Anthony Young,* Agroforestry for Soil Conservation, T. J. Hardwick Publisher, Wallingford, Oxon, United Kingdom, 1989, p. 102. By permission of CAB International.

forestry techniques used in the rural areas of the Philippines are illustrated in Figs. 6.7 and 6.8.

Figure 6.7 is the so-called SALT technology variously referred to as sloping agricultural land technology, and as Sustainable agroforestry land technology. It is typically implemented on hills or mountains that have been logged, farmed, and then abandoned to grassland or secondary scrub. The first step is for the farmer to lay out level contour lines, 4 to 6 m apart. After the lines are laid out, woody perennials, particularly nitrogen-fixing species, are planted to form hedgerows. Grass may be planted between the trees. Crops

Table 6.7 *Economic and Social Aspects of Agroforestry*

Possible Advantages
Increased income opportunities
Year-round distribution of income
Year-round distribution of employment
More evenly distributed requirement for labor
A variety of products (food, firewood, posts, poles, craft wood, fodder, fertilizer, medicinal products)
Potential for improved human nutrition
Reduced costs for establishing tree plantations (see photo-essay, Social Forestry in Thailand, this chapter)
Reduced weeding requirement
Reduced risk
Increased predictability of crop yield
Increased ability to manage for sustained yield
Increased economic stability
Possible Disadvantages
May have high labor cost
Requires transition costs to establish
Requires a long-term investment
Produces reduced yield of grain crops
Requires protection against fire and grazing (early stages)
Calls for land-use rights

SOCIAL FORESTRY IN THAILAND

Teak (*Tectona grandis*) has been an important export commodity for Thailand since the middle of the nineteenth century. The teak harvest increased until it peaked in 1971, when 298,800 m^3 were cut.

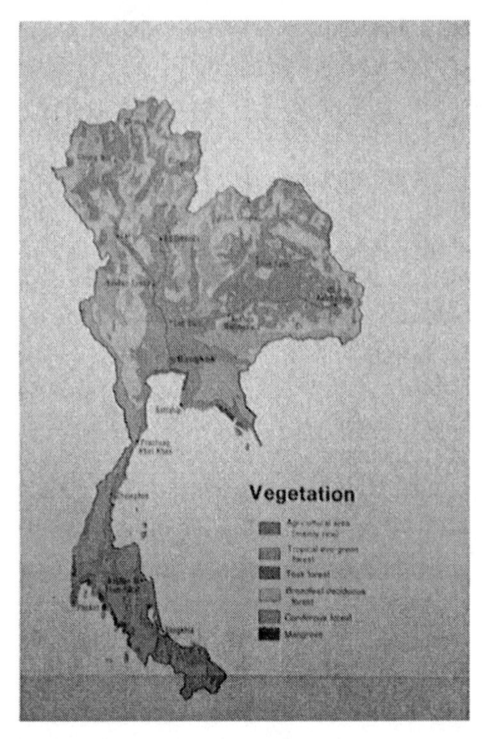

Fig. P.1 A map of Thailand. Teak forests occur in the northwestern highlands.

Fig. P.2 Teak occurs naturally in mixed deciduous forests and can make up 70 percent or more of forest biomass.

Fig. P.3 Fire often occurs in teak forests during the dry season. Teak is more resistant to fire than many other species, which may give it a competitive advantage.

Fig. P.4 Teak can attain huge girth and height, but such specimens are now rare in the wild. This one has been preserved in a park

Fig. P.5 Teak is harvested in the north and is floated down the Chao Phraya River to sawmills like this one in Bangkok.

Fig. P.6 In many areas, teak is still hauled from the forest and yarded by elephants. The elephant and its driver are a team that lasts throughout the life of both.

Fig. P.7 Since 1971, the teak harvest in Thailand has declined from close to 300,000 m^3 per year to 39,200 in 1985. Decline has occurred through both legal and illegal logging. Here a worker measures the diameter of a teak that has been illegally cut.

❧

Fig. P.8 The decline in the teak harvest has hurt Thailand economically. To reestablish the teak industry, Thailand has instituted a reforestation program called the Forest Village System which allows up to 100 families into each village. Family members are hired to work in reforestation and are also allowed to carry out subsistence agriculture. The headquarters of the village shown here is near Lampang, in the center of the Northwestern Highlands. (Fig. P. 1.)

Fig. P.9 The program encourages the enrollment of peasants such as the one shown here, who otherwise would practice shifting cultivation that degrades the remaining forestland.

Fig. P.10 The program provides subsidized housing for the villagers.

Fig. P.11 Schools, health care, and recreation are also provided in these villages.

Fig. P.12 Soccer games between Forest Villages stimulate morale.

Fig. P.13 The villagers maintain their own home gardens. This woman is spinning silk from cocoons grown from larvae fed on mulberry leaves from her garden.

Fig. P.14 Workers plant teak seeds and cultivate the young plants.

Fig. P.15 When the seedlings are about a foot tall, they are harvested, and the leaves and fine roots are removed.

Fig. P.16 These cuttings are then stored until the beginning of the planting season.

Fig. P.17 Meanwhile, other workers clear the site where the cuttings are to be planted, and they pile the brush.

Fig. P.18 The brush is then burned.

Fig. P.19 When the site is prepared and the rainy season begins, the cuttings are planted in a hole made with a dibble stick.

Fig. P.20 Between the teak seedlings, the farmer cultivates rice for his own use. Here he weeds the plot, which benefits both the rice and the teak. The system of planting agricultural crops among tree seedlings is called *taungya.*

Fig. P.21 The first year, the rice quickly overtops the teak.

Fig. P.22 The rice can be harvested without damaging the teak.

Fig. P.23 After two to three years, however, the teak becomes taller and cultivation of rice between the teak no longer is possible. The farmer no longer voluntarily weeds the plot.

Fig. P.24 Workers from the village are hired to weed the plantations.

Fig. P.25 Sometimes cattle are grazed on the weeds beneath the teak.

Fig. P.26 As the plantation matures, growth slows because of competition between individuals. Because the plantation is a monoculture, all the roots grow in the same zone and compete for the same nutrients and same water. The canopies overlap, and there is competition between leaves for light. However, diffuse light reaches the forest floor unused. The plantation is thinned to reduce competition. The trees removed during thinning have little value because they are so small.

Fig. P.27 To relieve the problems of monoculture plantations, experiments are beginning with mixed species plantations. Here a scientist is standing in front of a tamarind tree (*Tamarindus indica),* a legume whose fruits have high commercial value. This species has been interplanted between the teak. The fruit can give economic returns during the long interval before the teak is ready for harvest (30 to 60 years). In addition, the tamarind may improve the soil. Because it is not a tall tree, it will eventually be overtopped by the teak and shaded out. Thinning may not be necessary. There will be less need for weeding because the tamarind will use much of the light filtering through the canopy of the teak. The tamarind gives side-shading to the teak, an important factor in giving the teak a cylindrical-shaped bole.

Fig. P.28 Buddhism is an important factor in promoting the idea of reforestation because humans are viewed as part of nature and humans and nature are interdependent and equal. Here a monk from the village is revering a seedling.

Although the Forest Village program has been successful in establishing teak plantations, the system is not without problems. One is the desire of the farmer to own his own land. This cannot be done in the Forest Village System. Another is the long time period required for the teak to mature. Innovations such as interplanting fruit trees are helpful because they give economic income in the shorter term and allow the farmer to cultivate a single area for a longer time.

SUGGESTED READINGS

For a detailed discussion of the taungya system in Asia and the Forest Village System in Thailand:

Jordan, C. F., J. Gajaseni, and H. Watanabe, eds. 1992. *Taungya: Forest Plantations with Agriculture in Southeast Asia.* CAB International, Wallingford, Oxon, United Kingdom.

For the cause of declining teak harvests in Thailand:

Gajaseni, J., and C. F. Jordan. 1990. Decline of Teak Yield in Northern Thailand: Effects of Selective Logging on Forest structure. *BioTropica* 22: 114–118.

For an overview of economic development, forestry, and agriculture in northern Thailand:

Kunstadter, P., E. C. Chapman, and S. Sabhasri. 1978, *Farmers in the Forest: Economic Development and Marginal Agriculture in Northern Thailand.* University Press of Hawaii, Honolulu.

such as corn, upland rice, beans, and pineapple are planted in the alleys formed between the hedgerows. Sometimes permanent crops such as coffee, cacao, banana, and citrus, and fruit trees are planted in every third strip. After several rainfalls, the soil from the top part of each alley begins to wash toward the bottom, and gradually the strips turn into a system of living terraces. Such terraces are much easier to maintain than stone terraces. If a terrace springs a leak, it is only necessary to plug it with leaves and branches.

The hedgerow trees are allowed to grow up to around 2 m, and then they may be cut, leaving a half meter stump to sprout. The foliage may be left in the alleys as mulch, or removed and given as fodder to livestock.

Various species of trees are planted upslope from the terraces in order to provide firewood, poles, charcoal, construction materials, and furniture. Gmelina may be planted where there is a market for pulp. When the uppermost part of the mountain is still in natural

Table 6.8 *Agroforestry Practices*

Trees with Crops (agrosylvicultural)

- Rotational
  - Shifting cultivation
  - Improved tree fallow
  - Taungya
- Spatially mixed
  - Trees on cropland
  - Plantation crop combinations
  - Multistory tree gardens
- Spatially zoned
  - Hedgerow intercropping (barrier hedges, alley cropping)
  - Boundary planting
  - Trees on erosion-control structures
  - Windbreaks and shelterbelts
  - Biomass transfer (i.e., litter from forest used as mulch on crops)

Trees with Pastures and Livestock (sylvopastoral)

- Spatially mixed
  - Trees on rangeland or pastures
  - Plantation crops with pastures
- Spatially zoned
  - Living fences
  - Fodder banks

Tree Component Predominant

- Woodlots with multipurpose management
- Reclamation forestry
- Taungya (cf. Jordan et al 1992)

Other Components Present

- Entomoforestry (trees with insects)
- Aquaforestry (trees with fisheries)

Source: *Adapted from Anthony Young,* Agroforestry for Soil Conservation, T. J. Hardwick, Publisher, Wallingford, Oxon, United Kingdom, 1989, p. 12. By permission of CAB International.

FIGURE 6.7 "SALT" technology, as recommended for upland areas on the island of Mindanao, the Philippines.

forest, SALT can be used as a buffer zone to prevent further upslope encroachment. (See photo-essay, The Farmer First Approach to Agriculture in the Philippines.)

Figure 6.8 illustrates an integrated low-input rice production system used in parts of the Philippines. About 80 percent of the rice is for sales to market. The rest is small-grain rice used to feed the ducks, which also eat leftovers from the farmer's table. The ducks control snails that attack the rice. Some of the income from raising ducks is used to buy pig feed. Manure from pigs and ducks is a natural fertilizer for the plankton and the duck-weed eaten by the fish. Manure also fertilizes azolla and algae, which are then inoculated into rice fields where they increase the supply of nitrogen. Some of the pig manure and urine is used to fertilize vegetables and fruit trees. Fish raised in the flooded rice field eat weeds and insects and provide natural fertilizer for the rice. Mushrooms are grown in the shade of banana trees, in straw and earth pressed together. The moisture from the banana trees helps the mushrooms grow, while the decayed straw becomes fertilizer.

The area around the fish ponds and fields is used to grow more than 50 kinds of plants, both seasonal and perennial. Some of the plants contribute to livestock, whereas others are used as food. The plants help reduce erosion, and trees such as mango, co-conut, and jackfruit yield fruit and help to shade the fish ponds.

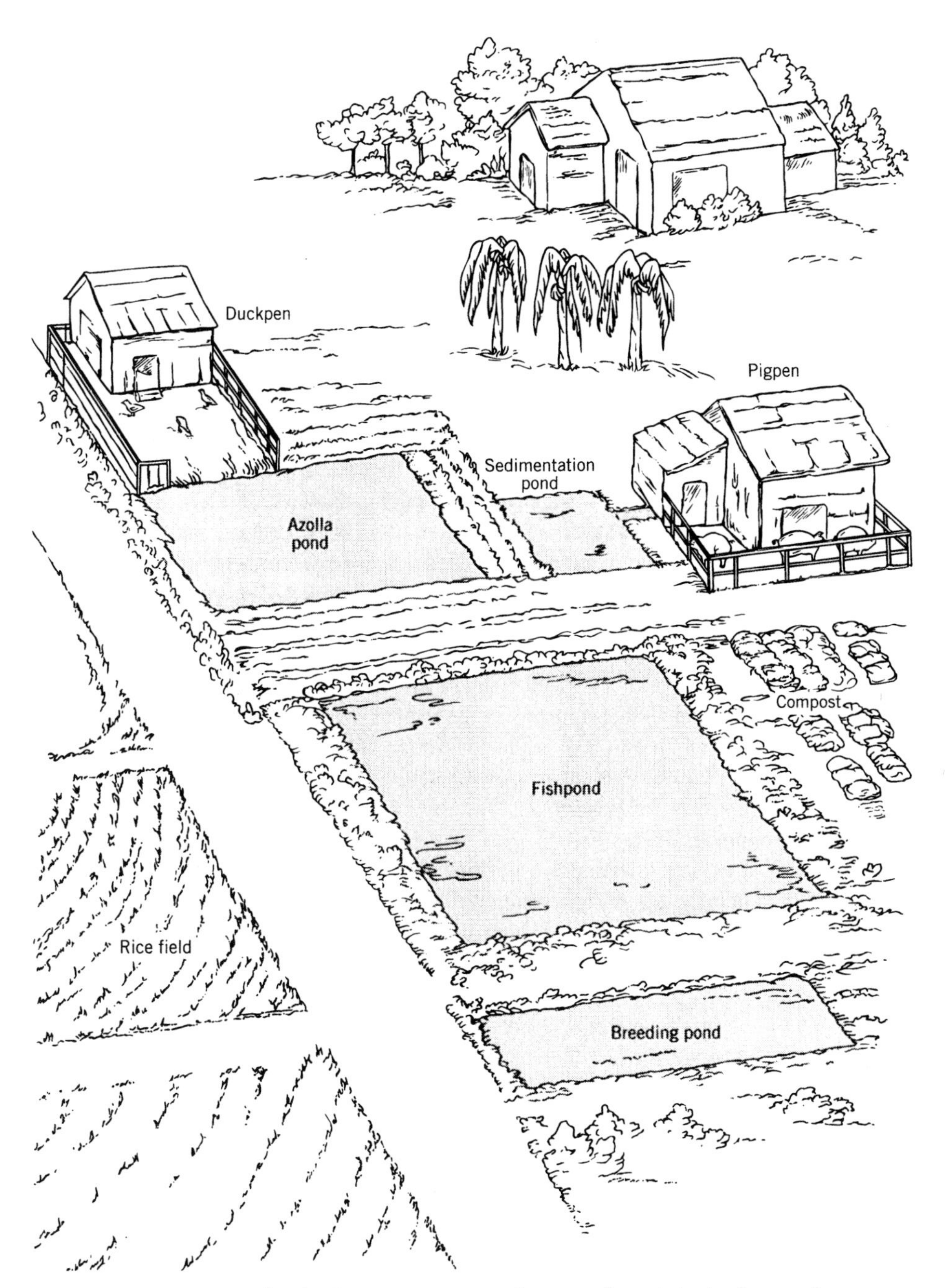

FIGURE 6.8 Integrated cycle of nutrients as practiced on some farms in lowland areas of the Asian tropics.

There are hundreds, if not thousands, of different combinations in agroforestry. The particular type found in a particular setting depends on the terrain and native ecosystems, economic conditions, social customs, and possibly political directives, and all these factors influence the system's success (Lal 1991).

Forestry

Temperate Zone

In forests managed for sustainable timber production, the volume harvested should not exceed the growth rate. Thus, if a forest stand puts on 5 m^3 of wood per year, then every 10 years, 50 m^3 could be logged from that stand. In hardwood stands, selection cutting can often produce sustainable yields. In selection cutting, only certain individuals are harvested. They usually are more mature individuals, and they are removed because of their relatively high value compared to smaller trees, and because their removal opens up a space in the forest that allows younger trees to respond with vigorous growth. A question sometimes arises as to which mature trees should be left as seed source. Those with the best shape are usually harvested, but by so doing, the best genetic stock is removed from the forest. Some foresters believe that the best trees should be left as seed sources, but the relationship between genotype and phenotype is not always straightforward. Sometimes a crooked tree may be crooked not because of genes, but because of damage due to windthrow or animals. Sometimes a straight tree is straight, not because of its genetic stock, but because of silvicultural management such as pruning or proper spacing.

Although selective cutting is ecologically desirable, it is seldom practiced because of logistical problems in harvesting. Much more common is clear-cutting. Although clear-cutting is often criticized, when forests are managed on a landscape basis clear-cut logging can be sustainable. If on a 200-acre tract, 5 acres are clear-cut every year and then allowed to regenerate for 40 years before the next cut, the practice can be sustainable. Sometimes, clear-cut stands are allowed to regenerate naturally. This can occur through stump sprouting of many hardwood species or through natural secondary succession. The clear-cut stand also can be replanted with seedlings. In the conifer stands of western United States, clear-cuts, or "patch cuts" as they are called, often are reseeded or replanted with seedlings.

Technically, there are essentially no obstacles to sustainable forestry for wood production in the temperate zone. The germination requirements of the seeds are known, as are the physiological characteristics of the seedlings. The seedlings of virtually all economically important species are readily available. Sometimes it is argued that nutrient loss owing to clear-cutting can limit the sustainability of a stand. Although it is true that clear-cut logging damages streams and disrupts fisheries and can also result in the siltation of reservoirs, rarely if ever are the amounts of nutrients removed sufficient to prevent the reestablishment of another stand (Jordan 1986).

The most important problems in temperate zone forestry do not include a lack of ability to manage for sustainable timber production. The serious problems arise from the competing nontimber demands being imposed on the forests. The increasingly urban population of the United States, Europe, and Japan is turning more and more to these forests for recreation, including hiking, camping, hunting, fishing, bird watching, wilderness experience, and scientific study. Logging almost always reduces the quality of recreational experiences. Logging also can threaten rare and endangered species. Thus there is increasing political pressure to stop logging in the publicly owned forests. It is the irreconcilability between logging and other uses, regardless of whether or not the logging is sustainable, which is causing conflict.

So although sustainable forestry for timber production is technically feasible in temperate regions, there is increasing pressure to reduce or phase out logging on public lands no matter how sustainable it may be. But pressure is resisted by those who say that the wood is needed to supply demand from housing construction and other wood-using industries. To restrict wood supplies, they say, would damage the economy.

Why don't private landowners take up the slack and start producing the timber that the world will clearly need? There are two reasons. One is that in the United States, the Forest Service, at least through 1993, has undercut the efforts of private landowners by heavily subsidizing logging on national forests (O'Toole 1988). The Forest Service manages the stands and puts in roads at taxpayer expense in order to enable loggers access to the timber. The price of the timber frequently does not cover the cost of the roads. This subsidy keeps timber prices low, but at the same time, it discourages private investment in timber production. If timber prices were to rise, the result could be a boon for private timber production.

A second problem with private investment in traditional forest plantations is the relatively long time between initial investment and the sale of the trees. However, initiatives that combine economics and ecology may reduce this problem. An example is the mixed plantation being tried by researchers in the Southeast. The experiments combine trees of the fast growing genus *Paulownia,* which is highly prized in Japan when the logs are straight and clear, and slower growing traditional hardwoods such as oak. The *Paulownia* serves as a "nurse" tree, mitigating extremes of microclimate and improving the soil with leaf litter. It can be harvested at 20 years at a good profit (if the trees have been well tended). The harvest serves as a "release" for the interplanted oaks, which by then have a well-established root system. The oaks can then reach marketable size in another 15 to 20 years. Other advantages of combinations such as this one are that the oaks provide side-shading that gives good form to the *Paulownia,* and a mixed plantation gives more ground cover, thus better suppressing weeds. A further innovation is the interplanting of pulp species that can be harvested at 5 to 8 years.

Because of large, privately owned forests in the southeastern United States, timber production is important in that region. It is important that managers there try to work with and not against, nature. In the Southeastern Piedmont, pine occurs naturally in recently abandoned fields, for reasons given in the preceding section under Succession. Management for pine during the early stages of succession requires little energy subsidy. However, once the first crop of pine is harvested and hardwoods have invaded, management for pine requires a large investment of energy for site preparation. Bulldozer-powered root-rakes, roller choppers, and gang plows are needed to return the soil to an early stage of succession so that the soil flora and fauna favor pine. Once the soil in the Southeastern Piedmont has evolved toward a condition that favors hardwoods over pines, it is expensive to try and fight nature by trying to cultivate pines by eliminating hardwoods. A more sustainable strategy is to encourage mixed plantations of hardwoods.

High Latitudes

Species diversity of trees decreases with increasing latitude. The forests of Canada, Scandinavia, and the taiga of the former USSR have a relatively low diversity of trees. Large, homogeneous stands of trees face particular problems with regard to management. In homogeneous stands, outbreaks of pests can be particularly devastating. The spruce budworm, for example, is a serious forest management problem in northern New England and eastern Canada. The pine-bark beetle is a serious pest in the pine stands of the southern United States. As a stand of pine or spruce matures, the trees have lowered resistance to pests. Outbreaks of pests, once they are established, can be devastating and render the timber over large areas almost worthless to the logger.

Under natural conditions, nature heals itself. With decreased vigor, the stands become more susceptible to fire. When a fire roars through, it kills the insects, sterilizes the soil, and the ash provides a fertile seedbed for a new crop of spruce or pine. The removal of cover provides plenty of sunlight to allow vigorous growth. However, when a stand is managed for timber or pulp, the owners cannot afford to have a fire. To keep trees healthy, they are sometimes sprayed with insecticides, but that is expensive and environmentally damaging. One solution may be managing on a landscape scale instead of on the scale of individual stands (Erdle and Baskerville 1986). The landscape is managed so that there are stands at all stages of growth. The most mature are cut just as they become susceptible. The number of stands is adjusted to meet the economic requirements of the market.

The Question of Fire

The use of fire in forest management has been controversial. In the early decades of the twentieth century, the policy of the U.S. Forest Service and of State Forestry departments was to suppress fire. Fire was seen as damaging to timber and therefore as damaging to economic interests. "Smoky Bear," the cartoon of a bear dressed in a forest ranger's uniform, was created to educate the public regarding the dangers of forest fires. He became an extraordinarily successful symbol of a fire prevention campaign (Hart 1990).

In some cases, however, suppression of fire did more damage than good. When fire is suppressed, fuel in certain ecosystem types such as the conifer forests of the semiarid West accumulates in dangerous proportions. The result is that when a fire does get going during a drought year, the results can be disastrous. The fires in Yellowstone Park in 1988 are a striking example. They whipped through the canopy and damaged or destroyed many tourist businesses in or near the park.

Periodic lightning fires prevent a dangerous accumulation of litter and duff on the forest floor. Such fires usually creep along the forest floor, do little permanent damage to the forest, and ensure a regeneration patchwork of trees and grass that supports a richer and more varied faunal population. Fire can provide other services desirable for ecosystem management. It is essential for the maintenance of particular types of communities and species. In the midwestern United States, from Wisconsin and Minnesota south to Oklahoma, fire rejuvenates the tall-grass prairie. When fire is suppressed, the ecosystems revert to forest. Efforts to reestablish prairie require the use of fire because it kills woody vegetation and allows more sunlight and moisture to penetrate to the roots and rhizomes of prairie species.

Fire favors the reproduction of the valuable long-leaf pine forests on the coastal plain of the southeastern United States, and jack pine in Michigan where it is the habitat of the endangered Kirtland warbler. Fire helps preserve the semi-open cover preferred by many game birds such as quail. The sequoia forests of California have an understory of fir, as a result of fire suppression in the stands (Biswell 1989). Without fire, the fir will replace the sequoia. In the 1960s and 1970s, the Tall Timber Research Station near Tallahassee, Florida, began a series of workshops focusing on the beneficial aspects of prescribed fire in a variety of ecosystems. Fires for ecosystem management are still controversial, but increasing research on their effects is increasing their acceptance (Dahl et al. 1978).

Tropical Forestry

Just as in the temperate zone, sustainable forest management for timber supply in the tropics is technically possible. It is rarely, if ever, achieved in practice, but the reasons are different from those in temperate areas.

Management of a forest stand in the tropics is much more difficult than management of a stand in the temperate zone. There are many more species in tropical forests, each with its own individual physiological characteristics and environmental requirements. Although only a few species may have a market value, all species must be considered in a management plan. And in contrast to temperate species, very little is usually known about the ecophysiology of tropical species.

Nevertheless, in the early 1900s, foresters of the colonial empires in Europe devised management plans for naturally occurring tropical forests. In Africa, the systems were often referred to as the tropical shelterwood system (Fox 1976). These systems were

FIGURE 6.9 Strip-cutting system designed to facilitate commercial exploitation, as well as to take advantage of the services of nature such as the downhill migration of seeds and nutrients needed for regeneration.

designed to promote the establishment, survival, and growth of the seedlings and saplings of desirable species by poisoning undesirable trees and removing vines and weeds. After several years, when surveys showed that the reproduction of desirable species was well established, the canopy trees of the timber species were harvested. The tract was then monitored, and weedings and thinnings took place as needed. The system was abandoned in the 1960s, partly because it did not make sufficiently intensive use of the land to compete with other forms of land use such as cocoa, oil palm, or agricultural crops (Lowe 1977).

Several demonstration plots of sustainable forest management are located in the tropics and subtropics, such as those established by the Institute of Tropical Forestry in Puerto Rico. However, such management is rarely practiced outside of experimental forests. In areas of substantial tropical forest, such as the Amazon basin, it is much cheaper and easier to log a stand of virgin timber, take out the valuable trees, and leave the rest than it is to undertake a sustained yield management plan, such as the shelterwood system. Timber from carefully managed stands cannot compete in the market with timber "mined" from virgin stands.

In some tropical forests experiments are being conducted with strip-cutting (Fig. 6.9). This system combines the logistic advantages of clear-cutting with the restorative ability of small cuts where seeds and saplings quickly recolonize. To be successful, however, the system must be adapted to the economic and cultural situation.

Wetlands

Often conservation management must extend beyond the bounds of the farm or the forest itself; management must be on the scale of the landscape. Preservation of wetlands is an integral part of conservation management on a landscape scale because of the vital life support services that wetlands carry out for a whole region.

Wetlands are coastal and inland ecosystems characterized by daily, seasonal, and long-term fluctuations in water levels. The U.S. Fish and Wildlife Service has defined wetlands as

> *. . . lands transitional between terrestrial and aquatic systems where the water table is usually at or near the surface or the land is covered by shallow water. . . . Wetlands must have one or more of the following attributes: 1. at least periodically, the land supports predominantly hydrophytes (aquatic vegetation), 2. the substrate is predominantly undrained hydric soil, and 3. the substrate is nonsoil and is saturated with water or covered by shallow water at some time during the growing season of the year.*

The U.S. Army Corps of Engineers and the U.S. Environmental Protection Agency use a more restricted, legal definition of wetlands in their regulations covering dredge and fill permits under Section 404 of the Clean Water Act.

> *The term wetlands means those areas that are inundated or saturated by surface or ground water at a frequency and duration sufficient to support and that do support, a prevalence of vegetation typically adapted for life in saturated soil conditions. Wetlands generally include swamps, marshes, bogs and similar areas.*

Wetlands have a high natural productivity and are important refuges for a wide variety of species, including important game birds such as migratory waterfowl. In some regions, they are important for groundwater recharge, and they also act as buffers against flooding. Salt-water wetlands are important biologically. Adult shrimp spawn offshore, and high tides carry the larva of their offspring into semi-enclosed estuaries. There they find the food and protection they need for rapid growth during their young and adolescent stages. When they mature, they move into deeper coastal waters and then into the open ocean.

Wetlands form a buffer that eases the impact of high water during floods (Kusler et al. 1994). Buffers have evolved through natural processes. The rapid flush of water from headwater areas during periods of heavy precipitation to more level terrain at lower elevations produces an interconnected wetland system. Silt deposition along the main channel forms natural levees that are stabilized by vegetation. Behind the natural levees are interconnected floodplain lakes that drain into the main channel. When water is high in the main river channel, the water flows into the wetlands behind the natural levees. As the water rises, they absorb the excess flow that cannot be carried by the main channel. After the peak passes, they gradually release the water back into the river. (Fredrickson and Reid 1990).

Many of the world's wetlands are being lost through developments such as dams for hydropower and flood control, diversions to speed water flow, levees for flood protection, wetland drainage for commercial districts and agriculture, and filling of wetlands for marinas.

The filling of wetlands along coastlines has produced economic and ecological problems. During heavy storms, salt-water coastal marshes absorb the waves that wash up over the beaches of barrier islands facing the open ocean. When the waves dissipate their energy across the islands, they deposit sand that replenishes the beach. When the marshes are filled or when there is construction on the sand beaches, rebuilding of the spits is replaced by erosion that destroys the beaches.

No single piece of legislation expresses federal goals for wetlands protection and management in the United States. Goals that relate directly or indirectly to wetlands management are embedded in several pieces of legislation reviewed by Bingham et al (1990). These include mitigation policy, that is, actions that must be carried out when federal activities damage wetlands.

Active political opposition has been launched against some of the laws and regulations that restrict the development permitted in areas designated as wetlands. The rules sometimes are seen as interfering with the rights of private landowners. However, recent natural disasters have demonstrated the importance of maintaining wetland functions. The floods that occurred on the upper Mississippi River in the summer of 1993 resulted in a tremendous economic loss, as flood waters breached the levees and destroyed whole towns and thousands of square miles of agricultural lands. If the natural wetlands along the river had been preserved, damage would have been much less severe because they could have absorbed part of the flood.

Management of Succession

Management of successional communities is an important part of conservation. An example is management of rights-of-way under power lines.

Forest growth is undesirable under power lines because the trees prevent access for maintenance and can grow up into the transmission lines and damage them. Periodic cutting of the trees once was the only available technique. Then when herbicides first became available, many thousands of miles of rights-of-way were blanket sprayed to achieve complete brush and tree control. In forested areas, these management practices created much bare ground that facilitated tree regeneration by seeds or sprouts. Rights-of-way communities developed more tree stems than before spraying. Management activities had to be repeated often to keep trees out of transmission lines.

A solution was proposed through selective herbicide application that encouraged growth of shrubs. Once a thick cover of shrubs is established, the establishment and growth of trees are sharply inhibited and the need for spraying or cutting is greatly reduced. Luken (1990) reviews details of the methods.

Restoration

The intensity with which humans have disturbed ecosystems of the world varies greatly. Plains Indians once lived in what is now the Dakotas and burned the prairie to increase the grass fed upon by bison. That was a relatively minor disturbance, and recovery of denuded grassland occurred within a matter of weeks. In contrast, surface mining in Kentucky where coal beds were uncovered by removing the tops of mountains with bulldozers proved a very severe disturbance. Because of barren rock and acid mine spoils, it might take hundreds, if not thousands, of years for forest to regenerate.

Restoration is a field encompassing many techniques suitable for rehabilitating a wide variety of degraded ecosystems (Berger 1990; W. Jordan et al. 1987). A common approach to restoring degraded areas is tree planting. Often the technique is to prepare an area, and then plant seedlings, usually at intervals of 2 to 4 m. A major problem with this approach is the expense. Often, only small areas can be reforested because of limited funds. Under certain conditions, however, much larger areas can be reforested at relatively low cost by taking advantage of the services of nature.

Such an approach has been explored by Kolb (1993) in an effort to expand the reserve of the golden lion tamarin (marmoset) in the Atlantic Coastal Forest area of Brazil. After examining "islands" of vegetation within the pastures surrounding the reserve, she found that these islands ranging in diameter from 2 to 10 m were strong attractants for birds, bats, and small mammals that dispersed tree seeds. In contrast, virtually no seeds were dispersed into treeless pasture. Kolb's work suggested that planting islands of vegetation instead of continuous plantations could be an economical approach to reforestation because of the seed-dispersal services of birds and mammals.

This is but one of many ways to take advantage of the services of nature. Others would include selecting nitrogen-fixing species for planting on degraded land or nutrient-poor pastures; using early successional species for improving microclimate and soil conditions, and thereby facilitate the later establishment of more valuable trees; and perhaps most important, being alert to possible roles for any and all species, even though they may be considered pests. As an illustration of the last point, fire ants are considered invasive pests in the southeastern United States. They *are* pests, if you stand on their nest, because their sting is bothersome and they can injure a young child. Yet it has been observed that in plantations on old degraded soils, saplings show improved growth when their roots are surrounded by a colony of fire ants. The improvement could come from the improved texture of the soil resulting from the nest and/or from greater phosphorus availability. Preliminary tests showed higher water-soluble phosphorus in soil under the nests of fire ants. The increase may be due to the effect of organic matter transported in by the ants. The bottom line regarding fire ants is not necessarily that they be encouraged. It is that one should not always rush to try and exterminate them, especially in an old, degraded field that one is trying to restore to forest.

Fisheries

The idea that resource systems should be managed to take advantage of the services of nature is relevant to farms, some forests, and rangelands, but it has limited relevance to fisheries. Maintaining free-flowing rivers or removing dams from spawning runs is one example of working with nature in the field of fisheries. But aside from this and from other relatively minor measures such as protecting and encouraging habitat including reefs and turtle grass, and from guarding wetlands that nourish nearshore populations, managing fisheries often consists of regulating the catch and, in certain special cases such as salmon, restocking.

Regulating the catch by imposing "bag limits" or quotas has proven difficult in marine fisheries because of the limitations of scientific knowledge on levels of sustainable yield. The standard approach to commercial fishery management relies almost universally on the idea that there is a strong relationship between the size of a fish stock today and the number of recruits, or young marketable fish, in the next generation of fish. The idea is the basis for long-term control of the fishery: if current catch can be regulated by controlling the level of effort in the fishery, the size of the catch in the future can also be controlled.

The presumed relationships between fish stock size and recruitment, between fishing effort and yield, and fishing effort, cost, and revenue are succinctly summarized by Townsend and Wilson (1987) (Figs 6.10, 6.11, 6.12). If or where these relationships hold, controls on harvest would be important in sustaining a fisheries resource.

These relationships often do not hold, however (Ludwig et al. 1993). For any individual species in a system, there appears to be almost no relationship between current and future stock sizes, when populations are above some critical minimum size (Hennemuth 1979, cited in Townsend and Wilson 1987). Above the minimum, recruitment is highly variable, with an annual average that is largely independent of stock size. Below the critical minimum, average annual recruitment falls dramatically. The likelihood of a high level of recruitment is much lower below than above the critical minimum.

A second problem associated with fisheries management is that fishing grounds often are accessible to everyone. Although Hardin's *Tragedy of the Commons* (1968) has been

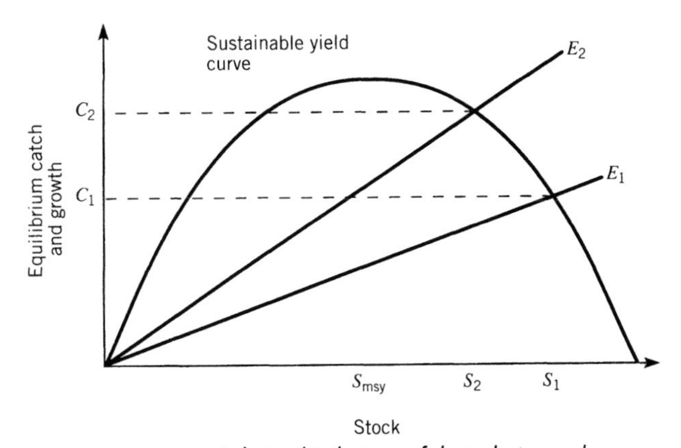

FIGURE 6.10 *Relationship between fish stock size and recruitment. Source:* Ralph Townsend et al., "An Economic View of the Tragedy of the Commons," *The Question of the Commons,* University of Arizona Press, Tucson, 1987, p. 314.

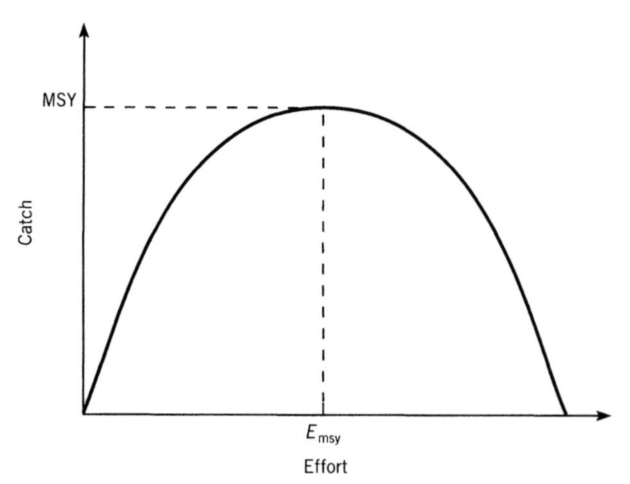

FIGURE 6.11 *Relationship between fishing effort and yield. Source:* Ralph Townsend et al., "An Economic View of the Tragedy of the Commons", *The Question of the Commons,* University of Arizona Press, Tucson, 1987, p. 315.

the most widely cited recent statement of the idea, the problems resulting from the exploitation of commonly owned resources has been recognized at least since the time of Aristotle (Ostrom 1990). Gordon (1954) described the problem of the commons in fisheries as follows:

> *There appears then, to be some truth in the conservative dictum that everybody's property is nobody's property. Wealth that is free for all is valued by one because he who is foolhardy enough to wait for its proper time of use will only find that it has been taken by another. The fish in the sea are valueless to the fisherman, because there is no assurance that they will be there for him tomorrow if they are left behind today. (p. 124)*

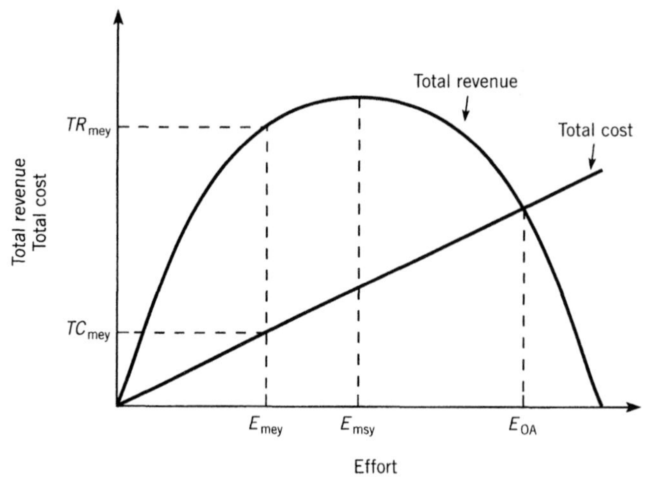

FIGURE 6.12 *Relationships among fishing effort, cost, and revenue. Source:* Ralph Townsend et al., "An Economic View of the Tragedy of the Commons", *The Question of the Commons,* University of Arizona Press, Tucson, 1987, p. 314.

Harvesting of irregular or fluctuating resources such as fisheries that are commonly owned are subject to a "ratchet effect" (Ludwig et al. 1993). During periods when the resource remains relatively stable, harvest rates tend to stabilize at levels predicted by steady-state bioeconomic theory (Figs. 6.10–6.12). When harvests drop owing to natural climatic phenomena or problems such as pollution, the industry appeals to the government for help because substantial investments and many jobs are at stake. The governmental response often is a direct or indirect subsidy. The subsidies may initially be regarded as temporary, but often they become permanent, and their effect is to encourage overharvesting. The ratchet effect is caused by the lack of inhibition on investments during good periods, and by strong pressure not to disinvest during poor periods. The long-term outcome is a heavily subsidized industry that overharvests the resource.

Some principles of effective management for commonly owned resources suggested by Ludwig et al. (1993) are as follows:

> Recognize human motivation, short-sightedness, and greed as a major cause of the problem.
>
> Act before scientific consensus is achieved. Calls for additional research often are mere delaying tactics.
>
> Rely on scientists to recognize problems but not to remedy them. Claims that basic research will (in an unspecified way) produce sustainable use of resources in the face of a growing human population will lead to false complacency and diversion of resources.
>
> Confront uncertainty. Once we are free from the illusion that science or technology (if lavishly funded) can provide a solution to resource or conservation problems, appropriate action becomes possible. Resource problems are not really scientific problems, but they are human problems that have been created at many times and places under a variety of political, social, and economic systems.

Are there any specific remedies that may ameliorate, at least in part, the problems of the commons? With specific reference to fisheries, Anderson (1987), said: "Ultimately, all answers to conservation problems can be reduced to one principle: make the users pay the costs. Passing on costs as 'externalities' to downstream users or to a diffuse public must be prevented." (Here "downstream users" refers to people living downstream who must clean up the water polluted by upstream users.) In the case of fisheries, the externalities would be a depleted stock, and the "diffuse public" would be the world that has lost another resource.

How can users be made to pay? Anderson (1987) has suggested privatization, public control, and localization.

Privatization

An example of privatization of a former commons is the establishment of private hunting and fishing reserves to replace public lands where fish and game are depleted. Privately held tracts of land and streams are leased to hunting and fishing clubs. To ensure the sustainability of income, landowners manage the tracts to ensure maintenance of fish and game populations desired by hunters and fishermen. The fees paid to the landowner by the clubs ensure that resources are available for economically sustainable management.

Public Control

Bag limits (limits on the take of fish and game) are common restrictions in sport fisheries and wildlife management areas. Regulations for commercial fisheries on inland and territorial waters are also common. However, public control involves many practical problems. Fishermen may misrepresent catches, and it is difficult to monitor all fishermen. When restrictions are placed on the method of fishing, the rules may result in an inefficient system of resource harvest. Furthermore, scientific knowledge on which to base bag limits may be inadequate.

Localization

Some authors have distinguished between marine fisheries in which there is open access to everyone and a commons in which access to resources is restricted to members of the commons. Such restriction is called localization.

Localization includes community control of local resources. The Maine lobster industry (Acheson 1987) is an example. This industry on the coast of Maine is kept sustainable through strong social mores and strong territoriality among the lobstermen. Although the coastal zone is a commons, there is a strong sense of territoriality. Although access to the coves and bays along the coast is free to anyone, access to lobstering is highly restricted by unofficial agreements among the lobstermen. The right to trap in a particular segment of the coast belongs to a particular lobsterman by tradition, and the right is inherited from one generation to the next. It is virtually impossible for a newcomer to gain access to a claimed territory. Claims are enforced by measures such as cutting loose the traps of anyone who infringes or trespasses on claimed territory. Sanctions are effective because they are enforced by (virtually) all members of the lobstering community.

The Maine lobster industry has features that are unique and amenable to local control. The resource itself is limited. The Maine lobster has a special taste and quality that gives it a value higher than lobsters from other regions. Moreover, the lobstermen are a relatively small and homogeneous group, thereby making agreements within an association possible.

Range Management

The principle that resource production systems that resemble natural ecosystems are ecologically and economically superior to those that are not is particularly exemplified in grazing systems. There is increasing evidence that on natural rangelands such as the Serengeti in Africa, moderate grazing can increase primary productivity and that some plants benefit from the presence of grazing animals (McNaughton 1992). The positive feedback from herbivores to producers may be chemical in that grazers induce a complex series of biological and physiological events that result in changes in plant function (Dyer et al. 1992). The feedback may be physical in that the impact of hooves may actually stimulate the production of range species in the long term, if the grazing pressure is low, as would occur with wild species (Savory 1988).

In contrast, pastures in areas that are naturally forest, especially tropical rain forest, are possible only when heavily subsidized. In undisturbed rain forests, decomposition is rapid. Heavy leaching of solubilized nutrients during rains could cause nutrient loss, but under forest cover, there is little loss because naturally occurring rain forest species have a whole host of nutrient-conserving mechanisms (Jordan 1989). Trees, with their deep roots, help prevent deep leaching and conserve some of the more readily leachable nutrients such as potassium in their trunks. Secondary plant chemicals prevent herbivory and protect plants from nutrient loss, an important characteristic in nutrient-poor conditions. Rain forests are rich in mutualisms that help sustain productivity in the face of nutrient-leaching stress. However, when the forest is destroyed and converted to pasture, the nutrient-conserving mechanisms are destroyed, and the productive capacity quickly deteriorates.

The adaptations of naturally occurring herbivores such as the wildebeest on the Serengeti give them a potential to be a more sustainable resource on their home range than domesticated substitutes. The idea of using wild species that are well adapted to local ecosystems instead of domesticated species that must be heavily subsidized is ecologically appealing. "Iguana ranching" in the canopies of tropical forest trees is much more sustainable than cattle ranching following the burning of trees. Sustainable harvest of native species such as North American bison should be less damaging to the environment than herding of cattle. However, the public is highly conditioned to eating beef, pork, and chicken, and social preferences are an important obstacle to implementing ecologically sustainable production of meat products. Furthermore, native species often have migration patterns that interfere with commercial production.

Costs of Shifting to Sustainable Systems

In this chapter, we have developed the idea that sustainable management of resources means moving away from fossil fuel subsidies and destructive use of resources such as shifting cultivation. However, changing to more sustainable methods of resource management may not be easy—or even possible in some cases—because of particular economic situations.

As an example, let us consider peasant farming in the tropics. Cultivating the sides of tropical mountains by slash and burn techniques is ecologically destructive and nonsustainable because soil erosion quickly depletes the ecosystem's base of production. The farmers do slash and burn anyhow because the only way they can survive is to exploit and destroy the forest.

The sustainable agroforestry land technology (Fig. 6.7) presents an example of a technically feasible agroforestry system that seems to have the potential to replace the more destructive slash and burn. However, the transition from hillslope monocultures of corn to terraced agroforestry systems is not quick and cheap. Often it will take several years. The costs that must be borne before the advantages of the sustainable system are realized are more than a peasant farmer can afford. In such cases, sustainable management techniques are not implemented because of the transition costs associated with abandoning the nonsustainable technique. If a poor farmer does not have the funds or resources to support the transition, he cannot make the change.

Almost all transitions from management for exploitation to management for sustainability require a transition cost. For example, Matta Machado (1993) found that it took three years for an alley cropping system to significantly improve the fertility of the soil. Farmers changing to the system would have to bear a cost as soon as the transition was begun because the hedges in the alley cropping system took 20 percent of the land area and thus reduced production by this amount.

One of the potentially most rewarding forms of international aid to developing countries would be a program of aid and assistance to farmers or other resource managers to help them in effecting a transition from a nonsustainable to a sustainable system. Once the system was established and profitable, however, the aid should be withdrawn because subsidies can be dangerous. When they begin, subsidies often seem like a good idea. They are put in place for a good reason. But subsidies build up a constituency that develops a political influence. When that occurs, the subsidies are very difficult to terminate, even if the national need for the subsidies has long since disappeared. This leads to a misuse of resources.

Subsidized programs should be initiated only when there is a clearly demarcated termination date for the subsidy. Everyone involved should understand that unless the system is able to sustain itself economically within the prescribed time limit, the system should be allowed to die.

TWENTY GREAT IDEAS IN ECOLOGY AND THEIR RELEVANCE TO THE RESOURCE MANAGEMENT PRINCIPLE

Throughout much of the history of resource management, guidelines were, to a large extent, empirical. Farmers, foresters, game managers, and the like managed their resources based on practical experience. A farmer would try something—for example, varying his crops and the amount of fertilizer he applied. If it increased his yield and was not too much trouble, he might continue the practice. If it did not, he would try something else.

Practical experience as well as common assumptions about the world is "common sense," and resource managers based their recommendations on common sense. This sometimes gave the desired results, but sometimes it did not. Killing wolves and cougars, for example, was assumed to be the logical thing to do. But hunting and eliminating top predators did not always prove the best thing to do. "Common sense" was not always an adequate guide for resource management when killing predators resulted in an uncontrolled deer herd that destroyed its own resource base.

The end of World War II brought an increased awareness of the need for better scientific understanding of the basic nature of resource systems such as forests, farms, and fisheries. Much of the new initiative fell within the purview of ecology. Previously, ecology had been concerned mainly with the description of naturally occurring populations and communities. But with two new developments, ecology emerged as a leading field in the investigation of the basic nature of fields, forests, and lakes. One development was the advent of the atomic age.

As a result of atmospheric testing of nuclear warheads, great concern arose over the threat to human health posed by radioactive fallout. The U.S. Atomic Energy Commission (AEC) initiated a program to study the fate of radioactive isotopes in the environment. This initiative gave impetus to what has become known as ecosystem ecology (Golley 1993b). A typical study would be to inject a small amount of radioactive material into the trees of a forest, and follow that activity through the herbivores, carnivores, decomposers, and mineral soil and back into the plant again. When studying the flow and cycles of radioactive isotopes such as Strontium-90 and Carbon-14 through a field, forest, or lake, one cannot help but focus on the flow of energy and the cycling of elements through the atmosphere, soils, water, and the biological populations and communities of an ecosystem.

A second important development in ecology relative to resource management was the emergence of population biology. This was a merging of old-style natural history with new advances in genetics and evolution. It has given insights into the role of species and communities in the functioning of natural and human-managed systems. It includes interactions between species and the critical role of such interactions in ecosystem function. Particularly important is how the evolution of mutualisms between species (coevolution) changes the structure and function of ecosystems ("evolution of ecosystems") and, conversely, how the evolution of ecosystems affects interactions between the species that compose it.

Empirical resource management is sometimes jokingly called the "Let's throw another ton of fertilizer on the corn field and see what happens" approach to research. In contrast, modern ecology is an inductive science based on studies of the processes and mechanisms of ecosystem function, or on analysis of ecosystem properties drawn from studies of whole systems. These process and system studies seek to understand the mechanisms that will help to predict whether or not a field of corn might respond to another ton of fertilizer. In this case of corn and fertilizers, process studies would look at factors in the soil that regulate nutrient availability and at factors in the plant that affect nutrient use. Systems studies analyze the dynamics of nutrients through the whole ecosystem.

As a result of the major funding initiatives in ecology during recent decades, a tremendous amount of basic information has accumulated on the structure and functioning of natural and human-modified ecosystems. The challenge has been to distill and synthesize the information in the tens of thousands of studies and reports that are available.

To consolidate some of the vast ecological information into concepts that could be useful in applications of many kinds, Odum (1992) synthesized what he has called twenty great ideas in ecology for the 1990s. Even though some of the ideas are controversial, and not proven in a strict scientific sense, they are extremely useful in that they condense a great deal of information into a comprehensible form. Here, these 20 "great ideas" are

Table 6.9 *Categories for "Great Ideas in Ecology"*[a]

A. To develop a sustainable system, we must understand that we are part of the system. (Idea 1)

B. To develop a sustainable system, we must analyze the system in which we are embedded. To analyze the system, we need a common currency. Energy is a convenient "currency" with which to analyze a system. Energy flows through all ecosystems, and the way it is used and stored determines the characteristics of each ecosystem. (Idea 2)

C. Energy flow through natural ecosystems is not random but is controlled and self-regulated by internal feedback interactions between organisms, or between organisms and environment. (Ideas 3–6)

D. Stability and sustainability of energy flow through ecosystems are enhanced by mutualistic functions. As ecosystems became large and complex, organisms within a natural community evolved toward cooperative functions. Evolution has selected many mutualistic species because of the higher efficiency of the functions they perform. (Ideas 7–12)

E. In managed systems, the cooperative functions and natural subsidies that lend stability and sustainability to natural systems are usually destroyed. For this reason, energy subsidies are usually required. (Ideas 13–15)

F. The stability and sustainability of a managed system can be increased by replacing external energy subsidies with cooperative functions of nature. (Idea 16)

G. The stability and sustainability of all systems can be increased by maintaining species and landscape diversity. (Ideas 17, 18)

H. The transition from nonsustainable to sustainable systems requires time and has a cost. (Idea 19)

I. Despite the cost, there is an urgent need to make the transition. (Idea 20)

[a] *The number(s) after each category refers to the number of the "Great Idea" in the text.*

used as a support for the resource management principle. They have been arranged in nine categories, "A" through "I" (Table 6.9), which form a framework in which the "great ideas" can support the resource management principle. The original literature on which the ideas are based is given by Odum (1992).[3]

A. To develop a sustainable system, we must understand that we are part of the system. (Idea 1)

1. "A parasite-host model for humans and the biosphere is a basis for turning from exploiting the earth to taking care of it." Man is an intermediate compartment through which energy flows on its way from the sunlight that impinges on the surface of the earth to the heat that diffuses away from it (see Fig. 6.5). In the biosphere, Mankind's functional role is similar to that of parasites.

A good parasite does not destroy its host. Through evolutionary selection, it coevolves to coexist with its host by not taking too much from it. To ensure its own sustainability, it must be certain that the host is sustainable. A good parasite functions as part of a system and may evolve to actually benefit its host or the entire system.

B. To develop a sustainable system, we must analyze the system in which we are embedded. To analyze the system, we need a common currency. Energy is a convenient "currency" with which to analyze a system. Energy flows through all ecosystems, and the way it is used and stored determines the characteristics of each ecosystem. (Idea 2)

2. "An ecosystem is a thermodynamically open, far from equilibrium system." To understand a system, all elements of the system must be expressed in common currency. Energy is a convenient currency, and the laws of thermodynamics are convenient tools.

"Thermodynamically open" means that the ecosystem depicted in Fig. 6.13 is receiving energy from the sun which is outside the system, and discharging heat back outside the system into the atmosphere. Because the energy flow is one way through the ecosystem, the ecosystem is not in equilibrium with its surroundings. The populations within the ecosystem are not in equilibrium either. As climatic conditions change, energy flow through the system changes, resulting in changes in populations of species that comprise the ecosystem."

C. Energy flow through natural ecosystems is not random but is controlled and self-regulated by internal feedback interactions between organisms, or between organisms and environment. (Ideas 3–6)

3. "Feedback in an ecosystem is internal and has no fixed goal." Feedback is information sent from one part of a system to another for the purpose of regulating the system. In contrast to a mechanical system such as a thermostat which has a fixed set point, naturally occurring ecosystems have internal feedback systems that keep the system functioning not at a fixed point, but within limits that are tolerable by life. The feedback system in rain forests described earlier in this chapter keeps enough nutrients mobile to keep the system functioning.

4. "Indirect effects may be as important as direct interactions in a food web and may contribute to network mutualism." Pirañhas, a carnivorous fish of South America, indirectly

[3] *In the list below the first sentence after each numeral is Odum's idea. The sentences that follow are an explanation.*

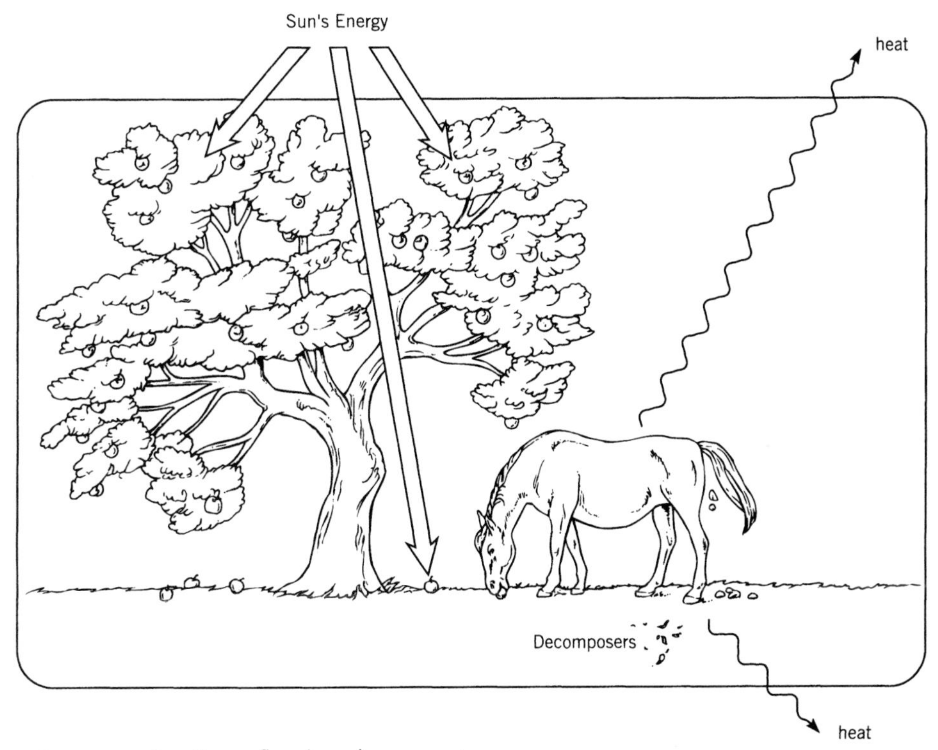

FIGURE 6.13 Energy flow through an ecosystem.

affect plant species in aquatic ecosystems in that they keep nutrients in soluble forms. When pirañhas eat herbivorous fish, they egest in a soluble form the nutrients bound in those fish. Nutrients can then be used by aquatic plants. When top carnivores are removed, nutrients accumulate in the bodies of a swelling population of herbivorous fish. When they die, they sink to the bottom and decompose anaerobically, and fewer nutrients become available to plants. The flow of phosphorus through this system is illustrated in Fig. 6.14. Pirañhas affect plant growth indirectly by regulating nutrient supply.

5. "Heterotrophs may control energy (and nutrient) flow in food webs." Organisms within the food chain that can switch among various food sources are more effective in controlling and stabilizing nutrient and energy flow than those that are limited to a single resource.

Heterotrophs switch from one food source to the other. When animal prey are in short supply, the heterotrophs switch to plant products, thereby removing pressure from the animal population and allowing them to recover.

6. "In a hierarchical organization of ecosystems, species interactions that tend to be unstable, nonequilibrium, or even chaotic are constrained by the slower interactions that characterize large systems."

This means that large complex systems such as oceans, the atmosphere, soils, and large forests tend to be more homeostatic than their lower level components. A higher level system is the environment for a lower level. As a result, ecosystems can remain stable in terms of energy and nutrient flow, while the populations that perform these functions change drastically. Individual populations can fluctuate widely, yet the total number of individuals in a community remains relatively stable, resulting in ecosystem properties (energy flow, nutrient cycling) that remain relatively stable.

D. Stability and sustainability of energy flow through ecosystems are enhanced by mutualistic functions. As ecosystems became large and complex, organisms within a natural community evolved toward cooperative functions. Evolution has selected many mutualistic species because of the higher efficiency of the functions they perform. (Ideas 7–12)

7. "Since the beginning of life on earth, organisms have not only adapted to physical conditions but have also modified the environment in ways that have proven to be beneficial to life in general." This is the Gaia hypothesis described in Chapter 3. The movement of energy and nutrients through an ecosystem occurs not merely as a result of the chance occurrence of species randomly associated. Rather, the individual species within an ecosystem have evolved simultaneously, each modifying the environment of the other, and adapting to an environment modified by other organisms. Each species in a natural system has an important role in sustaining the system.

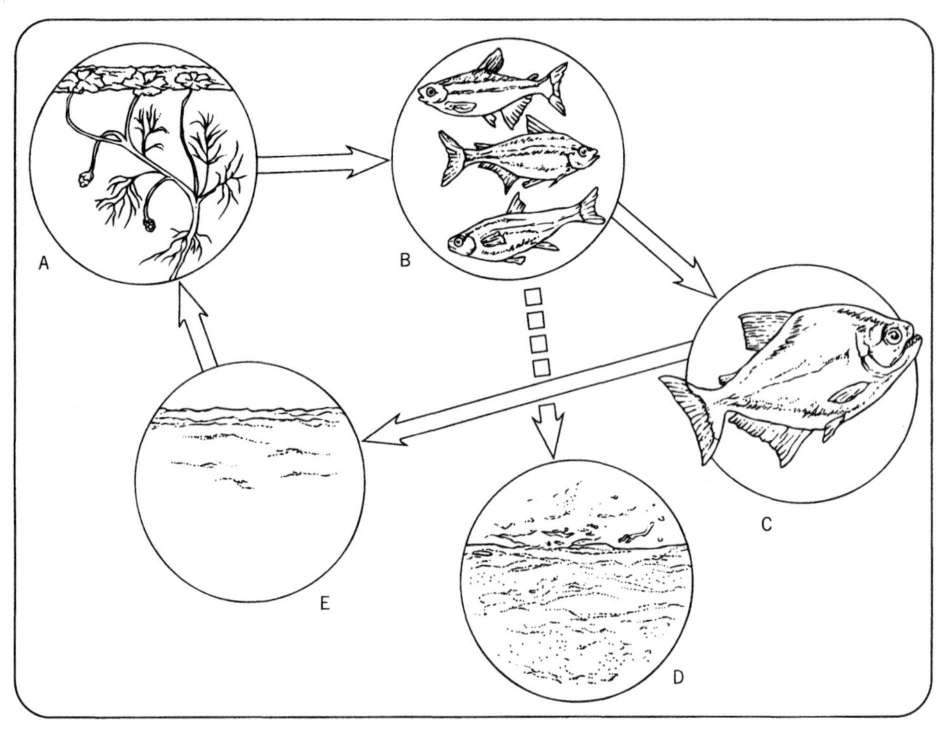

A = Aquatic plants
B = Herbivorous fish
C = Parañhas
D = Sediment
E = Water

FIGURE 6.14 Flow of phosphorus through an aquatic ecosystem.

8. "There are two kinds of natural selection, or two aspects of the struggle for existence: organism versus organism, which leads to competition, and organism versus environment, which leads to mutualism."

The emphasis on studies of competition as a mechanism for evolution reflects a bias. Capitalism and the Industrial Age were on the rise in the years when the evolution of species began to be accepted as a biological reality. Capitalism is based on competition, and the phrase "survival of the fittest" (best competitor) seemed to apply to biology as well as to business.

Only in recent decades has there been significant effort toward understanding mutualism as a factor in evolution. This may have been because of the recent increase in attention given to the environment and how species interact with the environment. Mutualisms are important mechanisms that provide sustainability to ecosystems and in some cases increase productivity.

9. "Evolution of mutualism increases when resources become scarce." When resources are plentiful, generalist species are often dominant. As resources become less plentiful, it becomes necessary to use them more efficiently. Specialist species evolve. But for specialist species to survive, they must rely on other specialist species to perform functions for which they are not well adapted. Thus evolve mutualisms.

Huston's (1993) evidence that, on a global scale, diversity is higher in regions where soils are less fertile supports this idea.

10. "Ecosystem development or autogenic succession is a two-phase process." The early phase is dominated by generalists adapted to exploit a resource-rich environment. As the structure of the ecosystem increases later in succession, resources become scarcer, and communities are taken over by specialists, many of which are mutualists. The same pattern of succession is followed by the human development of a resource

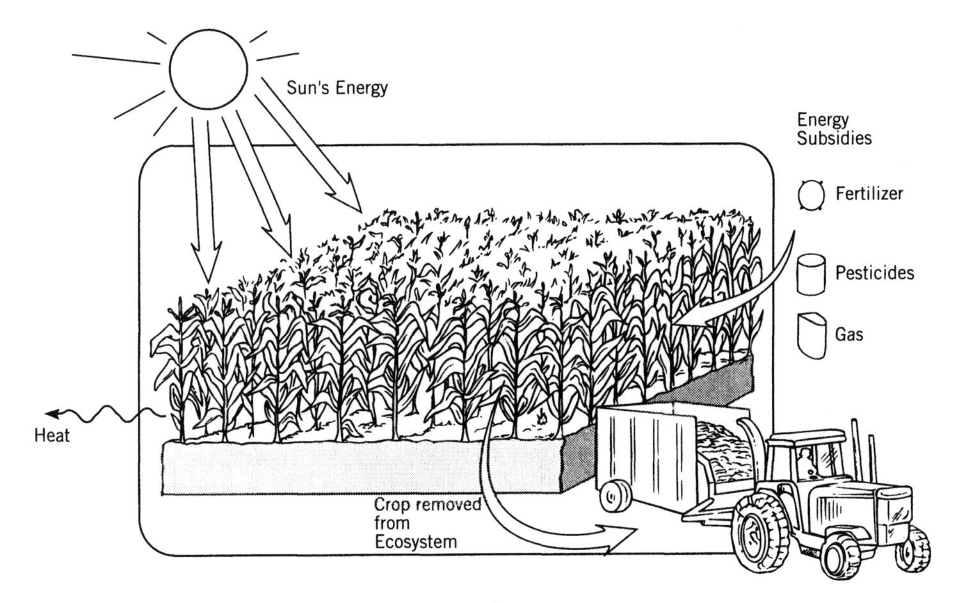

FIGURE 6.15 Energy flow through a crop production system.

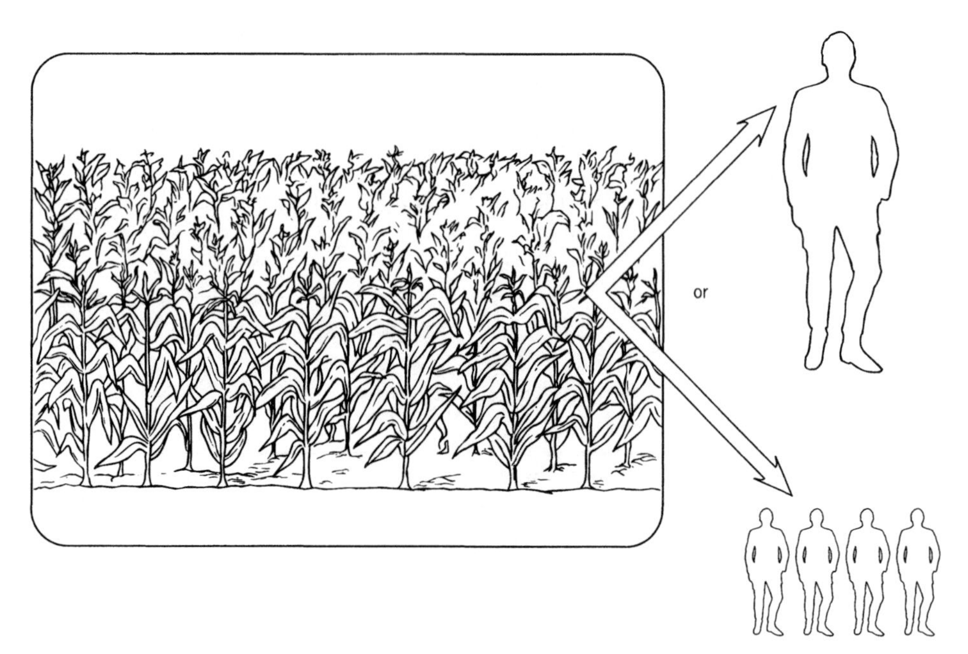

FIGURE 6.16 Carrying capacity depends on number of consumers and intensity of consumption.

frontier. Pioneers are generalists but can survive best when resources are abundant. Later, when resources become scarce, cooperation and mutualisms help humans use the decreased resources more efficiently.

11. "Competition may lead to diversity rather than to extinction." Competition in a completely stable environment can lead to extinction of one of the competing species. However, the environment is never stable; at one time, one species may have an advantage, and at another time, the other species. Species will evolve to exploit their advantage, and competition increases the pressure to differentiate.

12. "Natural selection may occur at more than one level." The classical view is that natural selection results in the survival of the best adapted species. A newer view is that selection also results in the survival of the best adapted ecosystem process. When a process is carried out more efficiently by groups of mutualistically interacting species, evolution will favor these groups of species.

E. In managed systems, the cooperative functions and natural subsidies that lend stability and sustainability to natural systems are usually destroyed. For this reason, energy subsidies are usually required. (Ideas 13–15)

13. "An expenditure of energy is always required to produce or maintain an energy flow or a material cycle." In undisturbed systems (Fig. 6.13), the sun's energy keeps food chains and nutrient cycles intact and functioning. When there is a harvest of materials, as in Fig. 6.15, there must be an energy subsidy to compensate for energy loss owing to harvest. The energy loss is the caloric value of the crop hauled out of the system. If the system were undisturbed, this energy would be used to fuel the species that perform functions that keep the system stable and functioning.

14. "The source-sink concept: one area or population (the source) exports to another area or population (the sink)." Sometimes, the energy source for a harvested ecosystem can be an adjacent system. A forest can be an energy subsidy for a field. Certain birds nest and reproduce in the forest but forage in the field. Birds are hunted and removed from the field, which is the "sink." The population of birds in the forest is the "source."

A forest can also be a subsidy for an agrarian society, as protection for a watershed, and a source of wood, leaf litter, and other products.

15. "Carrying capacity is a two-dimensional concept involving number of users and intensity of per capita use." The number of individuals of a species that can be supported by an ecosystem depends on the quantity of resources that each individual of that species uses. In the case of animals, the quantity of resources is more or less fixed. In the case of humans, the quantity varies tremendously (Fig. 6.16). Because of this variability, the resources required to support each person varies greatly. The higher the standard of living, the greater the resources required from the system on which an individual is dependent.

F. The stability and sustainability of a managed system can be increased by replacing external energy subsidies with cooperative functions of nature. (Idea 16)

16. "Input management is the only way to deal with nonpoint pollution." Here, Odum was referring to problems such as pollution in streams resulting from runoff of fertilizers and pesticides from agricultural fields throughout the landscape. It is almost impossible to control the runoff, so the only practical way to control the pollution is to limit the input chemicals to the farmer's field.

We can consider the idea in a larger sense in that energy subsidies are the generalized source of all pollution and ecosystem malfunction. It is what humans do to ecosystems with their earth-stirring, river-moving, mountain-shoveling, life-destroying petroleum-derived equipment and chemicals that is making the earth nonsustainable. The most practical way to control this damage is to reduce the amount of energy subsidies humans use in their management schemes and to replace these subsidies with the free services of nature.

G. The stability and sustainability of all systems can be increased by maintaining species and landscape diversity. (Ideas 17 and 18)

17. "The first signs of environmental stress usually occur at the population level, affecting especially sensitive species." When an ecosystem is stressed, for example, by pollution, the stress may not be apparent because it can affect different species differently. A stress that decimates one species may release resources for another less sensitive species, and total ecosystem function may not change. However, stress usually causes a decrease in species diversity because some species are more sensitive than others. Monitoring sensitive species and/or diversity can give an indication of ecosystem health.

18. "An expanded approach to biodiversity should include genetic and landscape diversity, not just species diversity." Preservation of species diversity is an important part of conservation. However, preservation of biodiversity per se is not enough. To ensure that the right combinations of species and communities remain viable and available to perform the services of nature necessary to substitute for petroleum energy, genetic and landscape diversity also must be preserved.

H. The transition from nonsustainable to sustainable systems requires time and has a cost. (Idea 19)

19. "Transition costs are always associated with major changes in nature and in human affairs." Management systems that rely on naturally occurring ecosystem functions are more desirable and more sustainable in the long run than those highly dependent on subsidies. However, there is a cost of switching to the more sustainable system in that it takes time, often years, to get the system operating. If there is no income or support during this period, the transition may not be possible. Lack of support for transition can hinder the efforts of subsistence farmers to adapt a more sustainable system. It can also hinder the efforts of developing countries to adapt a more sustainable national effort of resource management.

I. Despite the cost, there is an urgent need to make the transition. (Idea 20)

20. "There is an urgent need to bridge the gaps between human-made and natural life support goods and services, and between nonsustainable short-term and sustainable long-term management." Humans have moved too far in the direction of modifying resource ecosystems. They have too extensively substituted the services of petroleum for the services of nature. The results are pollution and resource degradation. Continued substitution of species adapted to survive and to maintain ecosystem function under natural conditions, by species genetically selected and managed for conditions highly dependent on technology ultimately leads to humans' complete dependence on technology. This might not be bad, were technology completely invincible, but as technological disasters from Chernobyl to the Valdez oil spill have proven, technology is not invincible. Thus, humans in their efforts to remove themselves from the system and from the vagaries of nature are replacing the vagaries of nature with the vagaries of technology—and the vagaries of the people who control the technology.

This is a call to action, to begin to replace destructive energy subsidization forms of resource management, which are sustainable only for as long as the energy supply (usually petroleum) holds out, with the services of nature—that is, with the mutualisms and functions that living organisms can carry out with little or no help from humans.

SUGGESTED READINGS

For the philosophy and scientific basis of nature as a model for sustainable agriculture:

Soule, J. D., and J. K. Piper. 1992. Farming in Nature's Image: An Ecological Approach to Agriculture. Island Press, Washington, D.C.

For an explanation of energy as a basis for understanding natural and human-influenced systems:

Odum, H. T., and E. C. Odum. 1981. Energy Basis for Man and Nature. McGraw-Hill Book Co., New York.

For recent advances in analysis of energy in agricultural systems:

Fluck, R. C., ed. 1992. Energy in Farm Production. Elsevier, Amsterdam.

Peart, R. M., and R. C. Brook, eds. 1992. Analysis of Agricultural Energy Systems. Elsevier, Amsterdam.

For an examination of mutualisms from a population and community perspective:

Boucher, D. H., ed. 1985. The Biology of Mutualism. Oxford University Press, New York.

For an examination of mutualisms from an ecosystem and resource management perspective:

Schulze, E., and H. A. Mooney, eds. 1993. Biodiversity and Ecosystem Function. Springer Verlag, New York.

For an analysis of species interactions in agriculture:

Vandermeer, J. 1989. The Ecology of Intercropping. Cambridge University Press, Cambridge.

For an explanation of the relation between stability of populations and diversity:

Pimm, S. L. 1982. Food Webs. Chapman and Hall, London.

For an overview of the role of alternative farming methods in the United States:

National Research Council. 1989. Alternative Agriculture. National Academy Press, Washington, D.C.

For the scientific basis of organic farming technology:

American Society of Agronomy. 1990. Organic Farming: Current Technology and Its Role in a Sustainable Agriculture. American Society of Agronomy, Madison, Wis.

For an overview of integrated pest management, with case studies:

Zalom, F. G., and W. E. Fry. 1992. Food, Crop Pests, and the Environment. American Phytopathological Society Press, St. Paul, Minn.

For an overview of agroforestry:

MacDicken, K. G., and N. T. Vergara. 1990. Agroforestry: Classification and Management. John Wiley, New York.

Young, A. 1989. Agroforestry for Soil Conservation. CAB International, Wallingford, Oxon, U.K.

For forest management in the temperate zone:

Hunter, M. L. 1990. Wildlife, Forests, and Forestry. Prentice Hall, Englewood Cliffs, N.J.

For forest management in the tropics:

Gomez-Pompa, A., T. C. Whitmore, and M. Hadley eds. 1991. Rain Forest Regeneration and Management. Man and the Biosphere Series, Vol. 6. UNESCO, and the Parthenon Publishing Group. Carnforth, U.K.

For a reference on nontimber products of tropical forests:

Smith, N. J. H., J. T. Williams, D. L. Plucknett, and J. P. Talbot. 1992. Tropical Forests and Their Crops. Comstock (Cornell University Press), Ithaca, N.Y.

For an introduction to wetlands protection:

Bingham, G., E. H. Clark, II, L. V. Haygood, and M. Leslie. 1990. Issues in Wetlands Protection: Background Papers Prepared for the National Wetlands Policy Forum. The Conservation Foundation, Washington, D.C.

For the view that restoration ecology is a test of ecological theory:

Jordan, W. R., III, M. E. Gilpin, and J. D. Aber. 1987. Restoration Ecology: A Synthetic Approach to Ecological Research. Cambridge University Press, Cambridge.

For management of commonly held resources:

McCay, B. J., and J. M. Acheson. 1990. The Question of the Commons: The Culture and Ecology of Communal Resources. University of Arizona Press, Tucson.

Ostrom, E. 1990. Governing the Commons: The Evolution of Institutions for Collective Action. Cambridge University Press, Cambridge, U.K.

CHAPTER

PRESERVATION OF BIODIVERSITY

CHAPTER CONTENTS

PRINCIPLE

Protection of species depends on protection of habitat. Because it is impossible to protect *all* habitats, we must choose those that will best contribute to maximizing global diversity.

CHAPTER OVERVIEW

Preserving biodiversity has been approached from two points of view. One perspective focuses on the species and then determines the habitat that must be preserved to save that species. The other focuses on habitat and is concerned with location, size, and shape of reserves to maximize biodiversity or to optimize environment for a species or group of species. Regardless of the approach, it is important to remember that the ultimate objective is to preserve the maximum global *diversity. In some cases, preserving existing biodiversity on a global scale requires a different approach from management to increase biodiversity on a local scale.*

Equally important for preserving diversity are social, political, and economic issues. Reserves will not be successful unless the local people, government, and economic conditions are incorporated into the conservation plan. ❧

THE PROBLEM

Loss of Species

An important part of conservation is preservation of the species of flora and fauna that inhabit the earth. Preservation is important from an aesthetic point of view: diversity is psychologically pleasing. It is also important practically: yet-undiscovered species can supply us with future foods and medicines; diversity provides redundancy of ecosystems' services such as nutrient cycling. Some conservationists argue that practicality is not the issue. They state that species have a right to exist apart from any benefit to humans and that humans have a moral obligation to preserve biodiversity.

The World Wildlife Fund (U.S.) estimates that at least 480 species of native plants and animals are known to have vanished in the United States in the last 200 years, including the ivory-billed woodpecker, the Florida black wolf, the passenger pigeon, the Arizona jaguar, and the Carolina parakeet. Currently, over 4000 species and subspecies are recognized as candidates for endangered species status within the United States. In many other nations, especially in the tropics, a greater number of extinctions have occurred. Biogeographers estimate that by the year 2100, 25 to 50 percent or more of tropical species will vanish (Soulé 1991).

A central objective in conservation biology is to stem the loss of biodiversity. Preserving biodiversity has been approached from two points of view. One focuses on the species, including the genetic diversity within species, and the other on habitat, including ecosystems and landscapes. Both viewpoints have advantages and disadvantages. The strategies are not mutually exclusive, and sometimes both are pursued simultaneously. This chapter reviews the strengths and weaknesses of these two approaches, and then discusses the practical aspects of management for preserving habitat and diversity.

POSSIBLE SOLUTIONS

Preservation of Species

International treaties and national laws protect species by making it illegal to kill them or trade them. On the international level, the most significant step has been the signing of the Convention on International Trade in Endangered Species (CITES) by the majority of the world's nations (see Chapter 5). In the United States, the Endangered Species Act has attacked the problem of species loss.

Endangered Species Act (ESA)

Within the United States, the major piece of legislation designed to preserve species has been the Endangered Species Act, passed in 1973. The act authorizes the National Marine

Fisheries Service (NMFS) of the Department of Commerce to identify and list endangered and threatened marine species. The Fish and Wildlife Service (FWS) identifies and lists all other endangered and threatened species. These species cannot be hunted, killed, collected, or harassed in any way. Once a species is listed as endangered or threatened in the United States, the FWS or the NMFS is supposed to prepare a plan to help it recover. In 1992 the federal list included 675 endangered or threatened species and nearly 900 "Candidate one" species that the FWS believes should be listed but has not acted on because of a lack of time, money, or personnel. In the "Candidate two" category—those considered likely to need listing—are more than 3000 other species.

For each listed species, a recovery plan must be drafted; This is an expensive and time-consuming process. Only about 50 species are added to the list each year. Of the 675 species actually on the list in 1992, recovery plans had been drafted for only about 370. By the time recovery plans are formulated, it can be too late (Rancourt 1992).

The Endangered Species Act also prohibits federal agencies from carrying out, funding, or authorizing projects that may jeopardize endangered or threatened species or destroy or modify their critical habitat; and bans the importation or trade in any product made from an endangered species unless it is used for an approved scientific purpose or to enhance the survival of the species.

The act has had a number of successes, for example:

- The number of nesting bald eagles in the lower 48 states has risen from less than 800 pairs in 1974 to more than 3000 today.
- Since the early 1970s, the number of nesting pairs of peregrine falcons has risen from 19 to 700.
- As a result of turtle excluder devices in shrimpers' nets, the number of sea turtle drownings has decreased.
- The population of gray whales has rebounded to the point where the species has been taken off the endangered list.
- The status of the Aleutian Canada goose has improved from endangered to threatened.
- Black-footed ferrets and California condors, successfully bred in captivity, were recently reintroduced into their former habitats.
- The red wolf, a species that has survived only in captivity, has been reintroduced into Great Smoky Mountains National Park.
- The gray wolf, once extirpated from most of its former range in the United States (except Alaska), is returning to Montana, Idaho, and Washington. (Bean 1993, p. 22.)

Problems with the Endangered Species Act

Despite these successes and others like them, the act has been criticized as being inadequate to the task of significantly reducing extinctions. Few species have recovered sufficiently that they can be removed from the list. Problems are difficult to solve because remedial action is not triggered until populations have already become dangerously low; funding is insufficient to gather the data necessary to determine the status of species whose populations are declining; recovery plans are developed only for a fraction of listed species; and many recovery plans are never activated (Orians 1993).

Many scientists believe that the species-by-species approach is inadequate. The goal of preserving biological diversity will never be achieved if efforts continue to focus primarily on species (Franklin 1993). The simple fact is that there are too many species to handle on

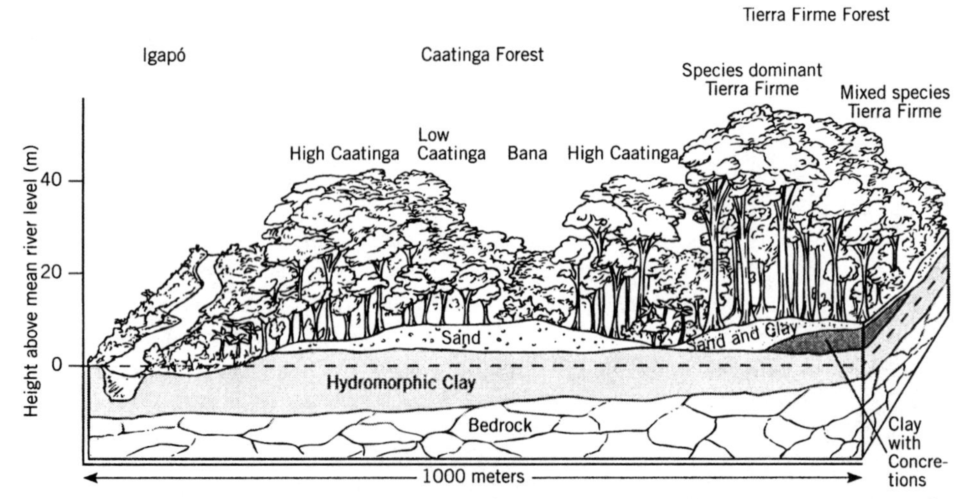

FIGURE 7.1 Idealized transect along a gradient from river to lateritic hill in the Rio Negro region of Venezuela. The igapó forest adjacent to the river is flooded for part of the year. Where coarse sand has been deposited, caatinga forest occurs with shallow water table, and bana occurs with deeper water table. Where bedrock forms hills, as toward the right of the diagram, granite grades into clay which comprises the subsoil. On the shoulders of the hills, there sometimes occurs a mixture of fine sand and clay which supports a forest dominated by one or a few species. *Source:* Jordan, 1985b.

a species-by-species approach. The procedures of the Endangered Species Act will quickly exhaust the time available, financial resources, societal patience, and scientific knowledge. Conservation at the habitat, ecosystem, landscape, or biome level may therefore be a more practical approach to saving the millions of species that constitute global biodiversity.

Supporters of the Endangered Species Act argue that Congress specifically stated that one purpose of the act was to conserve the ecosystems on which the endangered species depend, and that if the act were properly enforced, species would be used as tools to restore, monitor, and preserve ecosystems (Losos 1993). Nevertheless, action under the act is reactive rather than proactive.

Preservation of Habitat

Habitat destruction is the most prevalent cause of species endangerment (Wilson 1992), although there are exceptions. For example, some cacti and butterflies have become threatened because of the activity of collectors. Other species have been threatened because of their commercial value. Nevertheless, the most effective way to preserve the majority of the world's biodiversity is to protect habitats.

The habitat approach to conservation is appealing because only one listing would be required per habitat, regardless of the number of threatened species it contained. The difficulty is to define and delineate individual habitats. Classification systems based on life forms, climate, dominant plant genera, and physical and biological variables all have been tried, but no generally satisfactory or universally acceptable system has been found, except for biomes that are much too large to be useful as units to be protected by legislation.

Is there hope of devising a comprehensive system of habitat classification useful for conservation? Before answering that question, let us look at the relationship between biodiversity and habitat.

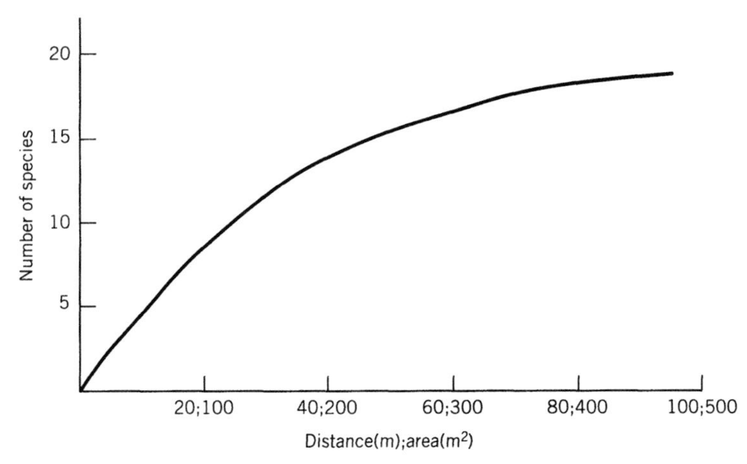

FIGURE 7.2 Cumulative number of species as a function of distance and area from main channel of river through Igapó forest in Fig. 7.1.

Habitat and Area

Habitats are relatively homogeneous areas that harbor fauna and flora that are more or less characteristic of that habitat and that differ significantly from other habitats. As an example, we can study a tropical rain forest near San Carlos de Rio Negro, Venezuela.

Figure 7.1 represents a cross section of the area's vegetation and underlying geology. Let us imagine ourselves at that site, delineating a transect 5 m wide along the cross section. We begin a reconnaissance of all species of vascular plants within that transect, beginning at the igapó (seasonally flooded) forest on the left of the figure and working away from the river. As we proceed, we examine and identify every vascular plant within the transect. Every time we encounter a new species, we write it down and keep a running tally of the number of species as a function of distance along the transect.

A graph of the number of species as a function of distance along the transect, for the first 100 m, might look like Fig. 7.2. At first, the number of species encountered as a

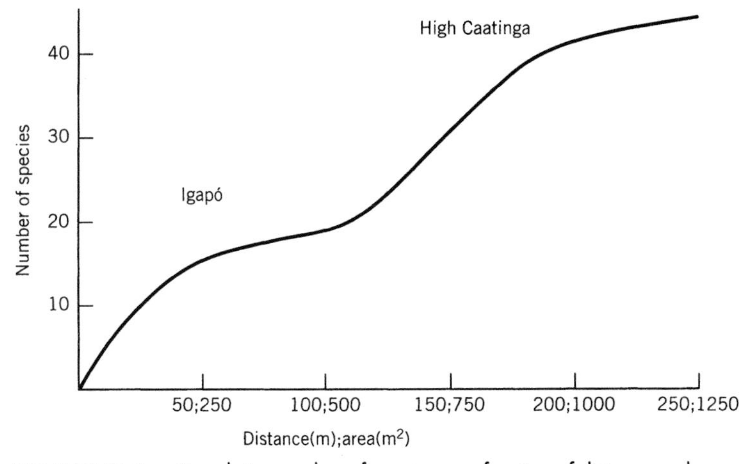

FIGURE 7.3 Cumulative number of species as a function of distance and area from main channel of river through High Caatinga Forest in Fig. 7.1.

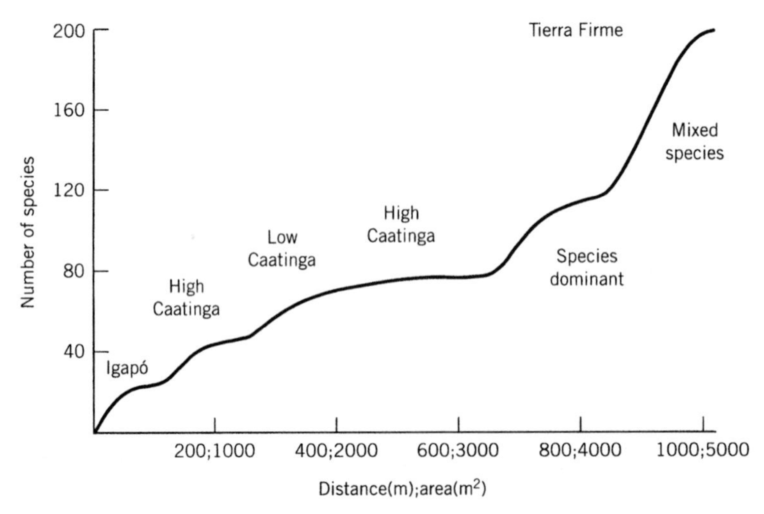

FIGURE 7.4 Cumulative number of species as a function of distance and area from main channel of river through Mixed Species Tierra Firme Forest in Fig. 7.1.

function of distance and area increases quickly, but then it decreases as more and more of the individuals found are members of species already tallied. The curve is an expression of alpha diversity, the number of species within a single type of habitat (Whittaker 1975). Alpha diversity is influenced by physical factors, food, predators, and competition.

If we continue our reconnaissance along the transect through the High Caatinga Forest, the diversity curve might resemble Fig. 7.3. At about 100 m begins another increase in the rate at which we encounter new species (rate = number of new species per meter). The reason is that we have crossed into a new habitat, that is, a site with soils and topography different enough from the riverbank site that a different suite of species occurs. The diversity that occurs as a result of including more than one habitat in the sample is called beta diversity. Beta diversity is dependent on extinction and speciation within a habitat, as well as immigration and emigration to and from a habitat.

A continuation of our transect along the entire cross section would result in a species-area curve like that shown in Fig. 7.4. Every time we cross into a new habitat, the rate of increase of species diversity increases. However, when we pass from Low Caatinga back into High Caatinga, we encounter almost no new species, because most of the High Caatinga species were recorded in the first area of High Caatinga. Upon passing into the Mixed Species forest on Tierra Firme, there is a big jump in rate of species increase, reflecting the relatively high diversity in that site and the large number of species that do not occur on other sites.

This procedure allows us to locate the limits of various types of habitats. Locating the limits of a habitat does not define what that habitat is, in terms of edaphic, climatic, and other factors, but for purposes of preserving biodiversity, such definition is not necessary. For this end, the limit of a habitat seems to be clear. It is that place along the transect where the number of species per unit area increases suddenly.

If we were to do another transect 10 km downstream from San Carlos, the shape of the species-area curve would resemble that of Fig. 7.4, but the species encountered would not be exactly the same. There might be an 80 percent overlap in species encountered. If we traveled 100 km, the overlap might be 40 percent. At 1000 km, there might not be any species in common. The total diversity in all habitats is called gamma diversity—the total number of species in all habitats of a region, or of the world.

Habitat and Structure

Within each of the habitats shown in Fig. 7.1 are many structural habitats. For example, one bird species may feed and nest on the forest floor, another in the understory, and a third in the canopy. Other species may live in the gradients between these habitats. A particular species of tree may comprise the habitat for a particular species of bird. Birds may nest in one habitat and feed in another.

The soil is a series of "micro-habitats." An abandoned termite nest will house a community of fungi and bacteria entirely different from that living in mineral soil a few centimeters away. A minor depression in the soil that collects a pool of water will have a complement of insects distinct from that living under an adjacent rock. Clearly, this level of habitat delineation is too fine for practical conservation purposes.

Habitat and Time

Not only are there many different habitats within an area, but a habitat can change into another habitat as a function of time. To illustrate, let us examine a species diversity curve along a time axis by taking a hectare (a plot 100 m on each side) of uniform habitat and counting the number of species in that hectare at different times during succession.

Assume that the plot was in the Piedmont of southeastern United States and that the first sampling occurred while the plot was under cotton cultivation. The second sampling occurred the first year after cultivation was abandoned. Following that there were yearly samples for 10 years, then samples at 10-year intervals for 200 years. A curve showing the cumulative number of species in the hectare as a function of time would look like Fig. 7.5.

Initially, the number of species would be low, and there would be only the crop species and a few weeds. In the first few years after cultivation was abandoned, the number of species would increase quickly, as a large variety of herbs, grasses, and pioneer

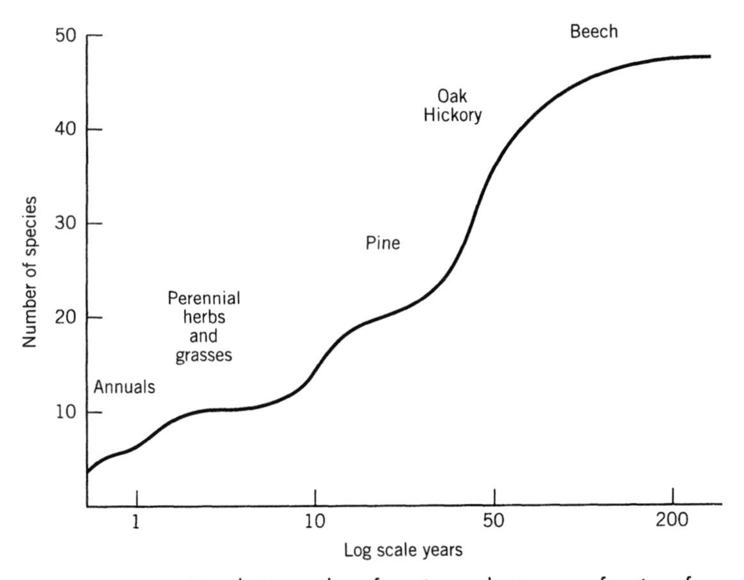

FIGURE 7.5 Cumulative number of species per hectare as a function of time during succession on the Appalachian Piedmont.

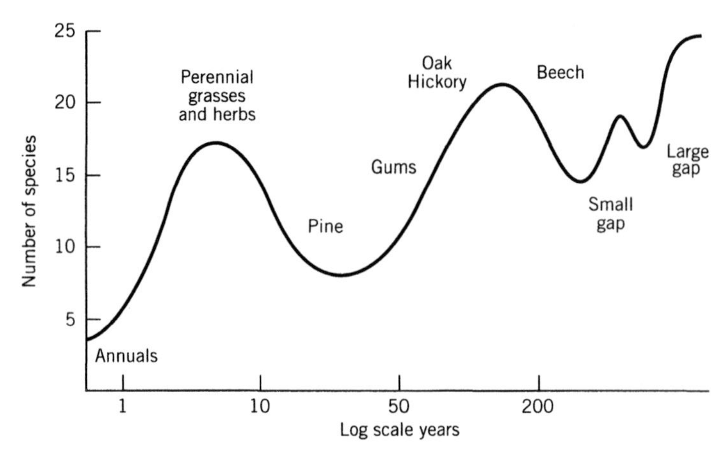

FIGURE 7.6 Total number of species per hectare as a function of time during succession on the Appalachian Piedmont.

species invaded. The occurrence of a plateau around five years in Fig. 7.5 might indicate the presence of fire which favored grass species and prevented establishment of later successional species. Gradually, however, midsuccessional trees would establish themselves, and the rate of accumulation of new species would increase. Eventually, the rate of appearance of new species would decrease, as the community reached "climax" or "equilibrium." In most cases, a static climax or equilibrium never really occurs because natural disturbances such as tree-falls, hurricanes, fires, and landslides open up part of the forest and set a portion of the community back to an earlier stage of succession. The species that occur in such disturbed areas, however, would generally be the same as those occurring earlier. Unless there was an invasion by an exotic species, the species-time curve would remain flat.

It is also important to look at the *total* number of species per hectare as a function of time during succession (Fig. 7.6). This is not cumulative species as in Fig. 7.5, but absolute totals found in the hectare at various times.

The first peak occurs between 5 and 10 years, with the mixed community of perennial grasses and herbs. As pines become established and shade out the grasses and herbs, diversity decreases. When the pines become mature, the stand is invaded by many species of oaks, hickories, and gums, and diversity increases rapidly. A peak of diversity often occurs in mid- to late succession. Diversity may then decrease as the shade-tolerant beech tends to dominate the stand. The decreased diversity often is reversed, when a gap opens up in the canopy. If the gap is small, as when a single tree falls over, the increase in diversity will be small. A few weedy species will come in but will be quickly shaded out again as saplings close the gap. If the gap is large, as when a hurricane knocks down many trees, the increase in diversity can be dramatic as representatives of all stages of succession enter the plot.

How Many Habitats Can We Protect?

For conservation purposes, a habitat is a place that houses a distinct community of species. In one forest biome alone, such as the tropical rain forest, there could easily be thousands or more distinct habitats. Is it possible to preserve all types of habitat, as a practical step toward preserving as many species as possible? In June 1993, the

Wildlands Project was proposed at the seventh annual meeting of the Society of Conservation Biology. Designed to protect biodiversity in North America, the project calls for a network of wilderness reserves, human buffer zones, and wildlife corridors stretching across hundreds of millions of acres and occupying as much as half the continent. On the Oregon coast, the Wildlands approach calls for 23 percent of the land to be returned to wilderness and another 26 percent to be severely restricted for use. Most roads would be closed, and some would be torn up. The plan basically calls for resettling the entire continent (Mann and Plummer 1993).

Critics of the project maintain that the program may be too much to ask of the people who will be affected. Asking the U.S. Congress even to consider such a plan, let alone adopt it, seems quixotic. In the quest to preserve global biodiversity, trying to conserve *all* individual habitats may not be any more practical than trying to conserve individual species.

Choosing Areas to Preserve

In light of the limitations and problems of the species-by-species and the habitat approaches to conservation of biodiversity, might another approach be more practical? Instead of trying to conserve *all* habitats, should we focus on a specially selected few?

Habitat as discussed here is too small a unit for conservation purposes. What is a reasonably sized unit? *Biome* is too large. *Landscape,* as generally used, includes human-influenced areas such as farms and thus is unsatisfactory for preservation of wild species. This is not to say that beautiful country landscapes should not be protected from developing into urban sprawl, but this kind of protection is not the issue here. *Ecosystem* in practical terms is defined by the interactions that occur within it. In Fig. 7.1, a species in the igapó habitat would be more likely to interact with another species in the igapó than with a species in the High Caatinga. If an ecosystem scientist were to begin a study of nutrient cycling, he or she would restrict plots to one of the habitats, unless a specific objective was to quantify nutrient flux between, say, the High Caatinga and the igapó. Thus *ecosystem,* as strictly defined, is no more useful a term than *habitat* as a unit for preservation.

Because it is impossible to save all types of habitats, another strategy must be employed to select sites for preservation. We must become selective in the habitats we preserve. We have begun by concentrating on areas that are exceptionally rich in species, that have particularly unique species, or that perform a particularly important ecosystem function.

Hotspots of Diversity

Myers (1988) identified 10 "hotspots" of biological diversity in tropical forests that contain 27 percent of the higher plant species in the tropics and 13.8 percent of the world's plants (Table 7.1). Wilson (1992) in his map of biodiversity hotspots also includes California, Hawaii, central Chile, the Ivory Coast, Tanzania, Cape Province, Madagascar, Western Ghats in India, Sri Lanka, and southwestern Australia.

A few countries require very special attention because they contain a very high percentage of the world's biodiversity, including marine, freshwater, and terrestrial diversity, or because of the large number of endemic species. These countries have been called megadiversity countries (McNeely et al. 1990). They include Brazil, Colombia, Indonesia, and Mexico, which are especially rich in species numbers. Madagascar and Australia, though not as high in total species numbers, are included because they have a

Table 7.1 *Areas of Unusually High Species Diversity and Endemism in Tropical Forests*

Area	No. of Plant Species in Original Forests	No. of Endemics in Original Forests
Madagascar	6,000	4,900
Atlantic Forest, Brazil	10,000	5,000
Western Ecuador	10,000	2,500
Colombian Chocó	10,000	2,500
W. Amazonian uplands	20,000	5,000
Eastern Himalayas	9,000	3,500
Peninsular Malaysia	8,500	2,400
Northern Borneo	9,000	3,500
Philippines	8,500	3,700
New Caledonia	1,580	1,400

Source: *Adapted from N. Myers, "Threatened Biotas: 'Hot Spots'".* The Environmentalist *8(3), 1988. Used with permission of Science & Technology Letters, Northwood, Middlesex, England.*

high number of endemic species. Other megadiversity countries include Ecuador, Peru, Zaire, China, India, and Malaysia.

Why are megadiversity countries often tropical or subtropical? Theories have included length of time for evolution to take place, higher spatial heterogeneity, competition, predation, productivity, and climatic stability (Pianka 1966). The refuge theory (Haffer 1982) postulates that isolated areas where environmental conditions were minimally affected by global events were a safe haven during episodes such as glaciation. During a favorable climatic era, the refuge population could disperse into a more continuous habitat.

Postulated forest refuges occur outside of the regions covered by glaciers during the Pleistocene (Fig. 7.7). Although there is evidence for the refuge theory (Prance 1982), it

FIGURE 7.7 Postulated forest refuges in the lowlands of the world during the peaks of glacial periods of the Pleistocene. *Source:* Haffer and Prance 1982. Reprinted by permission of the authors.

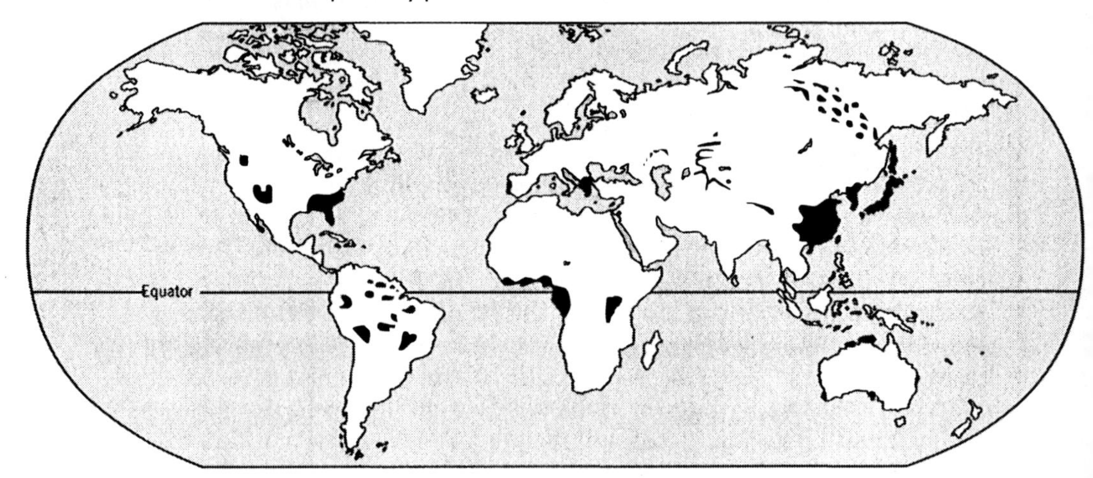

remains controversial. Although scientists may disagree on the *reason* why certain areas of the globe have relatively high numbers of species, they do agree that it is in these regions that conservation and preservation efforts should be concentrated.

Unique Ecosystems

The goal of conservation biology is to preserve as many of the earth's species as possible. Preserving areas rich in diversity is one step; another is preserving areas that have unique or rare fauna and flora, even though their biodiversity may not be high. These communities may be limited in extent because of high sensitivity to disturbance or because of endemism, that is, specificity to rare habitats (Carroll 1992). Their preservation contributes to the enrichment of global diversity.

Examples from shallow tropical marine environments include coral reefs, mangroves, and seagrass communities (Johannes and Hatcher 1986). Tropical continental aquatic habitats that deserve attention are lakes such as those of the Rift Valley in East Africa, tropical wetlands like the Pantanal in the Mato Grosso of Brazil, and tropical rivers (Sioli 1986). Caves constitute interesting problems for conservation biologists (Culver 1986). Arid lands are particularly sensitive to human disturbance. It has been estimated that about 9 million km^2 of the world's arid lands have been turned into humanmade deserts over the past half-century (Le Houérou and Gillet 1986).

Special Function Ecosystems

Certain areas should be selected for protection because of their particular ecological function. Wetlands are an example. They are particularly important because of their role in groundwater recharge. They are resting areas for migratory waterfowl and are spawning grounds for fish and crustaceans. They also serve as buffer areas along rivers to absorb water flow in times of storm. Their loss along the upper Mississippi River had a dramatic effect during the hundred-year flood of July 1993. Previous wetlands along the river that had been drained, diked, and converted to farmlands were flooded, causing millions of dollars of loss. Because the flood waters were not buffered and absorbed by wetlands, the flood crest peaked higher and faster, damaging towns, communities, and farms downstream.

Other special-function ecosystems are riparian and watershed ecosystems. Riparian communities should be left intact because of their role in stabilizing riverbanks and preventing erosion. In agricultural areas, they filter out fertilizers and pesticides that otherwise would drain from cropped fields into streams. Watershed areas behind dams for hydroelectric power or water supply must be managed to prevent erosion that could silt the reservoirs.

The Iterative Approach

The traditional method of reserve selection is to preserve sites with the largest list of species, the largest areas, or the highest values of some other indices for various criteria, the so-called scoring procedure. The disadvantage of scoring procedures is that the areas second or third on the priority list may duplicate the species in the first area, while other species or functions are missed. The iterative approach attempts to achieve maximum efficiency in nature reserve selection in order to preserve as many species as possible

(Saetersdal et al. 1993). This approach takes advantage of the computer's power to determine the maximum number of species that can be saved when a choice must be made between sites for purposes of establishing reserves. For example, if there are 50 potential sites for nature reserves, but there are funds to purchase only 12, which should be chosen to maximize the number of species preserved? The program will compare various combinations to arrive at a solution close to optimum. Because there are 1.4×10^{12} ways to choose 12 reserves from 60 sites, it is not possible to evaluate *all* possible combinations, but if the program is run 100 times, results show marked convergence. In the model of Saetersdal et al. (1993), selecting sites based on diversity or on endism of species produced similar lists, but selecting for birds produced a very different list than selecting for vascular plants.

THEORY OF NATURE RESERVES

Once an effort has been mounted or a decision has been made to establish a reserve in an area, the next step is to determine the size and shape of the reserve. Sometimes we must take whatever we can get, but at other times there is opportunity for negotiation. If a government is deciding on the fate of a tract of forest and a decision must be made on how much to allow to the loggers and how much must be set aside to preserve a species, it is important that decision makers be given a recommendation.

Size

In order to set aside enough area to protect a species, we must know how large a reserve is required to ensure preservation. What is the minimum area required (MAR) to preserve a species? It must be large enough to accommodate a minimum viable population (MVP) of that species. Minimum viable population, or the minimum genetically effective population, is the minimum number of individuals necessary for long-term survival of the population. It is determined by the characteristics of the landscape, by the population itself, and by the individuals that make up the population (Table 7.2).

A common rule of thumb has been that 50 individuals are required for short-term genetic fitness and 500 for long-term fitness. However, many scientists believe these numbers are too low. One recent model of minimum viable populations and the areas required to support them concluded that to allow small mammals a 95 percent chance of persistence for 1000 years, a million individuals and an area of tens of square kilometers are needed. For large mammals, hundreds of individuals may be sufficient, but they will require a habitat of a million square kilometers (Shafer 1990).

The MVP has been used to recommend the minimum areas required for keystone species. Keystone species are those that are so important in determining the ecological functioning of a community that they warrant special conservation efforts. Types of species that have been labeled keystone include predators, prey, mutualistic species such as mobile-link pollinators, hosts such as fruit-bearing plants, and landscape modifiers such as beavers. The keystone concept has been criticized on the basis that almost any species can be a keystone, depending on how the function and the ecosystem are defined (Mills et al. 1993).

The MVP approach to an MAR is useful for prescribing reserves for a particular species, for example, a bog pitcher plant that might be threatened by landfills. But since

Table 7.2 *Characteristics of Landscapes, Populations, and Individuals That Must Be Considered for Determining a Minimum Viable Population of a Species*

FACTOR	**CHARACTERISTICS**	
Landscape	Land use change Climatic change	Succession Disturbance
Population	Birthrates Death rates	Immigration Emigration
Individual	Fertility rates Growth Interactions Home range Host selection	Genetic selection

every species will have a different MAR, the concept is of limited usefulness for designing reserves where the objective is to preserve as many species as possible.

In preserves designed to maximum diversity, the only useful MAR is that for the top predator or predators. Presence of a top predator is the best insurance (though not a guarantee) that evolutionary processes can proceed in a natural way, with minimal skewing due to human activities. If an ecosystem lacks a top predator, food chain dynamics can quickly collapse, and mutually dependent species can face threats to their existence.

Top predators such as mountain lions may need a minimum of 13,000 km^2 to survive, and wolves may need between 39,000 and 78,000 km^2 (Frankel and Soulé 1981). Yellowstone National Park, with an area of approximately 9000 km^2, is too small to maintain an MVP of either species. However, because individuals may immigrate into the park from larger areas in Montana and Canada, the value of the park as a conservation reserve cannot be dismissed.

The World Wildlife Fund, the Brazilian Institute for Amazon Studies, and other organizations have launched one of the few field experiments to determine the effect of size of reserve on population dynamics of tropical species (Bierregaard et al. 1992). Forest fragments of 1, 10, 100, and 200 ha were isolated by cutting all the surrounding vegetation, and populations were surveyed and compared with those in "1000 ha reserves," which were actually part of the undisturbed forest. For virtually all classes of fauna, including insects, the 1- and 10-ha reserves showed diminished populations. Avian community structure was the most drastically altered. Immediately following cutting, the population in the reserves showed a dramatic increase, as birds from the cut area sought refuge. After 200 days, however, the number of birds fell sharply to levels below pre-isolation values. The rate and extent were greater in smaller than in larger reserves.

An important finding was the impact of edge—that is, the boundary between forest and clearing. Because of the fetch of wind impinging on the forest from the clearing, an increase in vapor pressure deficit was detected 40 m into the forest. This caused stress on trees, with the result that litterfall increased up to 50 m into the reserves. The amount of radiation reaching the forest floor increased, and the ultimate impact was a change in the structure and function of the biotic communities. Birds and mammals whose habitat was the undisturbed, low-light environment of the forest had to retreat from the forest edge, which was invaded by species adapted to higher light conditions. Species that lived in the forest but fed in openings had an advantage over those that used the forest for both nesting and feeding.

Studies of edge effect reveal that in the design of nature reserves, it is desirable to have a shape that minimizes edges. Thus, the ideal shape of a reserve is round because, of all geometric shapes, the circle has the maximum ratio between area and boundary. Squares are not bad, but the longer and thinner a reserve, the less the area/boundary ratio, the greater the edge effect, and the less effective will be the reserve.

Position

At one time, wilderness covered the earth, but as humans spread throughout the continents, they began to cut the forests, plow the valleys, and graze the plains. What was once a continuous expanse of natural vegetation became fragmented, as roads, farms, and cities were built. Relatively undisturbed areas of forest and prairie became smaller and smaller, until what remained in many parts of the world were patches that resembled islands in a sea of human-dominated landscape.

In 1967, MacArthur and Wilson published a book titled *The Theory of Island Biogeography* based on a theory dealing with the relationship between number of species on islands and size of the island, as well as distance of the island from the mainland. Large islands should contain more species than small islands because large islands have more habitats. In addition, because each habitat covers more area, the populations of each species are larger, thus reducing the chances of extinction.

Islands close to the mainland should have more species than those far away because of the greater immigration of species from the mainland to the close islands. The theoretical relationship between island size, distance to the mainland, and number of species on the island is shown in Fig. 7.8.

The applicability of island biogeography theory to design of nature reserves becomes apparent when remnant patches of undisturbed forest on a continent are seen as analogous to islands in the sea. For example, Yellowstone National Park can be considered a "large island" but distant from the "mainland," which would be the relatively continuous forest beginning in northern Montana and extending northward into Canada. A "small island" close to the "mainland" might be a one-acre patch of saw grass on the outskirts of Miami, separated from the Everglades by an urbanization. In retrospect, it seems obvious that bigger reserves should be better than small reserves and that reserves close to other

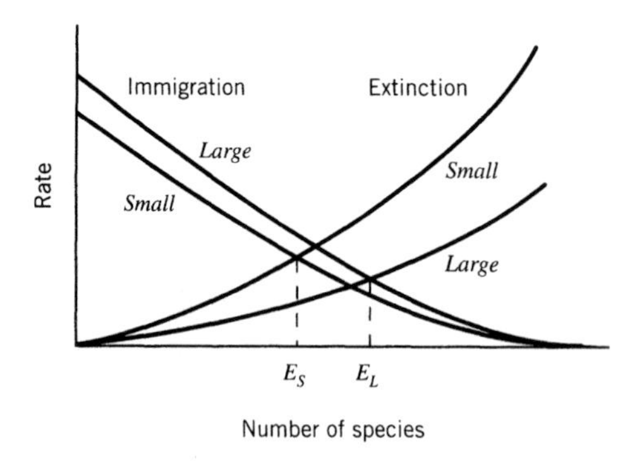

FIGURE 7.8 Relative rates of immigration and extinction on large versus small islands, the same distance from the mainland. A large island equilibrates with more species, E_L, than a small island, E_L. *Source:* Figure 5-14 from page 102 from *Geographical Ecology: Patterns in the Distribution of Species* by Robert MacArthur. Copyright © 1972 by Harper & Row, Publisher, Inc. Reprinted by permission of Harper Collins Publishers, Inc.

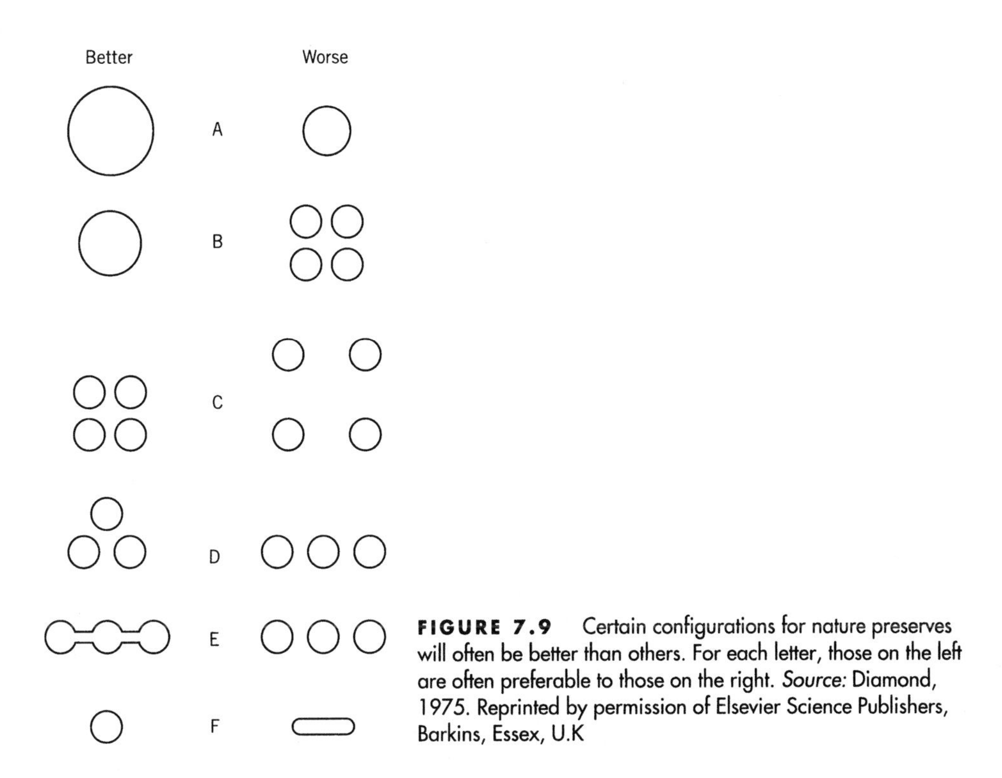

FIGURE 7.9 Certain configurations for nature preserves will often be better than others. For each letter, those on the left are often preferable to those on the right. *Source:* Diamond, 1975. Reprinted by permission of Elsevier Science Publishers, Barkins, Essex, U.K

reserves or to a "mainland" should be more effective than those far away. However, an understanding of the theoretical mechanisms underlying the seemingly obvious has marked an important scientific advance.

Another issue emerges when the planner of a biological reserve must choose between obtaining a single large reserve or several small reserves. The question has attracted so much interest among biologists that the debate has been given a nickname, "SLOSS," for single large or several small (Shafer 1990). A single large reserve is more likely than a small reserve to include minimum viable populations of a species, but several small reserves may each contain a subpopulation that will result in greater fitness of the population as a whole.

SLOSS must also be considered from the viewpoint of total diversity. Although a single large reserve will contain more species than a single small one, several small reserves are likely to contain more species than a single large one, because they are more likely to include a greater number of habitats. Diversity of smaller species—invertebrates, fungi, and bacteria—is better preserved through a large number of smaller reserves (Franklin 1993).

Another issue is that of corridors—that is, long narrow strips connecting islands of reserves. Corridors facilitate migration between islands and thus decrease the likelihood that a species will become extinct on any one island. However, corridors can facilitate the spread of disease and predators of endangered species.

Diamond (1975) incorporated these theoretical principles for design of reserves into practical guidelines for the conservationist faced with designing a network of conservation reserves (Fig. 7.9):

A. A large circular reserve is better than a small circular reserve.

B. A single circular reserve may or may not be better than several smaller reserves, when the area of the single reserve is equal to that of the sum of the smaller reserves. The advantage of the smaller reserves is that they will include more beta diversity. The disadvantage is that they pose a greater danger of extinction of a species.

C. Several reserves located close together may or may not be better than the same number of reserves with the same area spaced farther apart. The advantage of the closely spaced reserves is that migration between reserves will be greater, and thus gene pool in a reserve can be replenished more easily. The advantage of the further spaced reserves is that greater beta diversity is likely to be encountered.

D. Reserves distributed in a triangle may be better than reserves distributed in a line. In a triangle, the reserves are equidistant, and chances of dispersal from one island to the other are the same for all three reserves. In contrast, the chances of dispersal from island 1 to island 3 in the line are much less, and this would be critical if an endangered species occurred in 1 and 3 but not in 2.

E. Islands connected by a corridor may be better than islands unconnected, except in a case where an endangered species occurs in all islands and a predator on that species occurs in only one.

F. Round reserves are better than elongated reserves with the same area because the edge effect will be less. However, elongated reserves are more likely to encounter greater beta diversity and thus could be superior in that respect.

SOCIAL ISSUES

The Society for Conservation Biology came into being at the Second Conference on Conservation Biology, held in May 1985 at Ann Arbor, Michigan (Soulé 1986). During formal sessions, population and community ecologists gave presentations dealing with theoretical questions such as the minimum area required for a minimum viable population of a particular species. In the audience was a panel of applied managers, people who had to face practical problems of establishing and maintaining parks and reserves, in the United States as well as in developing countries. At the conclusion of each day's presentations, panel members were asked to give an evaluation of the day's talks. Their responses to the scientific presentations were almost unanimous: "What you told us is very interesting, but it will not help us one bit with our day to day problems."

In contrast to scientists who are interested in theoretical issues, land managers are concerned with the practical aspects of management of the site, stand, or landscape itself. They may be more deeply involved in economic, political, and social problems, and they have little time for scientific and technical issues. In real-world systems, economics and politics play a role that is more important than theory and more important sometimes than technical issues. Without a consideration of economic and social issues, habitat preservation is not possible.

Economic Issues

As in most human endeavors, the issue that ultimately determines the feasibility of a park or preserve is "Who pays?" Until this problem is addressed, nothing will happen. Some solutions at the national level suggested by McNeely et al. (1990) include:

Charge entry fees and user fees for visitors to parks.

Charge for ecological services. For example, part of the proceeds from hydroelectric power could be used for protection of the forest that prevents erosion in the watershed.

Collect special taxes on biological resources such as timber and wildlife extraction. This would include hunting and fishing licenses.

Include funds for conservation as part of the costs of large development projects.

Return profits from exploitation of biological resources. Trophy hunting of surplus game can bring in substantial revenues to African nations.

Negotiate concession agreements. When extractivists such as logging companies are granted a concession by a government, the concessionaires should be required to pay to restore the ecosystem.

Establish foundations for conservation. Foundations established for protection of a reserve can generate support if the reserve is well publicized.

Economic measures at the international level include the following.

Request support from international conservation organizations, such as the World Wide Fund for Nature, Conservation International, and the Nature Conservancy.

Use restricted currency holdings. In some countries, excess profits or local currency held by multinational corporations must be spent within the country.

Land Control Issues

Sometimes it is possible to buy areas that have potential conservation value. However, obtaining control of the land may be more than an economic issue. Overlapping and conflicting claims on land may occur. For example, it is desirable to establish buffer zones along the grassland/forest boundary in the uplands of Mindanao where conservation of high-altitude forests is critical (see photo-essay, The Farmer First Approach to Agricultural Development in the Uplands of Mindanao, The Philippines, Chapter 8). However, obtaining the rights to establish a conservation area there is not simple—for example:

Indigenous tribes living in the highlands have rights to the area through ancestral domain.

The upper slopes are under the jurisdiction of the Department of Natural Resources.

The national power company has an interest in controlling the land because it forms the headwaters of streams that feed hydroelectric dams.

Squatters who have lived on the land or who have worked it for a minimum number of years have legal claim to the land.

The Department of Agriculture is encouraging cultivation of potatoes on the area.

Loggers can obtain timber-cutting concessions through the right government contact.

The municipal government wants to run a pipeline through the area to provide uncontaminated water to the village.

It is not easy to reconcile the various groups with interest in the land in order to establish a conservation area.

Dealing with the Government

To protect and manage conservation areas, it is essential that the particular country's formal and informal legal system be understood. Different societies have fundamental differences in attitudes toward law and the extent to which a court system is used. Many different legal systems such as statutory or codified law, traditional or customary law, and religious law may operate simultaneously or at various levels. Each has its own authority structures that may include decision-making bodies separate from the formal courts. In some countries, informal negotiation may be the most feasible method for accomplishing protection goals. Frequently, this is through a strong persuasive role of key individuals or an agency that exercises ultimate authority. In such cases, the courts may play a less active role.

Conservation areas can be effectively established and safeguarded only if realistic and enforceable legislative and administrative arrangements are made for their protection and management. Legislative and institutional mechanisms that are not technically, socially, or economically feasible will not be effective. Too much law or authority can be just as harmful as too little and can lead to confusion and dissipation of effort. If laws are ignored or become redundant, the manager's authority and credibility are likely to be undermined.

MacKinnon et al. (1986) describe three general principles that must be heeded in drafting or strengthening legislation for conservation area management:

1. Conservation and management objectives must be ecologically sound and achievable with the technical and financial resources available.
2. Existing institutions should be used as much as possible so as to minimize the need for a new and expensive infrastructure.
3. A high level of national and local public participation should be encouraged to ensure broad social and political support.

Enforcing the protection of the conservation area is critical. Protected area legislation should name the enforcement agency and specify that agency's powers. Countries differ significantly regarding the amount of enforcement power they delegate to the protected area management authority. In some countries, protected area staff are armed and authorized to use firearms in carrying out their duties. In other countries, the staff may have no powers of arrest, confiscation of prohibited articles, or fining of trespassers. In cases where protected areas cannot be given staff policing powers, the management authority must develop particularly close working relationships with other law enforcement agencies.

Dealing with the Local People

Regardless of the laws or strength of the national or international authorities, a conservation area will not be successfully protected without the cooperation of the local people, or at least without provision for the needs of the local people. The photoessay on the Philippines in Chapter 8 depicts attempts to stabilize the agriculture in the uplands of Mindanao and protect the high-altitude forests by eliciting the cooperation and support of the local peoples. In contrast, in Thailand the strong national government feels less of a need to solicit the opinion of local peoples, but nevertheless realizes that it must provide for the needs of the local peoples if the teak reforestation effort is to be successful. (See photo-essay, "Social Forestry in Thailand," in Chapter 6).

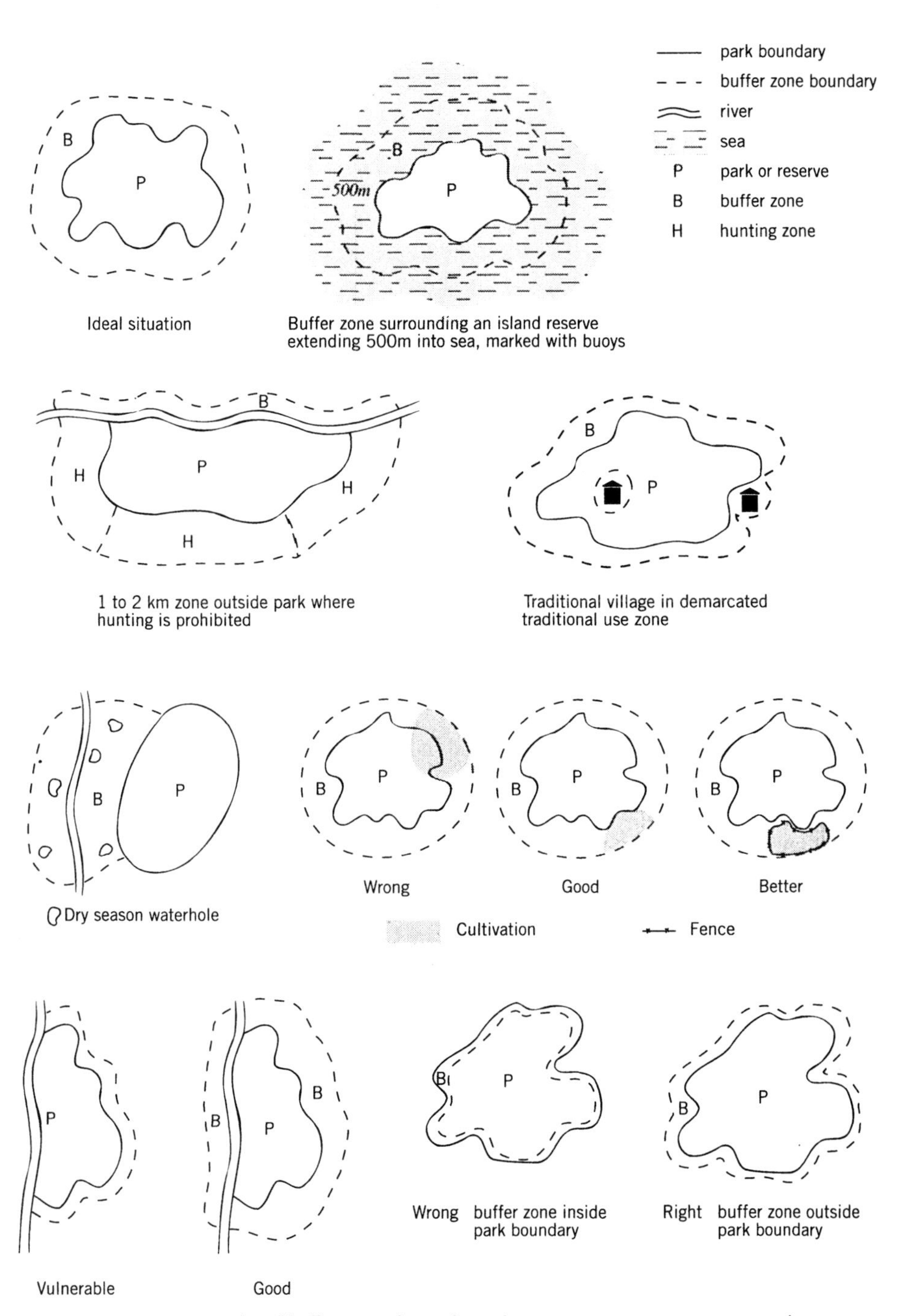

FIGURE 7.10 Examples of buffer zones for parks and reserves. *Source:* MacKinnon et al., 1986.

The local people's needs can be provided for throughout the landscape, as illustrated by the Philippines photo-essay, or in buffer zones where human activity is concentrated, as exemplified in the Thailand photo-essay. The merits of various configurations of buffer zones are illustrated in Fig. 7.10 (MacKinnon et al. 1986).

Case Studies

There are hundreds of examples of conservation projects specifically designed to integrate the needs of local peoples. Shafer (1990) provides the following.

1. In the Amboseli National Park in Kenya, large mammals disperse over 5000 km^2 of landscape during the rainy season. However, during the dry season, they concentrate in an area of 600 km^2, which is also used by approximately 6000 Maasai tribespeople, 48,000 cattle, and 18,000 sheep and goats. When it was realized that the economic benefits of the wildlife exceeded the costs to the Maasai, a program of monetary compensation was implemented to pay the Maasai for any losses suffered by their cattle from the migratory wildlife. Other forms of economic gain included plans for relocating tourist camps, payment to the Maasai to gather firewood, and employment by the park authorities. School and medical facilities also were made available to the Maasai.
2. Poaching in the Lupande Game Management Area in Zambia was a problem because the local economy was very poor, and often families lacked food. The experimental Lupande Development Project begun in 1981 was based on the hope that revenue derived from wildlife could meet the needs of both the local economies and of conservation. To achieve this goal, safari companies must hire a minimum number of local people. Safari concession fees are paid into a fund used by both the park service and the local chiefs for development projects.
3. The Sian Ka'an Biosphere Reserve is located on the Caribbean coast of the Yucatan Peninsula in Mexico. The 1.2 million acre reserve protects marshes, mangroves, lagoons, marine and reef environments, and tropical forests. A management plan bans disturbance in core zones but permits hunting, gathering, farming, and development for tourism in other parts of the reserve. The reserve has a lobster management program, a sustainable horticulture project, a palm research study, and experiments on sound cattle-raising techniques near the reserve.
4. The Gir Forest is a 1500 km^2 woodland in the Saurashtra Peninsula of Gujarat, India, and is a refuge for 200 of the last Asiatic lions in the world. The Gir ecosystem contains over a hundred settlements of Maldharis that tend approximately 16,000 buffaloes. Because of deforestation and land cultivation, loss of buffaloes to the lions became a problem. In 1972, the government of Gujarat established the Gir Sanctuary, and in 1974, the core area was made a national park. The Maldharis are being settled outside the sanctuary and being given land, monetary subsidies, and health services.
5. In the Royal Chitwan Park, Nepal, tigers were killing local people and livestock, and rhinoceroses were destroying crops. There was a cash flow into the local economy from tourism, but it was insufficient to motivate the local people to cooperate. Buffer zones have been added around the park, and in some places, local villages have been integrated into the management of the park. Though problems remain, allowing villagers to bring their cattle into a narrow buffer zone inside the park has promoted better relationships between the park and local people.

Not all of these projects were 100 percent successful in resolving conflicts between the needs of humans and nature, or among various groups of humans. Nevertheless, they are notable in that they have factored the human element into their design.

Miller (1982) summarizes the importance of relating the local people and their economic needs to plans for conservation areas:

> *As surrounding lands and waters become exploited for agriculture, grazing, forestry, fisheries and human settlements, the protected areas are called upon to relate in definable ways to human needs. Perhaps the most significant role protected areas can play in supporting adjacent lands is to contribute to the sustainability of the development process. . . . The salient point here is to question whether all possible efforts have been made to link protected areas economically and socially with the surrounding communities to provide tangible and visible support to the extent and in the form consistent with conservation goals. (p. 316)*

Dealing with the Tourists

The basic problem associated with tourists in conservation reserves is cultural clash. These clashes can occur on several levels. On one level, it could be a standard of living clash. Tourists might expect flush toilets, gourmet meals, air-conditioned rooms, and other amenities in luxury hotels. In this case, the guests would have to be told that this sort of development would not be compatible with the objectives of the reserve.

Tourist behavior can cause problems, as when skimpy bathing suits are worn in culturally sacred buildings and when demeaning gestures are directed toward porters, maids, and other service personnel, or when simple thoughtlessness occurs, such as littering with soft drink cans. Cultural clash can also cause a problem when management activities such as culling excess elephants because of overpopulation provokes complaints from animal rights advocates. These types of cultural clashes can usually be resolved by patient and friendly educational explanations.

Much more difficult to deal with are those tourists who may be more sophisticated regarding cultural differences but who have an agenda they are determined to achieve. Scientists are notoriously bad in this respect because the data they need to collect for publication in academically acceptable journals demand that sampling rigor take precedence over all other considerations. Adventure tourists such as mountain climbers and kayakers can cause problems when their activity poses a threat to the reserve or to the tourists themselves.

Even when individual tourists are respectful, sheer numbers can overwhelm a park. This lessens the aesthetic pleasure for each visitor, degrades the park, and endangers the wildlife.

A cultural problem that frequently confronts the ecotourist is how to take pictures of the local people without being offensive. Some cultures enjoy having their pictures taken, and the main problem there is snapping the shot before they start to ham it up. But more often than not, the local people seem to prefer not to have their pictures taken. In this case, it is helpful to take an interest in what the person is doing. A good example is the woman spinning silk in the photo-essay on social forestry in Thailand. Although she might have been shy about having to pose for a picture, she was more than happy to allow the photographer to take a picture of the silk spinning process, of which she was proud.

TECHNICAL ISSUES

A large amount of confusion exists regarding management techniques for biodiversity. The reason is that management can have one of two very different objectives, but which of the two different objectives desired is often not clearly specified. Thus far, this chapter has dealt with the objective of preserving global biodiversity. A different objective of management for biodiversity is to increase the stability of a managed ecosystem such as a farm or a forest plantation. It is important to distinguish between the two different objectives because in some cases, management to increase local biodiversity can cause a decrease in global biodiversity. When the objective is not clearly specified, actions may be counterproductive.

With the exception of management of buffer zones around preserves, little active management can be done when the goal is conserving global biodiversity. Management cannot increase global biodiversity. Global biodiversity can increase only through the process of evolution on a time scale much longer than the tenure of any land manager. Conserving global diversity is carried out through *preservation.* In contrast, the stability of a managed farm or tree plantation can be increased by *active management* for increased biodiversity. Chapter 6 discusses why a diverse ecosystem is more stable than a monoculture.

Management to increase biodiversity for ecosystem stability will often occur at the site level. Management to preserve global biodiversity is usually carried out with a global perspective. On a landscape scale, both preservation and management to increase biodiversity may be appropriate. The following three sections discuss management and/or preservation at these three levels.

Managing for Site Biodiversity

The set of resources used by a species is often called its niche. With greater complexity of ecosystem structure, there are a greater variety of niches that can be inhabited by plants and animals. Structurally more complicated ecosystems such as forests will have a higher diversity of fauna than structurally simple systems such as monocultures of corn. If you manage for structural diversity, you get biodiversity. With greater biodiversity, you may get greater stability through redundancy of pathways for nutrients and energy, and through better pest control by providing opportunities for predators of the pests.

Simple Ecosystems

An example of increasing diversity to promote stability is depicted in the photo-essay on social forestry in Thailand (Chapter 6). In the system of agroforestry called taungya, seedlings of teak are planted on government land by peasant workers, and families are allowed to cultivate subsistence crops among the trees for the first few years, until the canopy of the trees closes. In the northern region of Thailand, upland rice and sorghum are cultivated between the teak seedlings until the canopy closes, usually at about two years. Then the plantation becomes a monoculture. Because the teak is not ready for harvesting for 40 to 60 years, economic difficulty is experienced during this long period between the time when cultivation of grain crops ceases and the time when trees are ready to be harvested (Gajaseni 1992). There are also ecological problems in that the teak does not effectively use the light resource, and it may overexploit a portion of the soil resource.

To remedy this situation, an experimental system has been established in the Mae Moh basin, near Lumpang, northern Thailand, in which a variety of fruit trees are interspersed in the teak plantation. These trees begin to produce at about the time the cultivation of grains ceases. The fruit trees should increase the economic benefits of the system.

The trees are legumes and should enhance soil fertility. Their canopy shape is round, which complements the elongated structure of teak, and as a result, light is used more efficiently.

The increased heterogeneity that the fruit trees bring to the plantation may also provide a better reservoir for predators that prey on pest species. As plant species are added, the potential for adding animal species increases. For example, the region has over 100 species of nonflying mammals, and 65 percent of them can survive in disturbed areas, as long as there is at least some forest (Hoi-Sen 1978). Heterogeneous cropping systems such as agroforestry represent a way to help maintain the structural diversity required for high biodiversity, while at the same time providing for human needs.

The trend toward large-scale monocultures in both forestry and agriculture has reduced the complexity of managed ecosystems and resulted in the decreased diversity of their fauna. Several examples have been documented. For example, in Finland, biological communities in agricultural systems became simplified following changes from mixed farming early in this century to recent large-scale monocultures (Hanski and Tiainen 1988). Another example comes from Iowa, where researchers have shown that the disappearance of a patchwork of small diversified fields and their replacement by unbroken monocultures resulted in a sharp decline in the populations of ring-necked pheasants (Best 1990).

A monoculture of exotics such as eucalyptus provides very little variation in structure. Increasing the complexity of a stand by planting a variety of species or by allowing native species to recolonize will result in an increased diversity of fauna.

Heterogeneous landscape remnants of native vegetation on farms are often reservoirs for the disappearing biodiversity (Hobbs and Wallace 1991). Windbreaks and shelterbelts between agricultural fields in the United States shelter a greater biodiversity than surrounding fields (Capel 1988; Johnson and Beck 1988). Farmland hedges in England are important for the nesting success of birds (Lack 1988) and for the breeding of mammals. Of the 28 species of British lowland mammals, 14 are commonly found in hedgerows and breed there (Dowdeswell 1987). And in Germany, an entomologist found four times more species of arthropods in hedgerows and woodlots than in surrounding fields (Mader 1988). Many of the species were beneficial; for example, some spiders preyed on crop pests.

An increase in the amount of edge is a benefit when the landscape is a monotonous expanse of agricultural fields. Edges between forest and field are typically dominated by generalist species able to breed and forage in a variety of habitats (Wilson and Diver 1991). Patchy vegetation is effective as a reservoir for biodiversity in agricultural landscapes because of the edge effect. In contrast to mature stands where the edge effect diminishes biodiversity, in agricultural fields the edge effect increases biodiversity.

From the farmer's viewpoint, increased biodiversity can be bad as well as good. Although birds living in shelterbelts or hedges can be predators on insects in the fields, they can also be crop pests. Corridors also bear economic costs, in that they take land out of production, and other costs, in that they may channel disease, predators, and fire (Simberloff and Cox 1987).

Mature Sites

A number of reports have been issued showing that certain logging practices increase the species diversity of the forest. For example, Wang and Nyland (1993) showed that the richness of tree species was increased by the clear-cutting of northern hardwoods in central New York. Such data have sometimes been used to justify logging in areas that environmentalists would like to see preserved in pristine condition.

The problem is not in the data. Clear-cutting of a mature hardwood stand *does* increase the number of species in the area. This is because the regenerating forest will contain elements of several stages of succession. Many of the trees that are cut from the mature forest send out shoots that will grow into new trees. In addition, many early successional species such as black cherry invade the site. These species increase the biodiversity on the site, and if the weedy herbaceous species such as ragweed that usually invade disturbed sites are included, the diversity increases even more.

The problem lies in what the data represent. Adding a lot of weedy successional species to a forest plot because that increases the diversity index is not the goal. Rather, the goal is to maintain or increase diversity on a *global* scale by preserving intact habitats.

Logging an old-growth forest so that the local biodiversity is increased is a mistake. Such logging could very well reduce global biodiversity by causing the loss of a rare species—for example, a species of orchid that only occurs in old-growth forests. Such a loss might never be detected through compilations of indices of diversity. The disappearance of a rare orchid that occurs only in that site would be masked by the appearance of pokeweed, which occurs almost everywhere.

There is no need to manage an old-growth forest to conserve global biodiversity, nor is there a need to manage old-growth forest to increase local stability. Tropical rain forests, for example, are often remarkably stable. This is because the rain forest is very diverse in both species and structure. Because of the high biodiversity, there is little opportunity for an outbreak of pests or disease that could ravage the entire forest. Pests often specialize on one or a few host species. Tropical rain forests are also diverse in structure. Tree-fall gaps—that is, openings in the forest where large trees have fallen over—result in new resources becoming available for seedlings and saplings in the understory. The usually diverse structure of a mature forest also protects it against stresses such as hurricanes. For example, a hurricane that struck St. John, U.S. Virgin Islands, injured canopy emergents and small trees, but most intermediate-size trees remained intact (Reilly 1991).

Mature forests in the temperate zone also frequently have high diversity and stability, caused in part by tree-fall gaps. We seldom recognize this because there are so few mature forests left in the temperate zone.

Managing for Landscape Diversity

Management for diversity of landscape can be important for preservation of global biodiversity. In the wild, a variety of habitats can be critical for top predators such as eagles, lions, and wolves that roam across a variety of sites and prey upon a diversity of species in a diversity of areas. To a top predator, the complexity of a landscape may be very important. To be able to prey on a wide variety of resources helps ensure survival. A wide variety of resources might be all the different herbivores in all the different habitats to which the predator has access (Fig. 7.11). To the herbivore, a wide variety of resources might be all the different plants in all the different habitats to which the herbivore has access.

Management for landscape diversity also is important for stability and efficiency of resource production. It is important to manage landscape for diversity because if nature is not allowed diversity in space, diversity may occur in time. The spruce-fir forests of

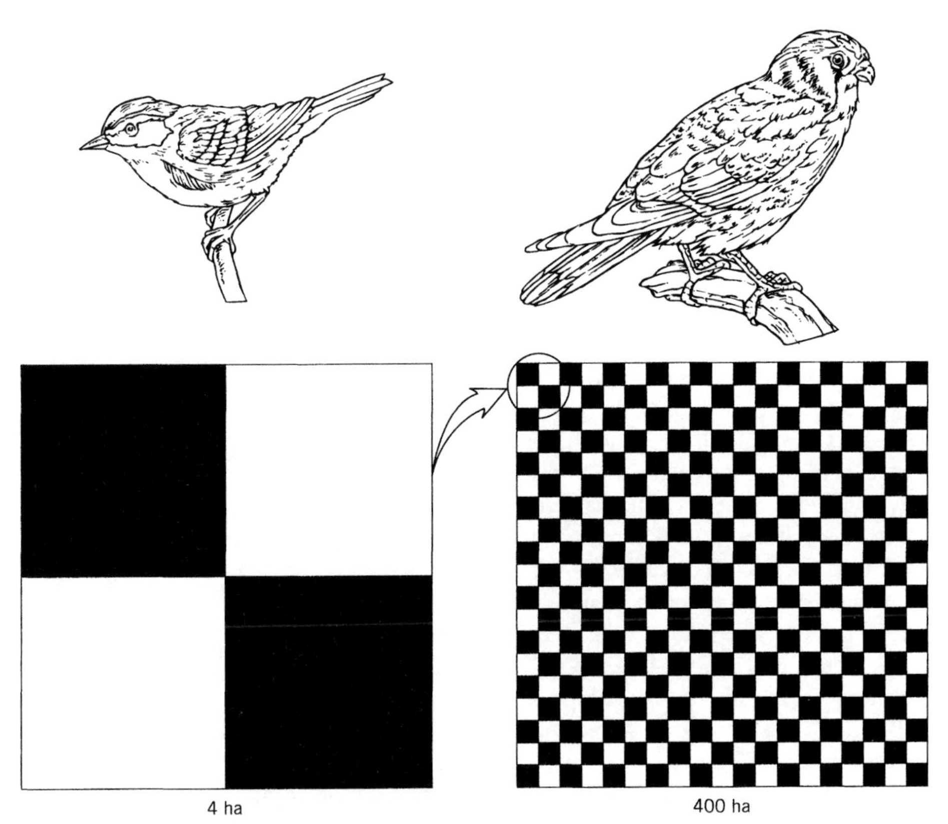

FIGURE 7.11 In each block, the individual cells represent 1 ha of 60-year-old forest (black) or 5-year-old forest (white). The left figure represents the scale of perception of a yellow-rumped warbler: the right represents an American kestrel's scale. *Source:* From M. L. Hunter, Jr., The Diversity of New England Forest Ecosystems. Pages 35–47 *in* Bissonette, J. A. (Ed.) 1986. *Is Good Forestry Good Wildlife Management?*, Maine Agricultural Experiment Station Miscellaneous Publication No. 689, 377 pp. Reprinted by permission.

northeastern Canada are an example. Under natural conditions, these low-diversity forests are kept heterogeneous by natural mechanisms. A stand will grow and mature, and as it grows older, some of the weaker and less hardy trees are invaded by the spruce budworm. The older the stand, the more trees are killed by the insects. Finally, such an accumulation of fuel occurs that a lightning storm sets a blaze, and a fire burns a patch that may vary in size but will often burn in an irregular pattern. The result is a new seedbed for germination of new reproduction intermixed with older patches that were vigorous enough to resist the insect damage.

In areas where these stands are managed for pulp production, vigorous campaigns of fire suppression have been launched. As a result, infestations of the spruce budworm have grown exponentially over large areas. Then, when a fire does break out, the damage to standing trees is much greater than would have occurred had the natural dynamics of fire been allowed to occur (Stocks 1987). Over time, the effort to maintain homogeneity in space resulted in increasing instability of the system. To counter the effect of homogeneity, managers are learning to deliberately maintain structural and spatial diversity (Erdle and Baskerville 1986).

Preservation of Global Diversity

We can conserve global diversity by preserving representatives of as many habitats as possible, some with high diversity and some with endemic species or rare species. There is no need to "manage" such areas in the sense of building roads, carrying out manipulations, or some other activity that impinges on the forest. Naturally occurring ecosystems have managed to keep intact for tens of thousands of years without the help of one or another agency. The only management that is required is to keep out the roads and keep out the machinery. By so doing, the most destructive elements of society will be avoided.

A common rationale for road-building in parks, national forests, or reserves is to enable better protection. Such reasons must be examined with extreme skepticism. The more roads in an area, the harder it is to keep out squatters, poachers, and timber rustlers. Furthermore, fire protection and insect control measures usually can be better carried out from the air than from the ground. If there is one phrase that will give the best guidance to agencies that strive to maintain biodiversity in conservation areas of developed as well as developing countries, it is the following: STOP BUILDING ROADS.

When habitat has already been degraded, restoration is an approach that is important. Efforts to restore forests, wetlands, and other ecosystems degraded by exploitation and development are critical for preventing extinctions. Restoration is also concerned with biologically unique areas that are not necessarily degraded but are nevertheless disappearing. The prairie east of the Mississippi River would not exist without fire. It evolved with the occurrence of naturally occurring fires, perhaps aided by fires set by Indians. It contained a flora that is now all but extinct, owing to agricultural development and to fire suppression. Unless prairie is periodically burned, a succession of woody plants will occur, eventually leading to an oak forest, or farther north, to a maple-basswood forest. Ecosystems in the Southeast United States that are dependent on fire are the long-leaf pine/wiregrass community and the broomsedge/goldenrod old fields.

Management is needed to compromise between competing interests in a tract of forest when conflicts between preservation and management for production are unavoidable. A case in point is the old-growth forests in the Pacific Northwest, which are valuable for timber but are also home to the Northern Spotted Owl. Because smaller areas are required for preservation of the owl when the sites to be preserved are surrounded by a mosaic of vegetation rather than a clear-cut, management of landscape for a mosaic of habitats may be a workable compromise (Franklin 1993).

Reintroductions of threatened species into former habitats is sometimes desirable. Captive-bred golden lion tamarins are being reintroduced into sites of the Atlantic Forest of Brazil, where a small, wild population was discovered. The objective is not to increase the diversity of the reserve, a relatively small patch of forest in the state of Rio de Janeiro. Rather, the value of the reintroduction efforts there is at the level of global biodiversity.

Sometimes, however, reintroduction or management for a rare or endangered species may require techniques that endanger or eliminate other species. For example, one species that has come under the protection of the Endangered Species Act is the snail kite, a small hawk found in the Everglades National Park of south Florida (Alper 1992). This species is a specialist that feeds almost exclusively on the apple snail which lives in areas of standing water. In the last century, much of the area surrounding the Everglades was drained and converted into agricultural fields, and the hydrology of the park was altered. In recent years, the Everglades has been struck by drought and an increasing demand for fresh water from nearby Miami and from the agricultural fields that surround the park. Restoring the hydrology of the park would benefit not only the 300,000 wading birds that inhabit the park during the summer, but also the entire ecosystem. However, by restoring the original waterflow in the ecosystem, one of the largest feeding grounds of the snail kite would be endangered because this area would be drained of its standing water. As a result, the snail kites that usually feed on apple snails that rest on sawgrass stems above the waterline would be forced to feed elsewhere.

What is the appropriate action in such a case? When management for preservation of a particular species such as the snail kite conflicts with management for biodiversity in an ecosystem such as the Everglades, the decision is difficult. If the law were changed to protect endangered ecosystems instead of endangered species, the problem would disappear.

Backup Measures

When preservation of site or landscape is inadequate to protect the species contained within them, *ex situ* mechanisms may be necessary. (*Ex situ* means places away from the site where the species naturally occur.) Examples of *ex situ* are botanical gardens, game farms, captive breeding programs in zoos, and gene banks. These facilities are particularly important for wild species whose populations are rare or endangered. They serve as a backup to *in situ* reserves as a source of material for reintroductions and as a major repository of genetic material for future breeding programs of domestic species. Even for wild species that are not threatened, *ex situ* collections are needed to make material readily available for breeding, so that the genetic base can be kept broad.

Through their research on captive populations, researchers in zoos have learned lessons about wildlife management that can be applied to protected areas that contain relatively small populations of certain species of particular concern. Methodologies and management techniques such as induced ovulation, transplantation of certain individuals between populations to ensure gene flow, and various veterinary procedures developed in zoos can be applied to species in protected areas that become isolated habitat islands.

In 1987, the International Union for the Conservation of Nature established a Botanic Gardens Conservation Secretariat to mobilize the world's botanic gardens into an effective force for conservation, particularly to monitor and coordinate *ex situ* collections of conservation-worthy plants. In the past, botanical gardens have been collections of species, focusing on particular taxonomic groups, particular regions of the world, or species of economic importance. Little attention was given to the interactions between the species within the botanical garden. Because interactions between species are important, conservation of threatened species might be more effective if entire communities of plants were established in safe locations rather than just individual species. Such measures would be most effective if the *ex situ* location had a similar environment to the native environment of the communities to be conserved. For example, the soils and climate of southeastern China are very similar to the soils and climate of southeastern United States. Botanical gardens in the Southeast United States would be the ideal location for establishing entire communities transplanted from southeastern China. Threatened species will survive better when their *ex situ* environment includes species that form part of the *in situ* natural environment. Costs will also be much less because special climate-control facilities are not necessary.

Storage of seeds in seed banks is frequently used because of the advantages it provides: ease of storage, economy of space, relatively low labor demands, and the consequent ability to maintain large samples at a reasonable cost. The disadvantages of seed storage include their dependence on secure power supplies, the need to monitor the viability of the seeds, and the need for periodic regeneration.

For many plants of global agricultural importance such as wheat, maize, oats, and potatoes, more than 90 percent of the variation has been preserved in collections. For the vast majority of the world's species, however, there are few collections, and there is little information on seed propagation methods.

CONCLUSION

The goal of conservation biology is to preserve species diversity, and ultimately, we are concerned with preserving *global* biodiversity. Although individual species may be exterminated because of their commercial value, the major threat to world biodiversity comes from habitat destruction. Thus, although a focus on species preservation may be most effective in conserving charming species like Pandas or commercially valuable fauna such as mahogany, conservation of biodiversity and ecosystem functions may be best achieved by preserving habitat.

Wilson (1992) has summarized the increasing importance of habitat preservation as a necessity for conservation:

> *From prehistory to the present time, the mindless horsemen of the environmental apocalypse have been overkill, habitat destruction, introduction of animals such as rats and goats, and diseases carried by these exotic animals. In prehistory the paramount agents were overkill and exotic animals. In recent centuries, and to an accelerating degree during our generation, habitat destruction is foremost among the lethal forces, followed by the invasion of exotic animals. Each agent strengthens the others in a tightening net of destruction. (p. 253)*

SUGGESTED READINGS

For the theory of conservation biology:

Fiedler, P. L., and S. K. Jain, eds. 1992. Conservation Biology: The Theory and Practice of Nature Conservation Preservation and Management. Chapman and Hall, New York and London.

Harris, L. D. 1984. The Fragmented Forest: Island Biogeography Theory and the Preservation of Biotic Diversity. University of Chicago Press, Chicago.

Meffe, G. K., and C. R. Carroll. 1994. Principles of Conservation Biology. Sinauer Assoc., Sunderland, Mass.

Shafer, C. L. 1990. Nature Reserves: Island Theory and Conservation Practice. Smithsonian Institution Press, Washington, D.C.

Soulé, M. E. 1986, ed. Conservation Biology: The Science of Scarcity and Diversity. Sinauer Assoc., Sunderland, Mass.

Soulé, M. E., and B. A. Wilcox, eds. 1980. Conservation Biology: An Evolutionary-Ecological Perspective. Sinauer Assoc., Sunderland, Mass.

For refuge theory:

Prance, G. T., ed. 1982. Biological Diversification in the Tropics. Columbia University Press, New York.

For the practice of conservation biology on a global basis:

McNeely, J. A., K. R. Miller, W. V. Reid, R. A. Mittermeier, and T. B. Werner. 1990. Conserving the World's Biological Diversity. International Union for Conservation of Nature and Natural Resources, World Resources Institute, Conservation International, World Wildlife Fund—U.S., World Bank, Gland, Switzerland and Washington, D.C.

Primak, R. B. 1993. Essentials of Conservation Biology. Sinauer Assoc., Sunderland, Mass.

For the practice of conservation biology on a local and regional basis:

Hunter, M. L. 1990. Wildlife, Forests, and Forestry: Principles of Managing Forests for Biological Diversity. Regents/Prentice Hall, Englewood Cliffs, N.J.

MacKinnon, J., K. MacKinnon, G. Child, and J. Thorsell. 1986. Managing Protected Areas in the Tropics. International Union for Conservation of Nature and Natural Resources, Gland, Switzerland.

For an overview of biodiversity:

Wilson, E. O. 1992. The Diversity of Life. Harvard University Press, Cambridge, Mass.

Wilson, E. O., and F. M. Peter, eds. 1988. Biodiversity. National Academy Press, Washington, D.C.

For a portrayal of diversity in the tropics:

Terborgh, J. 1992. Diversity and the Tropical Rain Forest. Freeman and Co., New York.

For photographs of tropical diversity:

Dalton, S., and G. Bernard. 1990. Vanishing Paradise: The Tropical Rainforest. Overlook Press, Woodstock, N.Y.

For an overview of environmental restoration:

Berger, J. J. 1990. Environmental Restoration. Science and Strategies for Restoring the Earth. Island Press, Washington, D.C.

CHAPTER

Culture and Development

Chapter Contents

PRINCIPLE

Development can be sustainable only in the context of culture.

CHAPTER OVERVIEW

Regardless of how ecologically benign and economically profitable a development plan might seem, it will be successful only if it is compatible with the local culture.

The cultural appropriateness of a development project was seldom considered in earlier days by the colonial powers, or recently, by international development agencies and banks. Lack of such recognition was an important reason why many development projects failed. Conservation efforts, too, have sometimes lacked cultural sensitivity, as when a nature reserve deprives local peoples of their subsistence.

Conservation and development projects must be appropriate to the people who will be most affected by the measure—the local people.

CULTURES

Primitive Cultures

The noble savage living in harmony with nature has long been a popular myth. This myth holds that among "primitive" peoples, culture is closely bound to the environment—to the cycles of sun and rain, to the fruiting and flowering of plants, and to the migration of animals and birds. Indigenous people have been viewed as having an innately pure and spiritual relationship with nature that ensures the sustainability of both the culture and nature.

How much truth this myth contains is a subject of considerable debate. An alternative view has been that indigenous tribes are stupid, indolent peoples who waste resources and destroy nature. Their worship of pantheistic gods instead of Jesus Christ has been taken as particularly strong evidence of their backwardness and moral turpitude.

Neither view is entirely correct, and each case is different. Nevertheless, at least *some* of the cultural traits of at least *some* indigenous tribes are now seen as contributing to sustainability of resource use. In recent years, as modern methods of resource exploitation have increasingly been seen to be unsustainable, increasing interest has been shown in those indigenous cultures that seem to be well adapted to their environment. An important question in these studies is how a well-adapted culture uses and cares for resources in a way that will ensure their sustainability.

Adaptations to the Environment

An adaptation is a biological or cultural trait that aids the biological functioning of a population in a given environment. It includes such aspects as a population's health, ability to feed itself adequately, functional capability in its physical environment, and reproductive performance. This definition encompasses "adaptation" as used in genetics (Baker 1984).

Hunter-Gatherers Important in modern anthropology is the study of adaptations of indigenous peoples to their environment. The work of Posey (1982) on the Kayapó is an example. Today the Kayapó live on a 5 million acre reserve in the Xingú River basin in the Amazon region of Brazil. Although the Kayapó are nomads for part of the year, cultivation of plants for food and medicine is an important part of their culture. The first step in preparing a new area for cultivation is clearing a circular-shaped field. Tree felling begins from the center and progresses outward. The fallen stems thus radiate outward like spokes of a wheel, and the bulk of the forest canopy biomass ends up near the perimeter of the circle. Corridors of relatively open areas lie between the tree stumps. Root crops such as yams, sweet potatoes, taro, and manioc are planted in the open corridors. The cultigens are already rooted and growing before burning occurs.

Burning is carefully managed. Tribal elders agree on an appropriate day when winds are minimal and the fields will burn thoroughly but not too quickly. The men begin burning the piles of dried debris one at a time. A protracted burn minimizes the heat so that the root crops will lose only their green tops, but not their viability. These pre-burn crops have a head start on weeds that will establish in the ash.

The Kayapó make the best use of the nutrients in the ash. Papaya, bananas, cotton, urucu, and tobacco, which require a high quantity of nutrients, are planted on the outer margins of the field, where ash concentrations are highest. A few weeks after the burn, men gather up unburned sticks and limbs, stack them in piles in various parts of the field, and set them on fire a second time. In the resulting piles of ash, other plants requiring high nutrients like beans, squash, and melons are planted.

The fields of the Kayapó last many years. Sweet potato and yam bear in fields that are four or five years old. Bananas and urucu, as well as domesticated varieties of a large vinelike plant called *kupa,* commonly continue to bear edible leaves and stalks for 8 to 12 years, and some fields that are 40 years old still yield edible kupa.

Many plants useful to the Kayapó establish naturally in the old fields. Some of these spontaneously colonizing plants have important medicinal values, and others provide seeds, berries, and roots for food. Some of the colonizing plants bear fruits that make excellent fish bait. Others attract animals or useful birds. The animals drawn to the leafy and bushy plants in these sites are easier to kill than those inhabiting the canopy of the high forest. Young boys construct blinds in the trees and quickly learn to kill birds for meat.

Because the Kayapó understand and can take advantage of the species that sequentially occupy a site as it progresses from open field to forest, they do not need to continually seek new forest to cut and burn. When, after many years, an old site changes again to closed forest, it can be cut again and used again, with no long-term degradation of the site.

Kayapó practices contrast with shifting cultivation in the Amazon as carried out by colonists from southern and northeastern Brazil. The colonists depend mainly on crops such as corn, rice, and cassava which grow well for only two or three years. When yields fall off, the colonists abandon the fields and move on to clear new forest.

Scientists once believed that these declines in yield were due to nutrient losses, but recently have found that clearing and burning in preparation for agriculture does *not* cause much of a loss of nutrients and a degradation of the site. The reason for the decline in crop production and the invasion of weeds in the fields of colonists is not loss of nutrients from the site but rather changes in the *form* of the nutrients. After burning, the nutrients are in the ash and are readily soluble, which means they can be easily taken up by annuals such as corn and rice. Within a year or two, however, the soluble nutrients become bound in the soil organic matter and in the inorganic clays. Here they are inaccessible to the annuals, but are still available to native perennial species that have the ability to extract nutrients held in chemically and organically bound forms (Jordan 1989).

Cultural Disintegration

The interactions of indigenous tribes with the outside culture leads to a gradual loss of traditional behavior and knowledge. Pressure on the tribe comes from both outside and inside the tribe. From outside the tribe, lack of understanding of indigenous cultural adaptations has lead to denigration and prejudice. For example, the Batak tribe in the Philippines have been forest dwellers, who have come into increasing contact with lowland Filipinos clearing for agriculture. The Batak have been characterized as "lazy farmers" and are looked down on by the settlers. The Batak, in fact, do not like sweating and toiling in the hot sun all day. They can obtain all the food they need in the forest with about four hours work per day (Eder 1987).

Cultural disintegration also comes from within, when tribal people are tempted to acquire goods such as knives and metal pots that make their lives easier. Once the process starts, work for cash takes priority over ritual and ceremony. When cash is owed or needed, tribesmen may shorten hunting trips or forego annual migrations to richer hunting regions. As tribespeople begin to intermingle with people of the national culture, traditions associated with kinship and mate selection are dropped, and often, young people prefer not to be associated with their parents' tribal group.

For example, the Yanomami or Yanomamo of Brazil who live near highways or Brazilian villages are losing their cultural identity as a result of their pursuit of manufactured goods. The introduction of Western goods and the practice of wage labor there is altering the Yanomamo social organization. Age, sex, and individual achievement in traditional skills no longer lend prestige in the highway villages. Traditional skills are devalued because they don't facilitate interactions with Brazilians. The young view the traditional knowledge and values held by elders as impediments to successful relations with Brazilians. As the Yanomamo become more dependent on Brazilians for economic resources, the bonds of kinship and marriage among them deteriorate. Moreover, modern medicine has eroded the position of the shaman. Many of the younger Yanomamo from the highway villages no longer wish to be considered Yanomamo (Saffirio and Hames 1983).

The indigenous group's resistance to outside culture takes different forms. The Penan in Sarawak have vigorously resisted the invasion of loggers into their homeland through direct confrontation (see the photo-essay "Trivializing Indigenous Resistance to Deforestation" in this chapter). Others, such as the "Jigalong Mob" described by Tonkinson (1974), resisted cultural breakdown because of high self-esteem. The Jigalong aborigines of western Australia exploited a desert environment by means of a seminomadic hunting and gathering way of life. In contrast to most aborigines, however, the tribe preserved many of their core cultural values and retained much of their traditional kinship system. Traditional religious life promoted self-esteem and ethnic pride, and helped the aborigines to maintain the belief that they alone controlled their own destinies.

Ancient Civilizations

It is not only hunters and gatherers who have lived within their environment in a sustainable way. Great civilizations of the past have existed for thousands of years practicing agriculture that provided not only subsistence for farmers, but also support for the development of a complex and highly developed culture. There is evidence that the Mayan civilization in lowland Guatemala, Honduras, southeastern Mexico, and Belize originated almost 3000 years ago and had origins 7000 years before that (Hammond 1982). The height of cultural development during the classic period occurred between A.D. 250 and 900, and remnants of the disintegrating empire still existed when the Spanish defeated the Aztecs in Mexico in 1521 (Graham et al. 1989).

The characteristics of Mayan agriculture suggest that the techniques they used were an important factor in the sustainability of their culture. One characteristic was "raised field agriculture" (Turner and Harrison 1981). Although there were variations in techniques, the basic idea was to create a network of canals. Between the canals in square or rectangular shapes that were the size of modern city blocks were fields used for agricultural crops or forage. Today, raised fields are still used in parts of Mexico and are called *chinampas.* Water in the canals is high in nutrients, and thus the production of aquatic weeds is prolific. Periodically, workers scoop the weeds onto the fields. This organic material acts as a fertilizer and also can improve the physical properties of the soil. Silt and clay that eroded from the fields into the canal also may be scooped out and applied to the fields. Trees such as willows may be planted along the edge of the banks to stabilize the fields. Fish can be an important resource from the canals.

Terracing may have been another important method for the Maya. Recent archaeological surveys of the central lowlands reveal large areas of terracing intersected by field walls (Turner 1974, cited in Hammond 1982). Until this discovery, it had been thought that the Maya had depended on slash and burn methods similar to those of modern colonists for much of the upland agricultural production.

The presence of native species of trees valued for fruit, fiber, bark, and resin within the classic Mayan cities indicates the close relationship that existed between the people

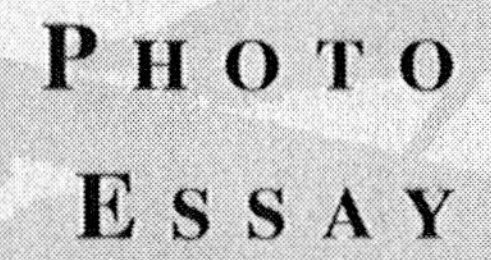

PHOTO ESSAY

TRIVIALIZING INDIGENOUS RESISTANCE TO DEFORESTATION: SARAWAK, MALAYSIA

Guest Photo-essay by J. Peter Brosius
Department of Anthropology, University of Georgia
Athens, Georgia

Fig. P.1 Throughout central Borneo live groups of hunter-gatherers known as Penan. Over the past decade, I conducted anthropological research among a group who inhabit the interior of the Malaysian state of Sarawak.

Fig. P.2 Though to all appearances a complete wilderness, the landscape is instead one that for Penan is imbued with cultural significance. For Penan, the landscape is more than simply a reservoir of detailed ecological knowledge or a setting in which they satisfy their nutritional needs. There exists for them a strong coherence between the physical landscape, history, genealogy, and the identities of individuals and communities.

Fig. P.3 Perhaps the factor that more than any other determines the movement of Penan bands is the availability of the sago palm, *Eugeissona utilis.*

Fig. P.4 Sago is the principal source of carbohydrate. Trunk sections are split, and the pith is then pounded as pictured here, rendering it soft and pliable.

Fig. P.5 It is then placed in baskets and trampled by women, while pouring water through. The starch is separated from the pith by trampling, and it washes through into the settling mat below.

Sago can be harvested sustainably. Sago reproduces vegetatively if the roots are left intact. When the sago in one area has been depleted, the people move to another area, leaving the previous stand to recover. The Penan ethos of resource use is one of explicit stewardship.

Fig. P.6 Penan use several kinds of palm. Here a Penan woman is preparing leaves of a *Licuala* palm, to be used for the roof of a hut.

Fig. P.7 Women are the exclusive weavers for the Penan. Their mats and baskets are in great demand by the "long-house peoples," who in turn trade with Chinese merchants.

Fig. P.8 Hunters use blowpipes for small game such as monkeys, squirrels, and barking deer. Here a hunter applies poison, extracted from a "tajem" tree, to a blowdart.

Fig. P.9 For larger game, dogs often are used to chase and corner the prey until it can be speared by a hunter.

Fig. P.10 The favored game is bearded pig, not only because of its large quantity of meat, but also because of substantial deposits of fat that can be rendered and stored for later use.

Fig. P.11 The first signs that the sustainable coexistence of the Penan and forest was ending were the survey markers for logging tracts.

Fig. P.12 In 1992, I returned to the Seping River Valley. A bridge and logging road had driven the game out and clogged the river with mud.

Fig. P.13 Rivers are the paradigm around which spatial, historical, and genealogical information is organized. Their names contain a wealth of cultural and environmental knowledge. Clearing of the forest presents an entirely different perspective of the rivers to the Penan. Here they discuss a previously familiar scene from the vantage point of a logging road.

Fig. P.14 Logging roads destroy hunting grounds, ancestral burial grounds, and the forest itself on which the Penan depend. Here a man looks ruefully at a species that is highly prized by the Penan for making blowpipes. It has been cut and then discarded by loggers.

Fig. P.15 The Penan responded to the logging by erecting symbolic barricades—

Fig. P.16 —and posting signs expressing their frustration.

Fig. P.17 The government's response has been to trivialize the issue, one of the most potent forms of dismissal. Trivialization is to act toward their complaints as one would act toward the complaints of wayward children. Authority knows best, and what it does is for the good of those disciplined. A slightly bemused and contemptuous attitude toward the Penan is displayed by loggers watching them demonstrate the use of the blowgun.

Fig. P.18 The persistence of the Penan has attracted the attention of the news media. A series of blockades was organized and galvanized global concern. (Photograph courtesy of Rainforest Action Network.)

Fig. P.19 The government has begun programs to help the Penan, such as giving them sheets of plywood for their shelters. Here, the Penan are rushing to unload the helicopter.

Fig. P.20 For the most part, however, the government has rejected emotional scenes such as this one, by claiming they are instigated by environmental imperialists.

Fig. P.21 The Sarawak chief minister recently summed up the government's attitude as follows. "How can we have an equal society when you allow a small group of people to behave like animals in the jungle. . . . I owe it to the Penans to get them gradually into the mainstream so that they can be like any other Sarawakian."

Development seems to be an inevitable substitute for sustainability in the lives of these Penan.

SUGGESTED READINGS

For case studies of other indigenous tribes threatened by deforestation:

Davis, S. H. 1977. Victims of the Miracle: Development and the Indians of Brazil. Cambridge University Press. Cambridge, U.K.

Denslow, J. S., and C. Padoch. 1988. People of the Tropical Rain Forest. University of California Press, Berkeley.

Eder, J. F. 1987. On the Road to Tribal Extinction: Depopulation and Adaptive Well-being among the Batak of the Philippines. University of California Press, Berkeley.

For a popularized presentation of tropical rain forests and their peoples:

Collins, M. 1990. The Last Rain Forests: A World Conservation Atlas. Oxford University Press, New York.

For an explanation of how other countries are dealing with the conflict between people and forests:

Gradwohl, J., and R. Greenberg. 1988. Saving the Tropical Forests. Earthscan Pubs., London.

Lugo, A., J. J. Ewel, S. B. Hecht, P. G. Murphy, C. Padoch, M. C. Schmink, and D. Stone. 1987. People and the Tropical Forest. U.S. Man and the Biosphere Program. Department of State, Washington, D.C.

Poffenberger, M., ed. 1990. Keepers of the Forest. Kumarian Press, West Hartford, Conn.

and their resources. The location of valuable species close to high-status vaulted structures suggests that control over these species was by the lords and priests (Folan et al. 1979). Common people, too, carefully cultivated their resources. One method was the "Pet Kot," a circle of stones in which garden crops were grown (Gomez-Pompa et al. 1987). The unusual height of trees found today growing within the Pet Kots suggests that the soil may have been enriched with organic debris.

Although the Mayan culture was sustainable for thousands of years, and the classic period for almost a millennium, it was not permanently sustainable. A wide range of theories have been advanced to explain the disintegration of the Mayan empire (Hammond 1982). One theory implicates social factors. Weakness in the framework of Mayan society led to internal quarreling and made invasion from outside tempting. A peasant revolt may have been partly responsible. The lower classes became disaffected by mounting demands for tribute for an increasingly esoteric and irrelevant priesthood. A gap opened between rulers and ruled, which led to a bloody revolution, the destruction of the elite and their cities, and a return to a less organized level of society.

Warfare between Mayan states may have been a factor in the collapse of the Maya. The struggle for supremacy diverted resources and energy into unproductive uses. Military intrusion from the Gulf Coast of Yucatan may also have contributed to the collapse.

Environmental deterioration as a result of overcultivation is yet another important possibility. Soil erosion filled the lakes, transforming them from a vital water supply into swamps and sources of disease. Soil fertility may have declined, and competition between food crops and weeds became too intense for the Maya to counter with their neolithic technology. Population outran resources. The carrying capacity of the land with the available technology had been exceeded.

Advancing Cultures

A primary characteristic of an advancing culture is its decreasing reliance on the patterns and rhythms of nature, and its dominance over the vagaries of winter and summer, drought, and flood. Before the Industrial Age, Europeans were, to a large degree, subject to the whims of nature. Irrigation was one of the few technologies that circumvented the climatic limitations to agriculture.

The turning point was the agricultural revolution. Some historians have argued that the agricultural revolution began in the sixteenth century, with the shift in Europe from subsistence to commercial farming (Bayliss-Smith and Wanmali 1984). The English agricultural revolution between 1760 and 1820 marked the most distinct change in the history of English farming. It was characterized by the introduction of cattle-feeds and artificial fertilizers, and investment in new buildings and drainage. Increasing control over nature continued early in the twentieth century with the introduction of the tractor and of labor-saving machinery, and later with the development of pesticides and herbicides.

The most significant shift during the agricultural revolution was from nature's domination of humans to the human conquest of nature (or at least apparent conquest). Humankind no longer appeared to be dependent on nature for survival. There was pride in the human triumph over nature. Many people now enjoyed a higher standard of living, and farmers became part of a middle class. The fact that culture became divorced from nature seemed not only inevitable but also desirable.

In recent years, however, some question has arisen as to whether nature really has been conquered. There have been some signals that it has not. One signal is the increasing resistance of some insects to pesticides; by 1992, this increased resistance had evolved in over 500 species of insects and mites. Resistance to multiple chemical insecticides has been detected in several species. Critical cases have been encountered in at least a dozen major economic pests, including the Colorado potato beetle, sweetpotato whitefly, and the diamondback moth. No class of insecticides has escaped the resistance syndrome, including the biological insecticide *Bacillus thuringiensis* (Duke et al. 1993). Other persistent problems such as soil erosion, stream pollution, and depletion of groundwater further suggest that human domination over resources is still far from complete.

Differences in Advanced Cultures

Cultures differ not only between developing and developed countries, but also between the developed countries themselves. The miracles of economic development in Japan, Korea, Taiwan, Hong Kong, and Singapore contrast with the lackluster performance in Spain and Brazil because of cultural differences (Harrison 1992) (Despite the vast rain forest in the north, Brazil is considered a developed economy because of the industry in the south.)

In recent decades the Spanish economy has grown at about one-half the East Asian rate, despite Spain's better resources, advantageous location, and, since 1986, membership in the European Community. Spain also saves less than East Asia, depends more on foreign investment and tourists, and has high unemployment. Between 1965 and 1980, Brazil grew as rapidly as the East Asian countries, but it saved less and borrowed more from abroad. It has been plagued by high unemployment and high inflation, and, since 1980, its economy has slowed greatly.

East Asia's success is attributable to an ethos containing elements of Confucianism, Taoism, and ancestor worship. These beliefs stress work, education, merit, the future, and frugality. In this respect, the beliefs resemble the Protestant ethic. The less spectacular economic performance of Spain and Brazil reflects the residues of traditional Iberian culture whereby the focus is on the present and past rather than on the future, and economic creativity is often perceived more as a threat than as a blessing. Harrison (1992) concludes that the many economists who argue that culture is irrelevant to development are missing a big part of the picture.

The interaction of culture and development in countries newly emerging over the past half century varied greatly, and an understanding of the differences is necessary to understanding the different success of development aid. Baker (1993) has summarized the differences:

> *Latin America struggled forward with a totally fossilized, centuries-old Iberian model of the worst form of semifeudal exploitation. Africa experimented with socialistic mind games and posturing, which bankrupted the entire continent, while Asia was divided among mind-numbing bureaucracy, ideological hot wars, and some of the most aggressively successful capitalism the world has yet seen, often achieved at terrible environmental cost. (p. 117)*

MISMATCH OF CULTURE AND ENVIRONMENT IN THE TROPICS: AGRICULTURE AND FORESTRY.

The Early Scientist Explorers in South America

The first European explorers of the American tropics were the sixteenth-century conquistadors who came in search of gold. For them, the rain forests of Central and South America were obstacles that interfered with the quest for riches. The great size, density, and lushness of the forests impeded them, and when their expeditions failed, they blamed the impenetrability of the forests.

Three centuries later another type of explorer, the explorer-scientist, traveled to the rain forests of the New World. These scientists, among whom were Alexander von Humboldt, Henry W. Bates, and Alfred R. Wallace, viewed the forests differently from the conquistadors. The explorer-scientists saw the forests as a potential source of economically important products. Although they sometimes encountered valuable species of plants, the qualities of the forests that impressed the scientist-explorers most were those that had most discouraged the conquistadors: the size and density of the forest.

These scientists knew that in Europe the size of the trees in a forest generally indicated the potential for agricultural use of the soil supporting the forest. Large trees usually meant the soil would be highly productive; small trees usually meant that crop yields from that soil would be low. The Europeans' cultural association between large trees and high agricultural productivity led them to conclude that the Amazon basin had extremely high potential for timber and crop yields.

This conclusion stimulated many attempts to exploit the area commercially. Sioli (1973) chronicled early efforts to colonize and develop the Amazon basin for agriculture and forestry. The most important point of Sioli's account is that all large-scale developments have failed, often because of the nutrient poverty of Amazon soils. Although some individual farmers have been able to carry out subsistence agriculture (Moran 1981) or cultivation of certain perennial crops (Alvim 1981), large-scale efforts to produce cash crops for export have not been successful. Even more striking than the failures is the depth of the cultural belief that agriculture *could* be tremendously successful if only things were done right. Failures are blamed on poor administration, an inadequate labor force, and insufficient official support; rarely, if ever, are they blamed on the productive potential of the soils (Sioli 1973).

How is it that a large rain forest growing on poor soil is capable of producing tons of leaf and wood biomass annually, but that the soils are incapable of sustaining anything more than a few years of agricultural production? The key is that the undisturbed tropical forest ecosystem is not merely a bunch of trees growing in the soil, but an integrated system of interacting organisms. Each group of species in a particular functional role such as producers, herbivores, carnivores, and decomposers is dependent on other groups or species. The mutual interdependence of the organisms of the rain forest is the key to the variety of species, large structure, high productivity, and long-term existence of the forest (Gomez-Pompa et al. 1991). The functioning ecosystem is just as much a system of interacting parts as an automobile. And like the automobile, it is subject to failure when some of the parts are removed.

The Jari Example

In 1967 one of the largest conversions of tropical forest to pulp plantation began near the junction of the Jari and Amazon rivers, in the state of Pará, Brazil. The project was initiated by Daniel K. Ludwig, one of the world's richest men and owner of numerous international corporations. Ludwig had anticipated a global shortage of wood fiber for pulp, and to meet this shortage, he and his advisers began looking for a site that could have high potential for pulp production. A site was selected in the Amazon because Ludwig believed that the Amazon region had a great and unlimited potential for high and sustained yield of commercial crops.

Ludwig wanted not only to raise trees for pulp, but also to develop an entire integrated pulp production project that included a mill suitable for producing high-quality pulp ready for delivery to overseas markets. The project included a port on the Jari River, a railroad system, and hundreds of miles of roads. A complete town called Monte Dourado was built, including schools, medical care units, a supermarket, bank, churches, houses, dormitories, central cafeteria, motor pools, machine shops, and a fuel depot. A series of outlying *silvavilas* was built to house and feed workers in remote areas of the plantation. Air service, complete with airport and planes, was established to ease the problem of slow access to and from Belem, 350 km to the east and the closest port of access. The single most expensive item was the pulp mill itself, built in Japan for $400 million and floated across two oceans, and up the Amazon River. By 1981, the total investment was approximately $1 billion.

In 1982, a majority interest of Jari was sold to a consortium of 27 Brazilian companies for $280 million. A question remains as to who finally paid for the $720 million loss sustained by Ludwig when he sold Jari to the Brazilian investors. Ludwig's parent corporation, International Bulk Carriers of New York, may have been able to take the loss as a tax deduction against U.S. income tax. If this was the case, then the U.S. taxpayers paid for part of the losses at Jari. If Ludwig defaulted on any of his debts to Brazilian banks, then the Brazilians themselves paid in part for the destruction of their own native forests.

No official reason has been given for the sale. However, Jari's low profitability would seem to be a reasonable explanation. Just before the sale, the yearly operating profit was $2.1 million, a rate of return of 0.2 percent on the original investment. And capital depreciation was not included in the calculation. Why was there so little profit? Reasons range from environmental to social and cultural inappropriateness of the project (Jordan and Russell 1989).

When Ludwig's advisers recommended melina (*Gmelina arborea*) as the best species to plant at Jari, he ordered the entire area planted to melina, without ever having his advisers examine the soils at Jari for their suitability for this species. Only after growth of melina was 40 percent below target were the soils examined. In half of the plantation, soils were found to be too infertile for this species. As a result, trials with pine and eucalyptus were begun. Studies of productivity and nutrient cycling at Jari (Russell 1983) indicated that stand productivity at Jari was relatively low compared to other plantations of the tropics, and lack of soil fertility was an important contributing factor. Other factors appeared to be inappropriate machinery, inadequate worker training, poor site preparation, inferior genetic stock, weeds, pests, diseases, and fire.

In an International Workshop of Forest Management in the Tropics, Palmer (1986) presented an independent analysis of the factors that caused the low profitability at Jari.

Loss of surface soil and compaction of subsoil owing to site clearing with heavy machinery.

Planting of melina on infertile soils, where it is not capable of good growth.

Machine removal at $500/ha of failed melina stands.

Less than optimal spacing of trees.

Delay in trials with pine and eucalyptus.

Failure to utilize native species to the extent possible.

Necessity for expensive fertilization.

High incidence of pests and diseases.

Fires set by disgruntled employees.

> Social and cultural factors such as a wholly owned foreign company in a relatively remote part of Amazonia; a company headquartered at an unfortunately named "Gold Hill" in a country fascinated by gold; and rigorous controls by guards on access to the estate.

These features aroused suspicion and dislike in increasingly nationalistic Brazil, which was seeking to distance itself from new forms of colonialism. Ludwig's prolonged reluctance to engage in dialogue with even the more responsible members of the press was apparently due to his belief in the power of his personal contacts with an increasingly tottering military regime.

Jari has not contributed to social stability in the region. In 1987, turnover of chainsaw operators averaged 90 days. These workers leave their families in other regions of Brazil, take a job long enough to make some money, and then return home. Turnover of technicians and engineers also was high, with 50 percent leaving every year.

In 1987, the problem of squatters and shifting cultivators seemed to be increasing. Slash and burn clearings were being farmed immediately adjacent to the plantations on the edge of the property. Hunters were frequently in the native forest reserve just north of the plantations and still on Jari property (Jordan and Russell 1989).

Despite the enormous financial losses and low return at Jari, the Brazilian government is pressing ahead with a program to deforest an area 10 times the size of Jari and establish eucalyptus plantations as part of the so-called Programa Grande Carajas. The plantations are to be used to supply 1.1 million metric tons annually of charcoal for seven projected pig-iron plants (Fearnside 1988).

The Impact of Culture

Meggars, in her 1971 anthropological study *Amazonia: Man and Culture in a Counterfeit Paradise,* sought an answer to the question of why developers never seem to learn from past mistakes. She stated: "The persistence of the myth of boundless productivity in spite of the ignominious failure of every large-scale effort to develop the region constitutes one of the most remarkable paradoxes of our time" (p. 5). The explanation for the persistence of the myth is a cultural one. In European and North American cultures, the frenzy of vegetative growth observed in the unmodified tropics is a signal of an environment with a highly productive potential. When temperate-zone-style agriculture and forestry fail in that tropical environment, the North American- or European-trained perpetrators cannot and do not admit a cultural misconception. Their belief in the absolute correctness of their conceptions is usually unshakable.

Scientists often are no more sensitive than others in appreciating the importance of culture in development. In 1992, the Soil Science Society of America and the American Society of Agronomy published *Myths and Science of Soils of the Tropics* (Lal and Sanchez 1992), a book that posits that while there are some infertile soils in the tropics, the key to using them to feed swelling populations is more soil studies, more use of more technology such as computerized Geographic Information Systems, and more fertilizers. Never mentioned are the economic costs of such technologies or the cultural acceptability of these solutions to the local populations. The persistence of capital-intensive, high-technology attempts at development, despite the history of their failure, is particularly unfortunate when advocated by science because of the scientist's traditional influence in development programs. Enthusiasm for scientific technology blinds many to the fact that development is more a cultural and an economic problem than a technical issue.

Frequently, it is the scientists of the developing countries who are the strongest advocates of high-tech solutions to their countries' agricultural problems. Many have been trained in agricultural colleges of state universities in the United States, where implementation of capital-intensive agriculture was deemed essential for achieving "progress." For these scientists, rejection of local traditions and indigenous knowledge is a prerequisite for achieving respectability in their chosen field.

Tropical Nature Reserves: Cultural Anomalies

Environmentalists often raise the issue of inappropriate cultural context when criticizing large-scale, energy-intensive schemes of development. Yet sometimes they, too, make the mistake of ignoring culture when prescribing solutions for the problems of tropical deforestation and tropical development. Advocacy of protected reserves in tropical rain forests exemplifies an environmentalist solution that does not consider the cultural implications.

Myers (1992) suggests that locking away hundreds of key ecosystems as parks and reserves will help solve the problem of extinction of species. In this regard, he cites Dr. Ira Rubinoff, director of the Smithsonian Tropical Research Center in Panama, who argues that tropical forests constitute critical areas that are unique to humankind, and thus merit the community of nations' collective effort to safeguard these resources for everyone's indefinite use. Rubinoff proposes a network of 1000 parks and reserves of an average of 1000 km^2 each. In principle, he says, the program would be financed by contributions from all nations. In practice, however, the burden would be borne mainly by those nations that could afford it. Rubinoff estimates the cost would be $3 billion, spread over 10 years. However, even if the nations of the world *do* contribute to the reserves, the proposal does not deal with the conflict between protected nature reserves and the use of the forest by the people who live there.

Unlike the parks in developed countries, tropical forests are full of people. In North America or Europe, with occasional exceptions, a park has discrete boundaries and rules for its use, and people respect the rules. Sometimes poaching or illegal logging occurs, but when the perpetrators are caught, they are usually punished. This is not the case in most tropical countries. Most parks are parks on paper only. Few if any guards are posted, and those few who are hired are severely underpaid, and thus amenable to bribes. Often, they will not report a squatter who has no other place to live. In some countries, squatters who have lived and cultivated a piece of land illegally for a period of time can eventually obtain legal rights to that land.

The establishment of a park in a rain forest produces a dilemma. If the park is truly protected, it will probably deny the local population a source of livelihood and create a great deal of hostility that may result in the deliberate trespass and taking of forest products. If the park is not protected, it is worthless as a sanctuary for threatened species.

The governor of Campeche State, Mexico, explained the problem of preserves in developing countries: "A decree doesn't preserve a natural protected area," he said. "It puts us in the position of having to send police or guards into too large an area. Well, we just don't have the resources to do it. And later, we'd have to send in guards to watch the guards. The only way we see of protecting this area is involving those who live within it" (Neumann 1993, p. A-11).

Here the governor was referring to a 2792 square mile biosphere reserve that had recently been transferred from the federal to the local government. Under the vision promoted by the reserve's director, the central nucleus of the area remains untouched. Just outside the nucleus, residents cultivate and market the trees they have planted, as well as breed game birds and other indigenous animals. Even with local involvement, it has proven difficult to stop illegal cutting in the forest. "But while it is difficult to provide management of the area on site," he said, "it is logistically impossible to manage the nation's protected areas from a desk in Mexico City. Without cooperation from local residents, there could be no reserve, only steady depredation of the area" (p. A-11).

MISMATCH OF CULTURE AND ENVIRONMENT IN THE TROPICS: DAMS, HIGHWAYS, AND RESETTLEMENT PROJECTS

Large-scale development projects in the tropics often follow a predictable pattern. A flamboyant initiative is launched because of its political attractiveness. The recipient government favors the project because it means large infusions of capital. Lending countries favor the project because it makes them appear altruistic. Lending banks and agencies favor large, dramatic, centralized projects because they are easier to monitor and control than small, diffuse, locally based ones.

Often national governments and international lending institutions collaborate on a project in an undeveloped region of a nation. Plans are made in the national capital and overseas, while the residents of the area affected may not hear about the project until plans are already finalized. As the initiative unfolds, serious flaws are revealed, especially in unanticipated side effects that escaped notice in project planning. Nonetheless, the initiative is maintained because of the political risk the leaders would face if it were abandoned or even modified.

Lack of early input on the local level denies the plan the necessary adaptiveness required for success. The central planning is often at odds with the coevolution of social systems to cope with changing situations. Centralization means that policymakers, who are typically far removed from the ecosystems under their control, lack intimate knowledge of the constitution and operation of these ecosystems. Decisions made at the center are prone to endorse plans that appear to be sustainable on paper but fail in the field, because sustainability depends on how people and resources mesh on the microlevel. When centralized government dominates decision making and controls financing, local organizing tends to languish. Thus, the communities in direct contact with the resource to be developed may fail to develop their own capacity to sustain their uses of the natural resources. The following case studies, condensed from Ascher and Healy (1990), illustrate the problem.

Indonesian Transmigration

Indonesia's transmigration program (*transmigrasi*) was unveiled by the independent Indonesian government in 1947. The objective was to relieve overcrowding in the country's "inner islands" of Java, Bali, Madura, and Lombok by giving the poor in these areas the opportunity to exploit the undeveloped outer islands of Sumatra, Kalimantan, and Sulawesi. Since 1969, transmigrasi has been part of the five-year development plans called Repelita I, II, and III. The Suharto administration's aims were to foster national unity, national security, equal distribution of population, national development, and preservation of nature, assist the farming classes, and improve the welfare of local tribespeople.

The short-run target of moving as many families as quickly as possible resulted in problems. Official statistics estimate that six out of eighteen receiving regions of the outer islands are already overcrowded, with a net overpopulation of over 1.6 million people. In 1987, the government estimated that the outer islands still had room for 8.3 million more people, but this figure has been strongly disputed in light of the fragility of the forests, the damage already done through slash and burn cultivation, and the abandonment of many farm sites. The most ominous environmental development occurred after Suharto reversed the 1979 ban on clearing virgin forest. This resulted in a loss of approximately 3.3 million ha of forest for the official migrants' first settlement plots alone. More was cleared by unofficial spontaneous migration and by new clearings that opened up when the original could no longer support agriculture.

The lack of sensitivity to specific conditions and the apparent lack of awareness of how the program was working in the field resulted from the policy-making mode adopted by the transmigrasi planners. All programs came directly from the central government, so that the formal program was originally conceived with hardly any input from the Indonesian public. In Indonesia, top government support of an initiative virtually assures the support of other officials and the silencing of criticism.

Whether or not the transmigrasi programs have been successful depends on which side evaluates them. From the perspective of the Indonesian government, the subsidies to official and spontaneous settlers have been worthwhile. Some of the settlers are truly better off, and their expenses have been borne by international loans or foreign aid. There would be fewer funds for other Indonesian projects if resettlement were discontinued.

From the perspective of project sustainability, however, the programs are deserving of increasing doubts. In 1984, the transmigration minister admitted that many settlements needed reconstruction, and 18 months later he recommended closing 18 of the 667 sites. Many sites have lost their fertility, especially when subjected to Java's wet rice cultivation methods which are inappropriate on fragile lands reclaimed from the forest. When environmental protection is used as a gauge, the programs can probably be classified as disastrous.

It might be expected that foreign donors and lenders would make transmigrasi officials more accountable to issues of sustainability, the rights of indigenous peoples, and environmental damage. Yet until very recently neither the Indonesian government nor the funding agencies have been very responsive to criticism from environmentalists and human rights activists. Official agencies had developed a vested interest in the progress of the resettlement program. Because an international financial institution like the World Bank has to fulfill its mandate to lend large volumes of money, the unimpeded progress of a mega-project like the Indonesian transmigration project becomes a high priority. However, when international criticism became too scathing, the World Bank's emphasis finally shifted from rapid resettlement to greater concern for finding better sites.

Brazil's Polonoroeste Project

After World War II, land consolidation due to mechanization in Brazil's south, combined with high birthrates, especially in the country's impoverished northeast, resulted in millions of landless rural people. Many turned to the Amazon region to obtain land and start a new life. The construction of the Trans-Amazon Highway project through the heart of the Amazon and the paving of the highway from Cuiabá at the edge of the frontier of developed Brazil to Pôrto Velho in the heart of Rondonia, a territory in western Amazonia, encouraged migration. Side roads from the highways led to planned villages. The government enticed settlers by enacting laws that in effect gave ownership to whoever "improved" the land—that is, cleared it of forest and used it for agriculture or pasture.

Most decisions on Amazonian projects were made at the highest level of government. These upper-level decisions generally were not based on significant input from local officials or specialists. Decisions on the Amazon region were regarded as being of the highest national security concern. Development of the Amazon has been a key element in Brazil's aspirations toward great nation status. Specialists on the scientific and environmental aspects of Amazonian development were relegated to troubleshooting the problems and damage caused by the projects.

By the 1970s, migration was rapid and largely uncontrolled. Colonists came whenever the highway was passable and settled on any land they found unoccupied. Violent conflicts occurred with Indians and with ranchers building huge estates. As land lost its fertility, the settlers moved on, clearing and burning new land. The Polonoroeste Project (Northwest Regional Development Plan) was launched to remedy these problems.

The specific objective of the project was to create regulated settlements for the 5000 migrants arriving each month in the northwest territory. The settlement sites were to be selected by the Brazilian colonization agency based on soil surveys conducted by the technical agency for agriculture. Further steps for establishing settlements were to include clarifying land tenure, providing credit, storing and transporting crops and fertilizer, reinforcing extension and research, improving education, and improving health conditions through clinics, clean water, and antimalaria programs.

By the time the soil and land surveys were completed in 1982, roads and settlement infrastructure had already been built on very poor agricultural land. Only 15 percent of the area under settlement had soils that did not need substantial fertilizer input.

The government policy of allotting each settler family 100 ha of land had an unexpected consequence. The intent was to allow each settler to clear 50 ha and reserve the other 50 in forest. However, even 50 ha was too much for a family to cultivate. Some sold the reserves to speculators or ranchers, and then cut all the trees on the remaining 50 and sold those too.

In late 1984, an assessment of the project revealed that road construction and other physical infrastructure were ahead of schedule, but Indian protection, environmental protection, health services, and agricultural extension were very poor. The negative reaction caused the World Bank to hold back a disbursement, but after promises of reform, it continued payments.

The Amazon settlement schemes were based on the notion of unlimited fertile soil in unoccupied land, but the notion was false. Ecologists had warned repeatedly that the initial soil surveys had been conducted in an unusually rich area and that most were of poor quality. Human rights groups had continually protested the incursions of settlers onto indigenous territories, but the protests were ignored. Trying to deal with the problems made the project too complicated. The issues could not be resolved at central planning headquarters, and so it was much easier to impose a simple single plan for everyone.

In 1987, part of the World Bank funding for Amazonian road construction was suspended on the grounds that the projects did not meet environmental standards. The Polonoroeste Project, at least under that name, was abandoned. Next, Brazil protested that its national sovereignty had been violated by the interference in domestic policy and so proposed a $100 million credit for an environmental protection plan. Whether a central governmental authority actually has the capability to balance migration with environmental protection remains to be seen.

Narmada Valley (India) Development

The Narmada Valley Project in India was designed to be a system of 30 major dams, 135 medium irrigation schemes, and approximately 1000 irrigation projects affecting 11.5

million people and costing 11 billion in U.S. dollars. The Narmada Valley Tribunal was formed in 1969, but only in 1979 did agreements among the affected states of Madhya Pradesh, Gujarat, Maharashtra, and Rajasthan clear the official obstacles to development.

The cost-benefit analyses of the projects were undertaken in a very limited way. The Narmada Sagar Project Report explicitly admits that it did not assess the loss of wildlife or other ecological effects of submergence. No geological studies of possible earthquakes were done, despite the fact that the Narmada Valley lies within a seismic zone. Although as many as 1 million people may be displaced by the entire project, no comprehensive studies of the implications of such displacements have been undertaken.

As the first steps were taken to begin construction in the early 1980s, nongovernmental organizations emerged to point out environmental problems and to defend the populations targeted for displacement. Rightly or wrongly, the groups have consistently assumed the worst of the governments' intentions and competence. They have mobilized communities, particularly among the tribals, to demonstrate against the construction projects.

In 1985, the World Bank approved a $500 million loan for the Sardar Reservoir irrigation project. One of the key stumbling blocks in its implementation had been the dispute between Madhya Pradesh and Gujarat over the fate of approximately 70,000 villagers, mostly tribals living in the submergence area. An agreement had been reached to give them land in Gujarat. However, when Gujarat villagers found out about the plan, opposition grew. Igniting much of the opposition was the lack of employment guarantees for the landless inhabitants who were to be displaced. According to government regulations, loss of private property is considered grounds for compensation, but loss of access to common property is not. The landless people who derived their livelihoods from gathering firewood and other forest products were not even formally considered as being affected by the Sardar Reservoir planners.

Kothari and Bhartari (1984) reported the conclusions of a study team that covered the entire Narmada Valley in mid-1983.

> *Nowhere during the planning have local people been involved. When asked about this, some of the officials we met seemed amused—their unstated attitude was obviously one of scorn for the abilities of the villagers; involving them in planning seemed quite absurd. Other officials admitted, however, that this was a serious fault in planning, and that this "we-know-best-for-them" attitude in the past had resulted in the failure of several projects. A case in point is the Tawa Project in Hoshangabad district (MP), where some planners who were quite unaware of the ground conditions decided to introduce canal irrigation into the area. If they had only asked the farmers, they would have told them that many of the black cotton soil areas do not need irrigation since this soil has considerable water retention capacity. But irrigation was brought in, and serious waterlogging resulted in some areas. (p. 909, cited in Ascher and Healy 1990, p. 119.)*

The project is ending in political impasse. This outcome is all the more tragic because the valley's promise had been so apparent. But it must be remembered that a government's capacity for effective action is eroded whenever

1. Nongovernmental groups exaggerate or otherwise misrepresent their circumstances and demands, out of disrespect or distrust of the government's willingness to use accurate information fairly.
2. The government loses the ability to distinguish between well-grounded criticism of its initiatives and criticism motivated by partisan political objectives.

3. Polarization prevents the government and its opponents from reaching reasonable accommodations for balancing growth, distribution, participation, and environmental conservation.

The objectives of development projects should include equity, environmental protection, and participation of the affected local peoples. These objectives are often overlooked when the issue is initially defined in terms of one simple combination of diagnosis and solution. In irrigation projects, for example, responsibility is invariably assigned to engineers who presumably know how to deliver the water to where it is needed. Soil specialists or extensionists who are essential for planning the actual agricultural use of the water are trivialized. As a result of single-problem definition and single-agency domination, the concern that first puts the issue onto the policy agenda tends to dominate, both conceptually and administratively, to the exclusion of other concerns vital to the success of the project.

WHY WE DON'T LEARN (OR LEARN VERY SLOWLY)

Why has it been so difficult to accept the idea that gigantic development projects are often inappropriate for less developed regions?

"Progress" as a Religion

Progress implies increasing control over nature, increasing the exploitation of nature, increasing the accumulation of material goods by humans, and increasing the amount of landscape dominated by human-made structures. The idea of progress assumes that the benefits of the process called progress should go to people. Humankind is at the center of the religion called progress. In "primitive" cultures that are subject to the forces of nature, the god or gods often are animals or natural forces like rain and wind. In "advanced" cultures, the god or gods usually are human figures. A correlation exists between the degree of control that a culture has over nature and the form of the god in that culture.

Another aspect of "progress" as religion has been the dedication of many developed countries to eliminating "poverty" in less developed regions. Poverty as defined in the United States and the West is culture-bound. For Americans, poverty is characterized by deprivation of material goods. In many so-called poor, traditional societies, however, people have homes, enough to eat, and relative security, and yet we regard them as poor because they don't generate much in the way of monetized gross national product. Before commerce and human-centered religions intruded into many of these societies, there is no evidence that the people thought of themselves as "poor."

Foreign Aid in the Reconstruction of Europe

A second reason why Western-style development continues to be viewed as desirable and achievable for everyone was the success of foreign aid in the reconstruction of Europe after World War II.

At war's end in 1945, the massive destruction resulting from the hostilities had left Europe unable to rebuild its economy. In the spring of 1947, General George Marshall, in a commencement address at Harvard University, announced a bold new program for European reconstruction. He appeared to impose only one condition for the aid: that it be used by the European countries in a coordinated way rather than be allocated individually to specific countries for specific purposes (Milward 1984). Whether the success of the plan was actually a result of increased international cooperation has been debated, but there is no doubt that the European economy recovered and that U.S. aid greatly assisted recovery.

Why was aid under the Marshall Plan so successful in comparison with the aid that was later poured into countries outside Western Europe, from Bangladesh to Burkina Faso to Bolivia? The reason has to do with North America and Europe's similar environment and culture. Europeans held many of the same values and the same goals, and adhered to the same principles of democracy basically as did Americans. The Europeans wanted what Americans thought was good for them. And it worked because Europe's resource management problems were similar to those in the United States.

The importance of culture and environment in Europe's recovery is seldom mentioned, however, and many people in the board rooms and bank offices where development aid is planned and charted still do not appreciate their significance for development in the tropics.

Success of the Green Revolution

Beginning in the 1950s, the United States and Europe became increasingly preoccupied with the problem of feeding rapidly growing populations in other regions of the world. Elimination of world hunger was the goal, and it was to be achieved by the green revolution, a research and extension program engineered by the international agricultural research centers, such as the International Rice Research Institute (IRRI) in Los Banos, the Philippines, and the International Maize and Wheat Improvement Center (CIMMYT) in Mexico (Arnon 1987). Work focused on three interrelated actions:

Breeding programs for staple cereals that produced early-maturing, day-length insensitive, and high-yielding varieties of grains (HYVs).

The organization and distribution of packages of high-payoff inputs, such as fertilizers, pesticides, and water regulation.

Implementation of these technical innovations in the most favorable agroclimatic regions and for those classes of farmers with the best expectations of realizing the potential yields.

The green revolution had a phenomenal impact on the less developed countries. Between one-third and one-half of the rice areas in the developing world were planted with HYVs. The greatest increases were in Bangladesh, Burma, China, India, Indonesia, the Philippines, Sri Lanka, and Thailand. The high-yield varieties, fertilizers, and irrigation increased production by nearly 100 million tons annually. Between 1964 and 1986, per capita food production in the developing countries rose by 7 percent, and in Asia, the increase was 27 percent. Only in Africa was there a decline (Conway and Barbier 1990).

The green revolution was successful only where cultural conditions matched those of the donor countries and where environmental conditions matched the ideal conditions at the experiment stations. Improvements benefited only those who accepted Western cultural practices and had sufficient capital to implement these practices. Those with insufficient capital to purchase high-yield varieties, fertilizers, and pesticides, and those who did not own fertile, well-watered lands—in other words, the majority of the farmers in the developing countries—did not benefit. However, in the worldwide plaudits to the success of the green revolution this was overlooked.

Arrogance of Western Culture

Ideas of progress and development in the multilateral banks and lending agencies are based on U.S. and European cultural values. Failure to appreciate the importance of environmental and cultural appropriateness in overseas development plans has often resulted in ecological disaster and social and economic disruption. But lack of appropriateness is seldom recognized. Blame for failures is placed on corruption, inefficiency, bureaucracy, laziness, red tape, and duplicity. Seldom, if ever, is failure of development aid ever attributed to the inappropriateness of the development project for the local culture, the local environment, or both.

Cultural imperialism and insensitivity, however, *are* reasons why many development projects have failed. Increasing the public's awareness of the problem is a difficult task. Although for anthropologists, the importance of culture is a basic tenet, the impact of culture is generally overlooked in scientific literature and in government reports on development. Newspaper and newsmagazine reports often focus on the technical aspects of development projects. When a project fails, a usual scapegoat is bureaucratic red tape. It is understandable that cultural imperialism is seldom mentioned. No modern culture likes to call itself imperialistic; rather, a nation prefers to see itself as benevolent. Foreign development aid is a mechanism by which a country can seem to be politically correct.

But is aid truly benevolent? The requirement that a nation receiving aid must conform to certain standards seems reasonable to most Americans. But a requirement that a recipient country must obey rules set down by the donor is really cultural imperialism masquerading as benevolence. "River of Rains" (Jordan 1992) is a story that illustrates two extremes of cultural imperialism which today are devastating the Amazon. The one extreme is the mentality that exalts conquering the frontier and bringing progress to the backward natives. Jordan's docudrama begins in 1968, when a North American forester brings his version of progress to the Amazon by overseeing jungle crushers in their wholesale destruction of the Brazilian rain forest and its replacement with monocultures of exotic pulp seedlings. During a vacation flight out to see his family, the foresters' plane is downed in a remote jungle. He survives hair-raising trials with aboriginal Yanomamis, who eventually accept him, cure him of malaria, and teach him how to survive in the jungle. In the process, he becomes aware that the intact rain forest has value. On his journey back to civilization, he stumbles on a mission controlled by a fanatical evangelist. Now appreciative of the wisdom of indigenous peoples, the forester foments a rebellion among the Indians. The enraged missionary hands him over to a renegade band of soldiers with instructions to arrange a fatal accident. With guile learned from the Indians, he escapes and returns to civilization, converted into a staunch environmentalist determined to preserve the rain forest in its primeval state.

His conversion is *to* the other extreme of cultural imperialism—the mentality that holds that North Americans have a responsibility to save the world by preserving intact the rain forests. The forester tries to fulfill this aim by overseeing a wildlife reserve for a U.S. environmental group. The reserve was thought to be pristine forest, but when he arrives, he discovers that it is inhabited by Brazil nut collectors. He lives with them, learns their culture, and sees how they use the forest without destroying it. In a second awakening precipitated by helping them battle the ranchers who want to turn the forest into pasture, he sees that the key to preserving the rain forest lies not in the excluding of people but in using the forest in a sustainable way.

After 20 years in the Amazon rain forest, the forester returns to the United States and is asked what he learned from his adventures. His reply summarizes the philosophy that is so important for Americans to understand. "North Americans have made a lot of mistakes," he said, "imposing their solutions on other peoples' problems. And in the end, they make things worse." The phrase "imposing their solutions on other peoples' problems" encompasses the cultural arrogance of many developers *and* preservationists. "We know what is best for you—just do what we tell you."

Not all Americans are blind to the importance of culture in sustainable development, but frequently, such people are not those in powerful governmental and business positions. Unfortunately, many of those who formulate development projects assume that *the American way* is the only suitable model to emulate. This assumption is part of the problem.

COLONIALISM AND NEOCOLONIALISM

European countries established colonies in Asia, Africa, and the Americas by planting on the shore the flag of the mother country and claiming the land for the king. When more than one country claimed a new land, disputes arose as to whose claim was valid. However, there never was any question as to whether these new lands could be "claimed." The thought that the lands might already be owned by the peoples who lived there never arose. The idea that the land belonged to the Indians in the Americas or the native tribes in Africa and Southeast Asia would have been just as silly as the idea that the land there belonged to the monkeys that swung from the treetops.

Modes of colonialism differed greatly, and the colonizers' underlying motivations have had a tremendous impact on development. Two extreme cases were the Puritans who came to New England and the Conquistadors who invaded much of Central and South America and the Caribbean.

Puritan Colonialism

The Puritans came to America because they rejected the elitist social privilege and ostentatious religious ceremony of Europe. They had a fundamental belief that rewards should be based on hard work and initiative and not on where one happened to be born in society. Another basic tenet was that elaborate and elegant ritual practiced in some European state churches subverted the intent of the Holy Scriptures.

The Puritans lived according to a philosophy that was more than a religion: it was an ethic that governed all aspects of their society. Their vision that hard work, initiative, and clean and simple living are virtues to be valued is known as the Puritan ethic, and this ethic has had an important influence in shaping North America.

Associated with the ethic of hard work was the ethic of saving and putting aside for the future. This included not only saving money, but also investing it in a way that could bring greater future rewards—building a bigger barn, buying more cattle, or investing in new farm machinery. The ethic also stressed efficiency. As the trees were removed from the land to clear space for farming, the logs were used to build cabins. When a deer was shot for food, the bones and the antlers were used for tool handles. A use was found for almost everything; little was thrown away. Efficiency was achieved in everyday life through simplicity. Citizens dressed plainly, and women did not wear make up. Resources were used for "important" things, and not on silly trivialities.

The Puritans' ability to make do with what they had gave rise to the North American frontier ethic that stressed rugged individualism and glorified the independent pioneers who could make it on their own in a hostile wilderness. An American ideal of independence was, "I can do it myself. I don't need nor want the government interfering in my business." This can-do attitude has carried over, in the twentieth century, into the United States' role in foreign affairs. For Americans, "doing" often is an end in itself rather than a means to a valued end. Americans scoff at Europeans as they dither about how to resolve a world crisis. In contrast, Europeans scorn Americans for their haste in rushing to police the world's trouble spots, often with no clearly defined objective.

Many immigrants came to "the new country" to sever bonds with "the old country"; America offered a new opportunity to build a new country. The country was for the people who occupied the land, not for a government or king back in Europe. Part of severing that bond was accomplished by the bringing along of wives. Through the generations, the Northern European settlers' wives were important in domesticating the North American frontiers and ultimately in developing a sense of societal obligation. Because community spirit and cooperation were necessary for taming the American frontier, these qualities also became part of North American culture.

Another manifestation of the break with European authority was the conviction that "The People," meaning the colonists, should govern themselves. This idea is reflected in the Preamble to the U.S. Constitution: "We the People . . . do order and establish this Constitution of the United States of America." Ninety years later, the purpose was reemphasized in Lincoln's Gettysburg Address: ". . . that government of the people, by the people, for the people shall not perish from the earth."

Much in the old ethic of Puritanism would be applicable to a new ethic of conservation. But although much in the old Puritan ethic was admirable, other aspects were less than commendable in a democratic society: intolerance; conventionality; sexual repression; and a materialism that grew from an overemphasis on hard work and savings.

The rebellion that simmered in the United States in the 1950s and emerged in the 1960s resulted in part from the oppression and conformity of middle-class Puritanism. The struggle for sexual and moral liberation has been too successful, however. The problem today, not only in the United States, but in most developed countries as well, is not repression, but not enough repression of cultural trends that are destroying the fabric of society: crime; drugs; sexual libertarianism. In sum, the society is afflicted by a lack of morality.

The 1960s rebellion also protested crass materialism, which seemed to obliterate spiritual values. But the struggle against materialism was less successful than that against sexual repression. Acquisition of material possessions has remained a dominant force in America.

Conquistador Colonialism

In contrast to the Puritans who came to the New World to build a new life and a new country, the Conquistadors came to the New World to enrich the old. The first Spanish explorers were looking for a new trade route to the East Indies, and South America was an obstacle to achieving this goal. But the disappointment soon changed when the Spaniards discovered the gold used by the rulers of the native American empires. Plundering for gold quickly became a primary motivation for the conquest of South and Central America. Infrastructure and development were undertaken to provide a basis for the exploitation of Mexico and Peru, whose boundaries then covered much of northern South America and Central America.

The benefits of colonialism were reserved for the mother countries. For the most part, the explorers came to the New World to enrich themselves or to be rewarded for enriching the mother country. The aristocrats who came to the New World to serve as governors and bureaucrats did so not because of a desire to leave behind the Old World, but to enrich themselves within the context and culture of the Old World.

The qualities of independence and self-sufficiency were not encouraged in the citizens of Spanish and Portuguese colonies. Initiative was discouraged. Paternalism was the norm, with the government or the owners of estates providing for the common people. People were expected to abide by the wishes of those in power. Provision for the lower classes was at a minimum level, but nevertheless there was *some* care. In return for the patronage, loyalty and subservience were demanded.

The benefits of development were for those in power, not for those who did the work. As a result, development of the resources in South and Central America stagnated. Although the Latin American countries became independent in the nineteenth century, the legacy of paternalism stifled economic development for a century. Most people still felt apart from government.

Prestige of career has had an important bearing on the development of Latin America. Traditional careers such as doctor and lawyer often have higher status than those of engineer or (heaven forbid) ecologist. In contrast to North America where the ability to do something practical was highly admired, ingenuity was given little importance in Latin America.

The religious differences between North and South America paralleled the secular differences between them. In Latin America, Catholicism was universal and orthodoxy was the rule. In the early years of colonialism, there was little rebellion against the dogma of traditional Catholicism. In contrast, religious freedom was a founding principle of North America.

In recent years religious changes have taken place in Latin America. Evangelical Protestant churches have now gained a strong foothold in much of South America, often by providing a spirituality that is more personal than that offered through the Catholic Church. Even within the Catholic Church there has been rebellion with the emergence of liberation theology, led by priests who believe that the Church should strive to help the needy and should occupy itself much less with traditional ceremony.

Despite these changes in church and state, much of the original culture remains, and failure to take that culture into consideration is an important reason why North American development initiatives often founder. These generalizations about culture are not invariably true, for a great deal of personal and regional variation exists. Nevertheless, it is important to remember that in Latin America, the people affected by a big development project are much less likely to be able to criticize the project, or provide input to the project, than are the people affected by a big North American development project. Local participation in designing a development project is essential for its success, but the traditional Latin deference and/or subjugation to authority often precludes such participation.

Neocolonialism

Economic Colonialism

Colonialism is the process of gaining control of a territory, and thereby its resources, by force. Neocolonialism is the process of gaining control of the resources of a territory or of another country by economic means.

Perhaps the worst type of colonization is an agreement between an industry within the economically colonizing country and the power elite of the colonized country. Exploitation of tropical timbers is an example. Officials of wood-using industries make an agreement with the government officials of a country rich in forest resources but poor in infrastructure. In return for payment, the government officials then grant concessions for the industry to exploit the forests. The destruction costs fall on the local people, while the profits from the sale go to the capital. A real tragedy is that the prices paid, whether legal or illegal, reflect nowhere near the true value of the timber, when the replacement costs and environmental services performed are considered. (For examples for Costa Rica and Indonesia, see Chapter 4.)

A somewhat less outrageous form of neocolonialism is the exploitation of a resource that the exploiting industry has helped develop, such as the banana plantations in Central America. Here at least, the company has made investment in the local resource and local people. Often, the people employed in the plantations are happy to have the work. Nevertheless, labor has been exploited. For example, the plantations seldom offered health benefits, despite worker illnesses resulting from fungicides used on the bananas. Exploitation of the land also took place in the sense that the services provided by the original forest were never reimbursed when it was cut to build the plantation. Because of the exploitation of labor and resources, bananas have been among the cheapest foods in U.S. supermarkets.

Where is the line between fair economic trade and neocolonialism? From a conservation point of view, when the purchasing country obtains resources from the less developed country or region at prices that do not reflect the costs of the services of nature that contribute to the formation of that resource, that is evidence of economic colonialism. For example, the price paid for tropical timber should approximate what the cost would be to raise the trees in a plantation, beginning with seedlings in a nursery and ending with a mature forest. Only then would the resource not be exploited.

A factor that drives down the prices paid for resources like tropical timbers is market manipulation by developed countries and competition among the sellers (Vincent 1992). Many underdeveloped countries want the cash exchange offered by bidders for their natural resources. In their desperation to obtain cash, the underdeveloped countries lower the price of their resources well beyond replacement costs. The true cost will be paid by future generations.

Foreign Aid

Because of the dependency foreign aid produces, nationalists in recipient countries sometimes view it as another form of neocolonialism. Foreign aid, they say, merely reflects the giver's enlightened self-interest designed to ensure that the recipient countries behave. With foreign aid comes an understanding, formally written or merely understood, that the recipient country must comply with the policy of the donor country. As a result of this obligation, recipient governments act in ways that may not be in the best long-term interest of the recipient country (e.g., selling resources at prices below cost of replacement).

Debt Forgiveness

Debt forgiveness is one way that economic exploitation of the past can be recompensed. Americans sometimes complain about Third World countries reneging on their international debt. The complainers should realize that the high standard of living in North America can partially be attributed to the exploitative low prices paid for resources from that country—low prices that have forced that country into debt.

The View from the Colonized Country

How do colonized people feel about their status? Puerto Rico is a good case study. Puerto Rico came under U.S. control in 1898 after the Spanish-American War. The value of this Caribbean island to the United States was largely strategic. The U.S. Army, Navy, and Air Force all had important bases on the island. Today, Puerto Rico's strategic value has diminished, but it remains an important link to the rest of Latin America.

In 1952, Puerto Rico was granted commonwealth status, which entitled its citizens to some but not all privileges of U.S. citizens. At present, they benefit from U.S. welfare programs; they are official citizens of the United States and can travel freely back and forth to the mainland; they serve in the U.S. military, and they were subject to the American military draft when it was in effect. However, Puerto Ricans cannot vote in presidential elections, and they have no voting representation in Congress. The commonwealth status is a compromise solution that allows Puerto Rico to retain aspects of nationalism—Spanish is an official language—while at the same time retaining the economic advantages associated with being part of the United States. A very important factor has been the provision of the U.S. tax code, which allows exemptions for U.S. corporations locating in Puerto Rico. This incentive played a major role in "Operation Bootstrap," which resulted in increased employment and an improved living standard for many Puerto Ricans.

How do the Puerto Ricans feel about being, in effect, an American colony? They seem to be ambivalent. On the one hand, nationalistic sentiment is strong, but on the other hand, they also fear independence would bring economic disaster. This ambivalence is reflected in the major political parties. The Statehood party advocates joining the union, the commonwealth party is for the status quo, and the Independence party supports Puerto Rican independence. In a plebiscite in the late 1960s, commonwealth won over statehood. It was feared that statehood would bring an end to industry's special tax status, and as a result the economy would deteriorate. Independence was not offered as an option.

In the most recent plebiscite, held on November 14, 1993, the Commonwealth party again won with almost 49 percent of the vote, and Statehood got close to 46 percent. This time, full independence was offered as an option, but it received less than 5 percent. Despite the low number of votes for the Independence party, the spirit of nationalism was reflected in the statements of supporters from both of the other parties. "Commonwealth is the best of both worlds" was the campaign slogan for the Commonwealth party. Governor Pedro Rossello, the island's most prominent statehood advocate, said, "Whatever we decide, we will always be Puerto Ricans" (O'Connor 1993).

RECOUPLING CULTURE AND ENVIRONMENT

If development throughout the world in both industrialized and agricultural countries is to be more sustainable than it presently appears to be, new paradigms of development must be adopted. These paradigms must incorporate more closely the culture and the environment of the region under development. In recent years, considerable thought and effort have been given to devising development methods that are more culturally and environmentally suitable and thus more sustainable. Initiatives have been carried out at many levels.

Indigenous Use of Resources

Neotropical Wildlife Use and Conservation (Robinson and Redford 1991) examines the sustainability of resource use by indigenous peoples in light of the decreasing natural habitat. The theme of the book is that the only hope for preserving the biological diversity of tropical forests is to allow the people who live in the forest to sustain themselves from that forest. The users of the forest then become its guardians. The scientific question asked is whether the resources are sustainable, given an expected increase in the demand for those resources.

Vickers' (1991) case study, for example, focuses on the Siona-Secoya Indian community of the Aguarico River in the Amazon region of Ecuador and their use of wildlife. Data were collected on ungulates, primates, rodents, edentates (armadillos), reptiles, and birds. The conclusion was that the native peoples' hunting of wildlife was in fact *not* sustainable. The problem was not so much an overharvest of the game, but a decrease in hunting territories as a result of the activities by the oil companies, agribusiness concerns, and non-Indian colonists. Their combined modifications of the environment pose a far greater threat to fauna than does the subsistence hunting of native people. "The issue," Vickers says, "is not so much one of debates about aboriginal adaptation, as it is one of developing strategies and policies to save neotropical flora and fauna from extinction" (p. 78).

Vickers asks whether native peoples should be subjected to the same hunting rules and regulations as other citizens. Although native people are "acculturated" in varying degrees, this does not negate their rights to resources within traditional homelands. Yet if unrestricted hunting and other developments lead to the extinction of fauna, no one will benefit, least of all the native people.

If the strategy is merely to unilaterally impose on the Indians limits or prohibitions on hunting, the regulations will be violated. A more effective strategy for conservation will include native people. Their cooperation must be enlisted, and they must be provided incentives and support to nurture and maintain it. Restrictions on the hunting of endangered species will be successful only if native people understand their purpose and help in their formulation. The greatest acceptance will occur, says Vickers, when native individuals are incorporated into park management systems as rangers, assistants, or administrators.

As one possible solution to the problem of how to enable indigenous peoples to maintain traditional uses of resources in the face of encroaching modernization, UNESCO's Man and the Biosphere program has proposed biosphere reserves. These reserves will include natural or core areas, manipulative or buffer areas, reclamation or restoration areas, and cultural areas. They may also incorporate existing protected areas within their boundaries, and the economic activities of traditional communities are to be included in the planning and management of these reserves. Essential for their success will be acceptance by politicians, scientists, administrators, and other decision makers that the indigenous presence is necessary, appropriate, and just.

How much can civilization encroach on such reserves without destroying the very element they are intended to preserve—the culture of the people and the function of the ecosystem? An important juncture occurs when indigenous peoples are urged to discard their beliefs in the gods of the forest and to take up a human-like image as their god. This transition changes their perspective of the forest. With the new point of view, it is no longer so urgent to preserve the forest because it is no longer holy. Rather, the placing of a human agent at the center of their religious faith means that all resources, including those of the forest, must be used for the benefit of humankind. Destroying the forest therefore becomes acceptable.

Extractive Reserves

Recently, Brazil has begun to protect areas of Amazon forest by designating "extractive reserves," which have the dual purpose of protecting the forest and providing a sustainable living for forest inhabitants. One of the best known is the Chico Mendes Extractive Reserve in the state of Acre, located in the southwestern part of the Brazilian Amazon (Perl et al 1991).

The settlement of Acre by nonindigenous people began a century ago, stimulated by the boom in extraction of natural rubber for world markets between 1850 and 1920 (Schmink 1992). Tens of thousands of migrants entered Acre as rubber tappers. A second wave of migrants was recruited during World War II, mostly from northeastern Brazil. The rubber tappers traditionally lived an isolated existence dispersed among the distant *colocaçôes.* These are areas of forest within a "rubber estate" for which individual rubber collectors had exclusive collection rights. The estates themselves were usually owned by absentee landlords.

As a result of synthetic rubber development and the intensely cultivated Asian rubber plantations, Acre's rubber economy declined. Many of the owners lost their rubber estates, which included all the coloca*çô*es in a region. Although the lands changed hands, the rubber tappers remained in the forest where they continued to tap rubber, collect Brazil nuts, hunt, fish, and cultivate subsistence gardens. On some estates rich in Brazil nut trees, nut collection was economically rewarding and the estates remained intact.

Some of those who obtained formal title to the old rubber estates established cattle ranches, an activity that destroys the forest and has low sustainability. Credit was readily available to most ranchers because ranching represented "development." The ranchers' first task was to get rid of the rubber tappers who occupied the land by, among other methods, buying the rights, threats, or violent expulsion. Many tappers who had few employment alternatives moved to town, where they quickly found they had few skills with which to make a living.

Many tappers then drifted back to the rural areas and, along with those who had remained in the forest, began to organize local unions. They sought to defend the forest, as well as their rights to collect and sell rubber and nuts. They developed nonviolent tactics to prevent the new landowners from clearing land, such as *empates,* or nonviolent confrontations. They also sought to provide literacy, health care, and marketing alternatives that would increase their autonomy from merchant intermediaries. The success of the tapper's resistance movement led to violence against its leaders. Chico Mendes, the head of the union based in the town of Xapuri, Acre, was murdered in 1988. The violence, however, often increased the collectors' resolve (Jordan 1992).

By the late 1980s, the rubber tappers movement began to receive international attention. The governor of Acre responded to this challenge by declaring his intention to pursue "forest-based development" for the state. But while the intent has been noble, problems have arisen. Continuing pressure has been mounted for forest clearing for agricultural and animal production. The tappers lack schools and basic health services, as well as technical assistance. Forest productivity must be increased in diversified and sustainable ways. The processing and marketing systems must be improved so that forest producers are able to earn more and raise their standard of living. A range of education and extension activities to accomplish these goals are now being tried.

Although extractivism is a use of the forest that can be harmonious with the culture and with the environment, it has been criticized on social grounds. The rubber tappers and Brazil nut collectors have a right, it is argued, to raise themselves up from a subsistence level of existence, which is surely what their existence would be were they to live sustainably within the forest. They should be allowed to participate in and enjoy the benefits of modern mainstream civilization. They should also have an opportunity to raise their standard of living through economic opportunism, just as any other citizen of the country in which they live.

But if these peoples are allowed or encouraged to benefit from modern civilization, they will have to leave a way of life that is sustainable and take up modes of life that are increasingly unsustainable, the degree of unsustainability being directly correlated with their success in the world of capitalism. If all Third World dwellers have the right to have at least the opportunity for a standard of living equal to that of the middle-class United States, then they indeed must be encouraged to leave the rain forest, for certainly they can't achieve such standards of living there.

On the other hand, should the world change course and begin to move toward a more sustainable mode of existence, then indeed the rain forest dwellers must be condemned to a life without two new cars in their garage.

These two positions, of course, are extremes, but they frame the issue. There can be compromises. Those living in folk cultures can be given vaccinations and health care. Iron pots and machetes don't seem too bad. But where are we to draw the line?

The issue is a basic one that separates a pure conservation ethic from an ethic of ruthless capitalistic development. In their purest form, the two ethics are incompatible. We must try to muddle through with a compromise.

Natural Forest Management

Selective forest cutting is ecologically superior to clear-cutting because in selective forest cutting, the basic structure and function of the forest remain intact. Processes disturbed by the harvest can recover quickly and easily, aided by spores, seeds, and animals from the surrounding intact forest. The problem with selective forest harvest is logistical. It is difficult and expensive to open up logging trails through the forest to harvest single trees, and even more difficult and expensive if care must be taken not to injure the remaining trees.

Strip-cutting is a compromise that combines the ecological advantages of small openings in the forest with the logistical advantages of clear-cutting. Strip-cutting resembles an elongated "tree-fall gap," an opening that occurs naturally in the forest when an old tree dies. A trial project in forest management utilizing strip-cutting was started in the early 1980s in the Palcazú Valley at the eastern base of the Peruvian Andes.

Approximately 3500 Amuesha (Yanesha) Indians inhabit the valley, where they practice traditional shifting cultivation of manioc, maize, and upland rice. Approximately 75 percent of the lower valley is still covered by primary forest. The Palcazú Valley is rich in native plant species, some of which are used for medicinal purposes. Hartshorn (1990) estimates that the forest contains 1000 species of native trees. Much of the remaining forests in the lower Palcazú Valley should be under permanent management for production forestry rather than for agriculture because of the highly erodible nature of the soils.

Timber exploitation under the plan is carried out in long, narrow clear-cuts interspersed in the natural forest. Each strip clear-cut is 30 to 40 m wide, and the length is determined by topography and logistics. Strips should be along the contour in order to minimize crossing ridges and streams (Fig. 8.1). Each strip is an elongated gap, bordered on each side by intact forest that supplies seeds for natural regeneration within each clear-cut strip. In successive years, new strips will be located far enough from recently cut strips, so as to ensure adequate stocks of reproductive trees to repopulate the harvested strips. Projected rotation time between successive harvests of a specific site in the strip shelterbelt system is 30 to 40 years.

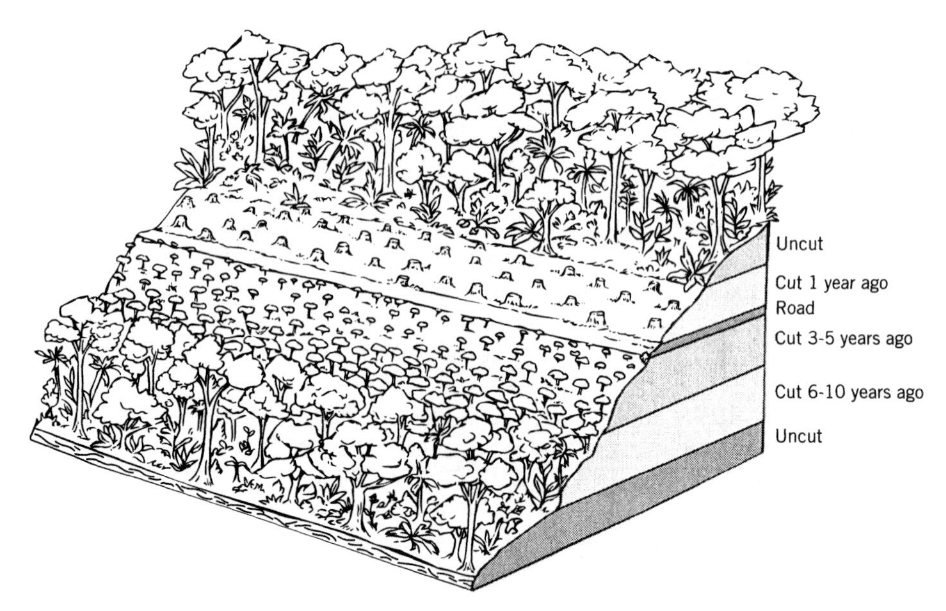

FIGURE 8.1 Strip-cutting scheme for tropical forests. After the strip is cut, stumps sprout and seeds move into the cut strip from upslope forest. Nutrients released by the cutting are used by downslope saplings. After several years, saplings have established in the cut strip, and it is possible to cut a strip further up the slope. The logistics of removing logs is easier when the forest is cut in strips, but the strips are narrow enough so that interference with the natural functioning of the forest is minimal. *Source:* Jordan 1982.

Natural regeneration of the cut strips has been striking (Hartshorn 1990). In addition to abundant regeneration from seed, many stumps, including the very dense, beautiful hardwoods, have vigorous sprouts. Fifteen months after harvest, there were 1500 regenerating individuals and 132 species in an area 20 × 75 m. The proximity of seed sources and the absence of burning and cropping probably contribute to the abundant natural regeneration on the strip clear-cuts.

The strip-cutting system is well adapted to the local environment both culturally and ecologically. Oxen or water buffaloes are used to extract the logs, poles, posts, and fuelwood. Extraction with draft animals is considerably cheaper, and logging and skidding damage to the soil is much less in evidence than with traction-driven tractors or skidders. Exceptionally large logs are sawn lengthwise in the forest with special portable saws.

The wood is processed in the valley at an integrated processing center organized cooperatively by the producers of the raw materials. This ensures that the added value from processing accrues to the local workers. The total harvested wood from the first demonstration strip was equivalent to 350 m^3/ha. The smaller logs are treated with a preservative and sold as utility poles and posts. Sawn timbers are sold in specialized markets based on their specific wood properties and workability, which are determined by laboratory tests and trials in a local carpentry shop. Wood that cannot be transformed into sawn products or preserved poles will be converted into charcoal, for which there is considerable demand.

Sustained yield management of tropical forests will depend on the replacement of traditional high grading of the forest with techniques such as strip harvesting and complete wood utilization. When integrated into the local economy, sustained forest management may have the potential to allow local communities to be independent of the pressures of outside exploiters of the forest.

Community Agroforestry

Farmers' economic cooperatives are common. When individual farmers band together to buy supplies or sell products, they often become better off economically than if they were to try to buy and sell as individuals. Most cooperatives focus on economic matters. However, surrounding the town of Tomé-Açu, in the eastern Amazon region of Brazil, is a union of farmers that is noteworthy not only for economic reasons, but for cultural and environmental reasons as well (Subler and Uhl 1990).

The town of Tomé-Açu located about 115 km south of Belém in the Brazilian state of Pará, has been a center of Japanese immigration to the Brazilian Amazon since 1929. A farming cooperative was formed, but during World War II the colony was designated as an enemy alien relocation center, and the cooperative was disbanded. Transport of produce to Belém was halted, and the Japanese farmers were cut off and had to be self-sufficient. The experience united the citizens and resulted in a restructuring of the cooperative system.

After the war, they discovered that the vine that produces black pepper could be profitably grown, and in 1949, the Tomé-Açu Agricultural Cooperative was founded to handle the marketing. The pepper boom ended in the early 1960s, owing to a drastic fall in world prices and to the spread of a *Fusarium* fungus which infected the roots of the vine. Some farmers left, but those who stayed began to diversify their farms and to experiment with innovative methods of agroforestry.

Individual farming practices evolved so that the sequence of crops planted in a given field mimicked the structure and function of the plants that would occur during natural succession. Although the specific species are different, the principle is the same as that employed by the Kayapó, described earlier in this chapter. After clearing old forest fallow, short-lived perennials such as passion fruit or papaya are generally planted. Between the rows of perennials, the farmers may plant annuals such as rice, beans, cucumbers, tomatoes, or leafy vegetables. At the same time, intermediate-lived perennials such as black pepper vines or fruit trees are interplanted. As harvest of the annual crops proceeds, long-lived plants such as cacao, rubber, cupuaçu, or trees valued for timber are interplanted.

In this system of relay planting, the short-lived crops grow and produce while the longer-lived crops are developing. When the short-lived crops are nearing the end of production, the longer-lived crops are ready to begin. The structural changes are the same as those in natural succession—from open, bushy, or vinelike vegetation to a closed, multi-layered forest.

Soil fertility is maintained in part by nitrogen-fixing leguminous plants. The nutrient-rich hulls of the cacao fruit are composted and burned, and then returned to cacao plantations to enrich the soil. Vanilla, a vine, is planted alongside the cacao which serves as support. Because native bees are lacking, the vanilla must be hand pollinated, a very painstaking operation.

Another notable aspect of the farmers' community in Tomé-Açu is their integrated co-operation (Jordan 1987b). Several farms concentrate on animal production. In one, some 13,000 chickens produce 30 tons of organic fertilizer per year. Husks from rice grown on another farm are spread on the floor of the chicken houses. Every few months the husk-manure is bagged and used to fertilize plantations of fruit trees on other farms. At neighboring farms, pigs are fed corn, rice, and manioc grown in adjacent areas and supplemented with minerals. Organic waste from the pigs is used to fertilize pepper plantations.

Particularly interesting in Tomé-Açu is the mix of culture and environment which has resulted in an apparently sustainable production system. The farming is labor intensive and tedious. A great deal of cooperation and self-sacrifice is required, at least early in the enterprise. The type of discipline that occurs in traditional Japanese society is required and has survived much more successfully in the conservative farming culture of Tomé-Açu than in many places in Japan itself. When this disciplined culture is applied in a coordinated way to carefully stewarded land, the result is sustainable production.

To achieve a successful and sustainable system, a culture must discipline its members to work hard and sacrifice. It must also emphasize cooperation between members rather than competition. Such cultural attributes are not common among groups that enter the rain forest. Some, such as miners, loggers, and ranchers, go in to get rich fast and then get out. Others, such as immigrant farmers, have little experience or communal support to establish cooperative systems.

Fishing Cultures

An important theme of anthropological research is the nature of ties between societies and the resources they exploit. In his analysis of fishing cultures, McGoodwin (1990) divided societies into two categories:

1. *Ecosystem people,* those who live within a single ecosystem, or at most two or three adjacent and closely related ecosystems. They tend to be aware that if they deplete their main subsistence resources they risk their own ruin.
2. *Biosphere people,* those who are tied in with global markets and the employment of more sophisticated and effective technologies. They subscribe to a "myth of superabundance"—the feeling that there are always other ecosystems and other resources to exploit should those they currently use run short.

Although McGoodwin's review concerns fishing cultures, the principle applies to all categories of resource managers. Traditional peoples who are dependent on local ecosystems get to know those systems intimately. Because they are less aware of the outside world and its multiplicity of other ecosystems, they do not behave as if important food resources were available in unlimited supply. Instead, their intimate association with the environment has been conducive to inventing measures that conserve the resource.

In contrast, large-scale commercial resource exploiters, after depleting a particular ecosystem of its resources, will merely redirect its efforts elsewhere. Commercial marine fisheries provide an example. Their unrestrained, global approach to fishing gained considerable momentum following World War II. By the late 1960s it had become so widespread that it was a prime factor in the leveling off of the world's fish catch. Huge fleets of factory trawlers roved the seas, wiping out whole stocks in some ocean regions and then moving on to new grounds. Some fleets consisted of more than a hundred vessels.

Cultural Mechanisms That Conserve Resources

A common strategy used by indigenous groups is the control of fishing space. Many fishing communities are markedly clannish and tend to exclude outsiders. In Japan, for example, coastal villages have for centuries asserted common property ownership claims over their traditional fishing grounds and striven to prevent entry by outsiders. Japan's contemporary sea-tenure systems have deep historical roots that antedate the feudal era.

Cultural traditions, customs, religious practices, superstitions, and taboos can be interpreted as passive means of restraint that result in resource conservation. In much of preindustrial Europe, for instance, fishing was forbidden on the Sabbath (McGoodwin 1990). Eliminating 52 days a year from the fishing calendar could conceivably have brought about a 14 percent reduction in overall fish mortality.

In the Hamito-Semitic cultures of North Africa, there are taboos against consuming nonscaled fish of all kinds, including shellfish. In many parts of Africa, there are special cultural proscriptions against women and children eating fish so as to reserve it for male adults. A taboo against or avoidance of seafood may help to conserve a reserve food supply that can be drawn on in desperate times, when sheer hunger could be expected to overwhelm the prevailing norms against its consumption.

The idea that cultural practices are conservationist just because they are traditional has sometimes been criticized. Northwestern American Indians were usually careful not to capture all the salmon that migrated upriver, and the practice has been interpreted as being conservationist. However, one of the motivations may have been the need to maintain alliances and forestall the aggression of peoples living upstream. Social considerations may often have been more important than resource conservation considerations. The elaborate "potlatch" ceremonies in which chiefs displayed and destroyed large quantities of surplus wealth including huge amounts of seafood may have been held to attract and hold allies, thereby solidifying their political power and prestige.

Just as indigenous practices are not always conservationist, industrial countries are not always blind to the importance of conservation. The Law of the Sea Treaty (See Chapter 5) is an example of a global effort at resource regulation for conservation purposes.

International Agricultural Development

Primitive Agriculture

The view that indigenous peoples and folk societies are often well adapted to the resource limitations of their environment has been embraced by many social scientists, especially anthropologists. In contrast, the view that resource use by native peoples is wasteful and that they must be educated to better ways of management has, until the recent past, been common among traditional resource managers such as agronomists and foresters. The view is a throwback to colonial times.

For example, the English colonial foresters who went to Burma in the early 1800s to supervise the exploitation of the highly valued teak forests were appalled at the apparent wastefulness of the Karen tribesmen who cut and burned the forest merely to provide themselves with subsistence (Takeda 1992). To counter this wastefulness, reserves were established where the Karen were forbidden to clear. This led to inevitable conflict between the foresters and the tribespeople who felt that their traditional lands had been unfairly confiscated. Refusal to abide by colonial law further reinforced the perception that native peoples were ignorant and uncivilized. The example has been repeated around the world, wherever developed cultures come in contact with indigenous peoples.

The colonial attitude of cultural disdain for local peoples was often accompanied by pity and compassion. Missionaries felt that the local peoples should be helped and should benefit from modern civilization. One benefit they offered was salvation. Other benefits, bestowed by scientists, included the weaning of indigenous peoples away from their primitive wasteful methods of resource management and introducing them to progressive techniques.

Beginning of Agricultural Development

A big spurt in international agricultural development occurred in the decades after the close of World War II. International aid for agricultural development was spurred by the increasing availability of preventatives and cures for diseases that had plagued the tropics. Malaria and yellow fever were brought under control in many regions, and health care was improved. As a result of vaccination programs, spraying for insect-borne diseases, and increased sanitation and health measures, populations began to increase exponentially. Increased populations brought increased demands for food. Traditional agricultural methods could no longer feed the burgeoning number of people. Spectres of hungry and starving children raised a sense of responsibility in the developed countries that something should be done to alleviate the suffering.

One of the most important steps was the establishment of a global agricultural research system (Plucknett 1993). The first organization in this system, the International Rice Research Institute (IRRI), was founded in 1961 in the Philippines. Its goal was to put an international team of scientists in well-equipped facilities, give it operational and intellectual freedom, and ask it to find ways to improve the yield of tropical rice. Since then, more than a dozen other centers have been added (Table 8.1).

A key ingredient in the international agricultural research effort is the Consultative Group on International Agricultural Research (CGIAR), an informal organization that supports and carries out research on major crops and commodities and problems related to agricultural development. CGIAR centers are supported by both industrialized and developing countries, development banks, private foundations, and international agencies.

Stages of Agricultural Development

The development of international agricultural research has passed through two phases, is in a third, and is heading for a fourth (Rhoades 1992). The phases coincide with four central issues: productivity, equitability, sustainability, and institutional viability.

Productivity Increasing productivity became the major goal of agricultural research after agriculture coalesced into a science. Consequently, the issue of increasing productivity dominated the first actions of the international development effort between 1960 and 1975. A major emphasis was on plant sciences, especially breeding. Adoption of high-yield varieties and hybrids was encouraged. The increase in yields, particularly in Asia, has been termed "an amazing success" (Plucknett 1993), and the phenomenon has been nicknamed the Green Revolution.

Despite the plaudits, the Green Revolution has had its shortcomings. Utilizing the hybrids often required highly technical infrastructure, including irrigation, fertilizers, pesticides, and mechanization of field operations. The farmers' economic ability to obtain capital that would finance the infrastructure was not considered. Farmers were not asked whether they wanted to use the hybrids, or if they could afford to implement the technologies. Researchers and administrators were the major policy setters. The farmers' only role was as recipients. As a result, only the rich farmers who could afford the technology were helped.

Equitability To remedy the problem of inequity that resulted from the Green Revolution, economic issues and constraints moved to the foreground between 1975 and 1985. Achievement of benefits more equitably distributed between rich and poor was the goal. In this phase, the researchers tried to understand the farmers' viewpoints. The main thrust was farming systems research.

Table 8.1 Centers Operated by the Consultative Group for International Agricultural Research (CGIAR)

CENTER NAME	LOCATION	SPECIALITY
International Center for Tropical Agriculture	Cali, Colombia	Rice, beans, cassava, beef
International Potato Center	Lima, Peru	Potatoes
International Wheat and Maize Improvement Center	Mexico City	Maize, wheat, barley, triticale
International Board for Plant Genetic Resources	Rome, Italy	Genetic research
International Center for Agricultural Research in Dry Areas	Aleppo, Syria	Dryland agriculture
International Crops Research Institute for Semi-Arid Tropics	Hyderbad, India	Food in the semiarid tropics
International Food Policy Research Institute	Washington, D.C.	Public policy
International Institute Tropical Agriculture	Ibadan, Nigeria	Tropical food crops
International Laboratory for Research on Animal Diseases	Nairobi, Kenya	Tick and tsetse control
International Livestock Center for Africa	Addis Ababa, Ethiopia	Livestock production
International Rice Research Institute	Los Baños Philippines	Rice production
International Service for National Agricultural Research	The Hague The Netherlands	Research systems
West African Development Association	Monrovia, Liberia	Rice self-sufficiency
International Centre for Research in Agroforestry	Nairobi, Kenya	Agroforestry
Center for International Forestry Research	Bogor, Indonesia	Tropical forests
International Irrigation Management Institute	(Being established)	Irrigation
International Network for the Improvement of Banana and Plantain	(Being established)	Banana and plantain
International Center for Living Aquatic Resources	(Being established)	Fish, shellfish, etc.

Source: *From R. Baker,* Environmental Management in the Tropics, *p. 134, Lewis Publishers, 1993, a subsidiary at CRC Press, Boca Raton, Fla. With permission.*

Some researchers focused on on-farm research in which new technologies such as alley-cropping or disease-resistant varieties were tried out in farmers' fields in the hope of better appreciating the environmental and socioeconomic constraints to adoption. A frequent problem was that in the researchers' desire for success, genuine farmer participation was reduced.

Another approach was to concentrate more on the farmers' needs. The starting point is not new technology, but the analysis of existing farming systems to determine needs, problems, and constraints to which subsequent technological innovation is directed. On-farm research with a farming systems perspective consists of the following steps (Conway and Barbier 1990):

1. Diagnostic survey
2. Identification of farmers' needs
3. Search for appropriate technology
4. Testing via on-farm trials
5. Recommendation for adoption

The farmers' role was to be a source of information for the researchers. Researchers interacted with extension specialists, who then interacted with farmers. There was a great deal of optimism that once technologies that considered the needs of the farmer had been developed, the need for these technologies would result in swift adaptation by the farmers. This idea was called the self-propulsion theory.

The most notable achievement of the farming systems approach has been "appropriate technology," and it includes both the farming methods and the tools used on the farm. Much of it has been broadly labeled agroforestry, a few examples of which have been given here in the resource management section. But appropriate technology wasn't always culturally appropriate. For example, the farmers did not accept an efficient stove introduced into the highlands of Peru to save fuelwood because the new stoves did not smoke sufficiently. Smoke from traditional cooking stoves helped preserve potatoes stored on overhead racks.

The major weakness of farming systems research was its tendency to carry out activities at the experiment stations under ideal conditions where the terrain was flat, soils were good, infrastructure was available and workers were educated. The results were often not transferable to farms, especially to poor farms. In addition, research was heavily biased toward economics and agronomy. Often ignored were cultural, social, and political factors, not the least of which was the issue of land tenure. It was not truly a systems approach in the sense that all factors and their interactions are considered.

Sustainability The next phase of international agricultural research began in about 1985 and can be called the ecological, anthropological stage. Sustainability is the main issue, and the operating slogan is "Farmer first and farmer last." The farmer is to be a full participant, identifying the problem, working to solve the problem, and benefiting from its solution.

Basic to this farmer-to-farmer approach is the rapid rural appraisal (RRA). "The RRA may be defined as a systematic but semi-structured activity carried out in the field by a multidisciplinary team, and designed to acquire quickly new information on, and new hypotheses about, rural life" (Conway and Barbier 1990). (See the photo-essay, "The Farmer First Approach to Agricultural Development in the Uplands of Mindanao, The Philippines" in this chapter.) Interviews with farmers are often less structured than the formal written questionnaires used in the farming systems era. To get an interview going, spade technology is used; that is, the interviewer takes a soil or plant sample from the field and asks the farmer a pertinent question.

The strengths and weaknesses of RRA are exemplified in the experiences of the group involved in the Mindanao effort. Because the issue was sustainability, an initial question was why the farmers abandoned their land and moved on to cut and burn new forest. The answer turned out to be the same as that for peasant agriculture almost anywhere in the world—declining crop production resulting from decreasing soil fertility and increasing weeds and crop disease.

The farmers' common solution to the problem was to ask for more credit so that they could buy more fertilizers, more pesticides, more water buffalos, and so on. Only later

THE FARMER FIRST APPROACH TO AGRICULTURAL DEVELOPMENT IN THE UPLANDS OF MINDANAO, THE PHILIPPINES

Traditionally, decisions on international aid for development of agriculture in tropical countries were made by economists and bankers in board offices in London and Washington. The type of project to be funded often depended on an analysis of international debt relations and the securing of resources for developed countries. When the development project failed, as most of them did, the usual excuses were government red tape and bureaucracy; lazy or stupid workers; lack of telephones; a mail system that did not work; the general indolence of the tropics; and bad luck. Rarely, if ever, was the idea proposed that the type of aid offered was inappropriate either for the local ecosystems or the local cultures.

About a decade ago, several anthropologists working with the CGIAR group (the Consultative Group on International Agricultural Resources) came up with a revolutionary suggestion: Why not ask the local people who would be most affected by the aid what *they* thought the problems were, and what *they* thought might be the best solution. This "farmer first" idea was at first rejected. Gradually, however, a few agencies began to listen sympathetically. In 1992 the U.S. Agency for International Development funded a project to develop sustainable agriculture in several tropical areas based on the idea that the farmer (or tribesman, etc., who would be affected by development) should not only be consulted, but should also have a say in the type of development undertaken. One site chosen was the Manupali watershed in central Mindanao, the Philippines. This photoessay documents the "rapid rural survey" through which project personnel began to understand the local agricultural, environmental, and social problems, and to begin working toward a solution.

Fig. P.1 The survey was carried out in the watershed of the Manupali River, which has its headwaters on Mount Kitanglad rising above the central plateau of Mindanao. Only 20 years ago, the entire valley was covered with tropical moist forest. Now, the valley and lower slopes are in agriculture or fallow grassland.

Fig. P.2 Survey participants were from North American universities, Philippine universities, nongovernmental organizations, and Philippine governmental agencies. Language was a problem, because even some of the Filipinos did not speak Cebuano, the language used in the region.

Fig. P.3 For the survey, the group split into teams of two or three, each with one member who spoke Cebuano. Every team was assigned an area to survey. The technique was simple. Teams walked through the countryside or village, and when they encountered a likely prospect, they asked a friendly but leading question. For example, this team might have addressed the farmer with a question like: "Your corn crop here seems to be doing a lot better than that of the farmer downslope. What are you doing differently?" Gradually, the conversation moved to the farmer's problems, and his ideas for solutions.

Fig. P.4 Often the farmer would invite the team into his house, as did the young migrant farmer on the left. He cultivated potatoes, drying on the rack in the background. The tobacco leaves drying on the line were for the farmer's personal use. Hanging on the post is a spray can for pesticides. Heavy spraying contaminated downslope drinking water, but there seemed to be no alternative for fighting insect pests.

Fig. P.5 At higher elevations, the farmers were often indigenous Tala-andig tribesmen cultivating lands claimed as ancestral heritage. Their method of drying corn seemed to protect it from insects.

Fig. P.6 Teams consisted of ecologists, agronomists, sociologists, anthropologists, and economists. Here an anthropologist holds a child while asking his mother about health conditions.

Fig. P.7 The team did not hesitate to explain to villagers what it was trying to do. When the interviewer spoke the language and understood the culture, as did the one shown here, the experience was enjoyable for all.

Fig. P.8 The teams also observed the landscape and tried to relate what they saw to what the people told them. These grasslands once were croplands. They were abandoned because of decreasing productivity owing to loss of soil fertility and invasion by highly competitive weeds.

Fig. P.9 Sometimes the grasslands are burned, purportedly to clear the area at the end of a fallow. Often the fire escapes upslope and burns into the old-growth forest. Such "escape" is often encouraged, because the burned forest is easier to invade, in order to clear and prepare for cultivation. Although invasion is illegal, it is usually overlooked.

Fig. P.10 Agricultural expansion into the uplands is driven by a population explosion, caused in part by the high birthrate of local residents and in part by rapid migration into the area from other Philippine islands.

Fig. P.11 Agricultural expansion is also driven by the high prices for vegetables, especially potatoes, in the markets of major Philippine cities.

Fig. P.12 Potatoes are often the first crop planted after clearing the forest. They do well only for about a year, and then they suffer from wilt. Cabbages and other vegetables follow for about two years before the area is abandoned.

Fig. P.13 Even the steepest slopes are cleared for cultivation.

Fig. P.14 Here a resource specialist has plucked two orchids from the crowns of trees that were about to be burned in a clearing high on the mountain. His point was that these orchids which will be destroyed could have a market value higher than the potatoes that will replace them.

Fig. P.15 The reservoir behind the hydroelectric dam on the Manupali River has never been full. During the dry season, the water is so low that the turbines run at less than 30 percent capacity. Low water during the dry season is due to deforestation which results in low infiltration of water into the soil.

Fig. P.16 After the survey, the team held a workshop in a village hall to explain the findings to local officials, farmers, and others. Here a team member explains the effects of deforestation and the social and economic factors that cause it.

Fig. P.17 The councilmen ("Baranggay captains") listened intently, then voiced their own opinions and questions.

Fig. P.18 Immigrant farmers made strong points about land rights.

Fig. P.19 And were often contradicted by Tala-andig tribesmen.

Fig. P.20 In the end, most agreed that a technical solution was already known, as illustrated by this poster from a "Sustainable Agricultural Fair" in Cagayan de Oro. It depicts a child's interpretation of sloping Agricultural Land Technology, discussed in Chapter 6.

The problem is basically not technical, but political and economic. How is the land for such terraces going to be controlled? Who will have the authority to control it? The answers to those questions are difficult. Nevertheless, there is a feeling of progress. Even though there are no ready answers, at least, the team is asking the right questions—a major advance over previous projects.

SUGGESTED READINGS

For a conservative view of agricultural development in the tropics:

Committee on Sustainable Agriculture and the Environment in the Humid Tropics. 1993. Sustainable Agriculture and the Environment in the Humid Tropics. National Academy Press, Washington, D.C.

For a liberal view of agricultural development in the tropics:

Conway, G. R., and E. B. Barbier. 1990. After the Green Revolution. Earthscan Pubs., London.

Many nongovernmental organizations and small rural colleges in developing countries have published how-to-do-it manuals for people-centered agricultural development. One that I happen to have is:

Bunch, R. 1982. Two Ears of Corn: A Guide to People-Centered Agricultural Improvement. World Neighbors, Oklahoma City, Okla.

during the workshops did long-term solutions, such as adopting more sustainable agricultural practices, emerge. These solutions are exemplified by sloping agricultural land technology (SALT; see Fig. 6.7).

Farmers in other RRAs often identified problems involving marketing, land tenure, transport, postharvest, pesticides, livestock, and lack of construction materials. Other problems often identified by the interviewers, though not always by the farmers, included lack of technology, ethnicity, and gender issues. Ethnicity is a recognition of how membership in a particular ethnic group affects political, economic, and social relations within the community. Failure to recognize these issues has often limited the effectiveness of aid programs targeted at lower economic classes.

Among gender issues is the recognition of women's role in both the farm and the village community. Their role in decision making and in carrying out tasks has frequently been unrecognized or undervalued. When development programs, workshops, and other training sessions focus only on men, they frequently are less effective than they could be. Including women in development training programs and soliciting their opinion on an equal basis with that of men will go far toward improving the effectiveness of development activities.

Although environmentally desirable technologies such as SALT are recognized, they often are not adopted. Fujisaka (1991) discusses the reasons, including adverse off-site effects; cost; incorrect demonstration by extension agents; insecure land tenure; the mining of resources by farmers; and inappropriate innovation. To address these problems, he makes the following recommendations.

> The problems addressed by research scientists must really be the same problems faced by the farmer.
>
> Innovations must work under the target conditions and must be better than or equal to the farmer's practices in terms of solving the problems. In addition, they must not work against other of the farmers' problem-solving practices and must not create additional serious problems.
>
> Near-term costs and actual and opportunity costs of adoption cannot be too high. Present survival should be assured before potential adoptors invest in possible long-term gains.
>
> Institutional development is needed to the degree that innovations are correctly communicated from research to extension and from extension to farmers.
>
> Policies are needed that encourage investments in innovations by land users. Such policies must balance the adoptors' needs against the potential for systems-destructive resource grabbing and mining by segments of the target population.
>
> Adoption may be facilitated by farmer participation and must be examined in terms of whether "real or spurious." Adoption can take many forms. Sometimes an effort is needed to recognize adoption when it is not in the form promoted by the extension agent.

Also important in the sustainable agricultural phase of development is the so-called landscape approach or watershed approach. Because a farm is not an isolated entity but is connected physically to the surrounding terrain, and economically, socially, and politically to the surrounding villages, problems and solutions must reach beyond the farm. The farm is a piece of the landscape, and farm problems cannot be solved without placing them in the context of the surrounding landscape. Because of discrete physical boundaries, a watershed is often the most convenient landscape unit used in development activities.

But even a watershed approach is not sufficient to effectively address the farmer's problems. Agricultural development in Mindanao is an example. Although deforestation in the uplands is being driven in part by problems within the watershed, other factors from outside the watershed are so important that no sustainable solution can be made without taking them into consideration. One is the market forces from major cities in the Philippines for the produce, especially potatoes, raised in the newly cleared land. These forces are increased by the Agriculture Department, which assists and encourages potato farming. A second factor is the influx of population from other regions of the Philippines to take advantage of the last morsel of undegraded soil in the uplands.

Institutional Viability The coming phase of international agricultural development, institutional viability, requires thinking beyond the boundaries of the farm, the watershed, and the landscape. Not only is a national strategy required, but also a global strategy in which farmers are full cooperators. Holistic management of the farm is not enough. A farm cannot be sustainable within the context of a national economy that is not sustainable, and a national economy cannot be sustainable within a global economy that is not sustainable. What is required is sustainable management of the globe. Highly developed nations cannot succeed in helping less developed countries to achieve sustainable development if the developed countries do not have a sustainable economy of their own—and if the global economy does not become sustainable.

Achieving global sustainability is what conservation is about. Earlier in this book, we have discussed various aspects, including ethics, economics, policies, and techniques of resource management. This chapter concentrates on the importance of culture in sustainable development.

Development cannot be sustainable unless it is compatible with culture. But culture is an adaptation to environmental conditions, and environmental conditions include the forces of development. It is a "chicken-and-egg" situation. Although trying to match development to culture might seem to be a hopelessly complex task, it turns out that certain values and goals are basic to all peoples and cultures, indigenous and developed alike. This congruence of values makes it easier to set goals for development.

CULTURE AND HOLISTIC MANAGEMENT

The approach to managing resources for goals such as sustainability and preservation of a pleasant environment has been called holistic management and holistic thinking. It is called holistic because it focuses on values and goals, and considers all factors that contribute to attaining those goals. The specific means of value and goal achievement are considered only after a decision is made on values and goals.

The focus in development plans is too often on means—getting rid of pests, raising crop production, increasing gross income—instead of goals and values. As a result, we often forget why we are trying to get rid of pests, raise crop production, or increase gross income. Thus, the means have replaced the ends as a management goal, and so it is that controversy over means rather than ends has usually fueled the arguments between environmentalists and resource managers.

In retrospect, it is clear how resource management has fallen into the trap of substituting means for ends. In the United States' early years, increasing production was essential to improving the quality of life. Foresters, agronomists, range managers, and fishery managers were all concerned first and foremost with increasing the output of their particular resource, and the effects that such increases had on other aspects of human life were considered secondary. In the last half-century, however, issues other than quantity of production have become more critical to achieving quality of life. Clean and healthful air and water and pleasant surroundings, including a diversity of plants and animals, have assumed greater importance. But despite the increasing realization that the emphasis on production contributes to the degradation of these qualities, most resource agencies and departments have continued to direct their attention primarily to greater production.

Holistic Management of Natural Resources

Small farmers may be considered holistic managers when their objective is to maintain and enjoy the quality of life on a family farm. These farmers are usually beset by a bewildering set of influences: the farm machinery dealer tells him he should buy a new harvester to increase his farm capacity; the extension agent tells him he should spray with more insecticide to combat an increasing infestation of grub worms; the banker who holds his mortgage tells him he should start raising soybeans because they are more profitable; the local environmentalist tells him he should use organic farming; the veterinarian tells him his cows need more vitamins; the slaughterhouse tells him his pigs have too much fat; the family wants a new car.

If the farmer were to follow the advice of any one of these, he would not have a sustainable farm. Instead, he focuses on his goal of keeping the farm and his marriage alive, and he accepts advice from others only to the extent that it will help him achieve this goal.

Aside from the family farm, resources are rarely managed holistically. There are exceptions, but frequently they are experimental. An example is the Instituto Biodinâmico, near Botucatu, São Paulo, Brazil, which is particularly concerned with the holistic approach in agriculture and is a member of the International Federation of Agricultural Movements (IFOAM) with headquarters in Switzerland.

The Rural Development Forestry Network based at the Overseas Development Institute in London, England, exemplifies those organizations that are experimenting with achieving sustainable forestry in developing countries by considering social and economic as well as technical factors.

In Albuquerque, New Mexico, the Center for Holistic Resource Management, known through its workshops and through *Holistic Resource Management* (Savory 1988), uses a near-holistic approach to resolving resource management problems. It bases its work on the recognition that resource management is a social as well as a technical activity, and that to achieve social goals values must be explicitly recognized. Goals can be set at the individual, family, or community level through consensus building. People are asked to think about and to discuss values as goals to be achieved. Setting goals is so important to the process that Savory begins his chapter on getting started with the words: "DON'T EVEN *THINK* OF MANAGING RESOURCES WITHOUT A GOAL."

When individuals within a community are brought together in a workshop and are asked what their goals may be, the uniformity of their responses is surprising, considering that community members are from such often-conflicting groups as ranchers and bird watchers.

One of the most frequently mentioned goals is *quality of life*. For example, a ranch family might set out to "create a warm and stable environment for our family and our staff, and to enable our children to return to the ranch when their schooling is over; to see our community revitalized; to enjoy life in the process of achieving our goals" (Savory

1988, pp. 55–56). Frequently, goals also include maintenance of a landscape that can be enjoyed. The rancher might describe it as "open grassland community at a high successional level, with scattered trees and shrubs, and a mosaic of brush thickets and grassland along the river bottoms giving us high complexity, including diversity of birds, insects, and other animals, and stability, an efficient mineral cycle, sound water cycle, and high energy flow" (Savory 1988, p. 56). The members of the local Audubon Society could well have the same goals.

The first step in holistic resource management is consensus building. When confrontational groups realize that they have the same values and goals, progress becomes much more feasible. When values and goals remain unstated, compromise is more difficult because arguments then erupt over means. Examples of means that have sparked controversy in the American West are: eradication of brush; flood control; eradication of wolves; and preservation of wolves.

A particularly unfortunate example of a means that often is seen as an end is the Endangered Species Act. Most environmentalists realize that this bill is merely a tool for achieving their real goal—maintaining quality of life. In the popular media, the relationship between species protection and quality of life is often lost. As a consequence, the defense of a seemingly insignificant creature has made environmentalists sometimes appear ridiculous.

The bitter confrontations between environmentalists and developers usually occur over means and rarely over goals. When asked why they want to continue logging in the Northwest, most loggers reply, "To preserve a way of life that includes living close to the wilderness." They do not answer "because clear-cutting is fun."

Sometimes a logger might answer, "To make some money." But if the logger is asked why he wants to make some money, then he is forced to consider answers such as "To enable me to continue living in the forest, because I like it there." Then the logger can be confronted with the choice between living in the forest and clear-cutting the forest. If the logger's answer were merely "To make a living," then the logger should be advised to pack up, move out, and get a job in the city where the pay is better.

With goals and values clearly stated, loggers and environmentalists can stop arguing about whether or not the forest should be clear-cut and begin to look at how they can achieve the goal of living in the forest without destroying it. Such goal setting could help reverse the ecological and social tragedies of international development frequently associated with projects such as huge dams. The problem is that the international lending agencies begin with assumptions that developing nations want dams or with conclusions that they should have dams whether they want them or not. Both local peoples and the government usually answer yes when asked if they want a dam, a road, or a canal. (Indigenous peoples sometimes say no, but their opinion has counted for little; see photo-essay entitled Trivializing Indigenous Resistance to Deforestation, Sarawak, Malaysia in this chapter). People in the mainstream culture see new dams and roads as a means of quickly increasing their short-term economic income. For the local peoples, however, greater economic income is often accompanied by a decreased quality of life. Usually, it is the local people who pay the costs of the dam (in environmental deterioration), and it is people far away who reap the benefits. No cost/benefit analysis of development projects can be complete without a clear statement of who pays the costs and who gets the benefits.

Theoretically, the lending bank or the development agency should begin by inquiring into the values and goals of the peoples who will be affected by a project. Only then should plans be made for achieving these goals. If the goal is to continue a traditional way of life, for example, then clearly building a hydroelectric dam will not help achieve this goal.

The politically correct approach does not always result in the politically correct response. For example, when members of the rapid rural appraisal team asked the mayor of the local Mindanao village what his values and goals were, he responded that his values were to improve the life of the villagers, and his goal was to persuade the United States Agency for International Development to finance a system of water pipes that would run from the top of the mountain because local springs were drying up owing to deforestation. The politically correct answer—stabilization of agriculture and establishment of a buffer strip to protect the remaining forest—had to be teased out of the farmers and politicians at the workshop.

On a national level, the approach can also backfire. When a government official is asked what he would value for development, he might respond that his country would like a dam because it would help repay foreign debt.

Appropriate Technology

Appropriate technology for developing countries is technology that is relatively compatible with the environment and the culture. Use of botanical pesticides or the natural enemies of insect pests instead of synthetic chemical pesticides and use of draft animals instead of tractors are two examples. When costs and benefits include social and environmental factors and not just money, appropriate technologies have lower costs and higher benefits to the local people.

To increase the sustainability of local resource management, the techniques must be appropriate—that is, they must enhance rather than degrade the environment and the culture. As has already been pointed out, however, local resource management cannot be sustainable within a global economy that is unsustainable. So what we must search for is technology that is appropriate for the global economy and modern advanced culture. These will be technologies that help achieve and then sustain cultures and environments of a high quality. They must not be technologies such as highways that once were means to that end but have become ends in themselves, and in the process increased pollution and traffic jams, while at the same time destroying the landscape.

Appropriate technology for the global village does not necessarily mean low technology; in fact, it usually means high tech. For example, the easiest, fastest, and cheapest ways to reduce reliance on oil and coal is to use energy more efficiently. By using new technologies to make homes more weather-tight, automobiles more economical in their use of fuel, and stoves more efficient, energy needs can be reduced while the growing needs of people are met. The incandescent light bulbs may soon be replaced by a variety of new lighting systems, including halogen and sodium lights. The most important new light source may be compact fluorescent bulbs that need only 18 watts to produce the same light as 75 watt incandescent bulbs.

Some of the greatest savings can come in refrigeration. Commercial models on the market today can reduce use of electricity from 1500 kilowatt hours per year to 750. Also under development are models that could reduce the figure to an astonishingly low 240. Nearly as great gains are possible in air conditioners, water heaters, and clothes dryers.

Large opportunities for increased efficiency are now available in industry. In the future steel-making is likely to rely heavily on electric arc furnaces that require half the energy of the open hearth furnaces. Aluminum, which is energy-intensive to refine, may be replaced in many applications by less energy-intensive synthetics.

Transportation systems in a sustainable world must be more energy efficient than they are today. During the past 15 years, automobile fuel economy has more than doubled, and even more efficient technologies are now being developed. More efficient appliances, housing, and transportation are appropriate technologies because they help us attain our goals of comfort, convenience, and pleasant surroundings by minimizing the impact on the environment.

An important direction in the development of appropriate (sustainable) sources of power generation is away from oil because of the finiteness of the supply, away from coal because of the pollution it causes, and away from nuclear power because of the impossibility of disposing of long-term radioactivity. Costs of renewable electricity based on wind and photovoltaic sources have decreased by an order of magnitude within the last decade. Building a solar-based economy is fundamental to achieving an environmentally sustainable global economy. Examples of solar-based technologies are described in more detail by Brown et al. (1991).

In *Making Peace with the Planet* (1975), Commoner points out that, although solar power would benefit industrialized countries, the initial expense would be prohibitive to many developing countries. But the argument does not diminish the importance of solar power for developed countries. It merely means that it should be used where cultural conditions are appropriate for the technology.

Recycling and reuse of the products of modern civilization, from aluminum cans to railroad tracks, are also making technology more sustainable and thus more appropriate.

The Worldwatch Institute in Washington, D.C., publishes an annual report on progress toward a sustainable society entitled *State of the World* which focuses on a series of issues. In 1992, the major issues were: conserving biological diversity; building a bridge to sustainable energy; confronting nuclear waste; reforming the livestock economy; improving women's reproductive health; mining the earth; shaping cities; creating sustainable jobs in industrial countries; strengthening global environmental governance; and launching the environmental revolution. In 1993, the issues were: facing water scarcity; reviving coral reefs; closing the gender gap in development; supporting indigenous peoples; providing energy in developing countries; rediscovering rail; preparing for peace; reconciling trade and the environment; and shaping the next industrial revolution.

The Worldwatch series examines the state of technology in today's global culture, ways to modify the technology to achieve sustainability, and the economic and political factors relevant to each technology.

Population and Culture

A thousand years ago and more, humans belonged to a fragile species whose existence was threatened by disease, predators, and starvation. To ensure the survival of the human species, it was extremely important that human life be considered sacred and that population be increased. To accomplish this goal, cultures had to develop an ethic whose sacred base was the preservation and increase of human life.

Today, over most of the globe, humankind is threatened by neither predators nor starvation; the threat rather has become crime, pollution, and diseases such as AIDS that may result in part from overpopulation. Conditions have changed, but the ethics have not. Overpopulation is the world's greatest threat to everyone's quality of life. Yet there is no ethic of population control. Population control, even through such means as taxes on children, is a subject that is still somewhat of a cultural taboo.

CONCLUSION

Cultures adapt to their environment, but it is a process that may take many generations. When technology changes the environment more rapidly than the culture can adapt to the environmental changes, trouble begins. The environmental problems that plague the world today—and the social problems as well—can in large part be attributed to technology outstripping culture.

It is tempting to propose that all science and all technological innovation stop for a generation or two so that culture can adapt to the changes that have already occurred. But we are caught in a trap. The only way we can solve many of our environmental problems such as pollution is through more and better science and technology. And it is not possible to specify that the only permissible science and technology is that which helps abate pollution. We either have science or we don't, and it is pretty clear that we will have science.

But more science, more facts, and more technology by themselves only widen the vicious circle. What is needed is a philosophy that will temper the facts with wisdom. What is needed is a philosophy of conservation, and developing that philosophy is the aim of this book.

SUGGESTED READINGS

For a documentary account of the cultural changes that occurred in an indigenous tribe as they came into increasing contact with Western culture:

Eder, J. F. 1987. On the Road to Tribal Extinction. Depopulation, Deculturalization, and Adaptive Well-being among the Batak of the Philippines. University of California Press, Berkeley.

For the historical roots of the culture-environment misunderstanding that is leading to the destruction of Amazonia:

Jordan, C. F. 1982. "Amazon Rain Forests." American Scientist 70: 394–401.

For an anthropological study of aboriginal adaptation in the Amazon rain forest:

Meggars, B. J. 1971. Amazonia: Man and Culture in a Counterfeit Paradise. Aldine Atherton, Chicago.

For a perspective on how colonial culture has impeded sustainable development in the tropics:

Baker, R. 1993. Environmental Management in the Tropics. An Historical Perspective. CRC Press, Boca Raton, Fla.

For case studies that examine the sustainability of hunting by indigenous peoples:

Robinson, J. G., and K. H. Redford, eds. 1991. Neotropical Wildlife Use and Conservation. University of Chicago Press, Chicago.

For conservation of rain forests through sustainable use by indigenous peoples and folk societies:

Redford, K. H., and C. Padoch, eds. 1992. Conservation of Neotropical Forests: Working from Traditional Resource Use. Columbia University Press, New York.

For a docudrama depicting the struggle of Brazil nut collectors in their efforts to save their extractive reserve:

Jordan, C. F. 1992. River of Rains. Ms.

For development of more sustainable agriculture and forestry in the Amazon region:

Anderson, A. B. ed. 1990. Alternatives to Deforestation: Steps Toward Sustainable Use of the Amazon Rain Forest. Columbia University Press, New York.

For the culture and ecology of communal resources:

McCay, B. J., and J. M. Acheson, eds. 1987. The Question of the Commons: The Culture and Ecology of Communal Resources. University of Arizona Press, Tucson.

For the culture and ecology of traditional and modern fisheries:

McGoodwin, J. R. 1990. Crisis in the World's Fisheries. Stanford University Press, Stanford, Calif.

For an examination of the interface between agriculture and anthropology in international development:

Rhoades, R. E. 1984. Breaking New Ground: Agricultural Anthropology. International Potato Center, Lima, Peru.

For holistic resource management with an emphasis on range management:

Savory, A. 1988. Holistic Resource Management. Island Press, Covelo, Calif.

For appropriate technologies necessary to sustain the modern world:

Brown, L. R., C. Flavin, and S. Postel. 1991. Saving the Planet: How to Shape an Environmentally Sustainable Economy. W. W. Norton, New York.

Commoner, B. 1975. Making Peace with the Planet. New Press, New York.

The "State of the World" Series, updated annually by the Worldwatch Institute. W. W. Norton, New York.

For case histories of appropriate technology in less developed countries:

Evans, D. D., and L. N. Adler, eds. 1979. Appropriate Technology for Development: A Discussion and Case Histories. Westview Press, Boulder, Colo.

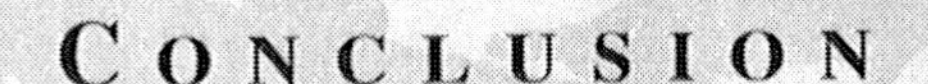

Conclusion

Conservation is the philosophy of managing the environment in a way that does not despoil, exhaust, or extinguish. The most effective way to begin such management is to replace quantity with quality as a goal for global management.

Pat Waak, director of Audobon Society's population program makes an eloquent plea for quality of life to replace quantity of material goods as a measure of our well-being in her "Letter to My Granddaughter," reprinted here from the *Audobon Activist*, January 1993, with permission from the publisher.

LETTER TO MY GRANDDAUGHTER

Dear Nisi,

I wanted to write you this letter because you are my only granddaughter. You bring me love and joy and remind me to look at all the beauty in everyday life.

This year is a passage for both of us. My birthday in February will mark 50 years on this planet, and your birthday in August will be your fifth. And I wonder: What kind of life will you have tomorrow, or in 10 years, or even in 45 years, in 2038, when you reach the age I am now?

Will you still be able to walk along a riverbank in the early morning and hear the birds sing, smell the flowers in the air, see the clouds floating overhead, feel the wind in your hair, touch the dew on a leaf? Or will we who are responsible for the Earth today have obliterated all those experiences in the next decade or two?

For many years I have been talking to people about population, and asking questions: How did we get here? How have we created so many problems? How have we managed to keep so much of the beauty that we have been given? But always it comes down to this question: What kind of world will our children and grandchildren—like you—live in?

The answer to that question lies in the choices we make today. Most people in this country have lived beyond their needs during the past several decades, but it was not always so. Your great-great-grandparents, who came to this country from many lands, had to struggle to make a good life for themselves and their families.

Somewhere along the way, the definition of a good life has been twisted out of shape. We want more and acquire more. You see commercials for newer, better toys, and you want them all.

I am grateful that you can have what you want, and above all you will have your health. But I fear that we—who have come so far and are so removed in time and space from the daily struggle—have forgotten that not everyone can afford what they want, or even need. Not everyone has the choices we do in our own lives.

How do we ensure that children around the world—and you and your children—have meaningful choices and healthy lives, and that the beauty and plenty of the Earth do not disappear? How do we make the year 2038, when you are 50, one of hope and promise?

First we should dream of the world through your eyes, a world of clean air and water, blue skies and drifting clouds, birds, animals, and laughing children. Only then can we plan the steps to make that dream come true.

Second, we must learn a big word—stewardship. The meaning of that word is to care for someone's property. For us grown-ups it means living in the world as if it were yours, Nisi, not ours—so when you are my age you can do the same for your granddaughter.

Third, we must make sure that there are only as many people as can share the Earth safely with all other creatures. To be truly fair, there should be enough for every child in every country.

Fourth, we must protect the special places that keep the Earth healthy—from the trees where we hang birdfeeders to the forests that are so important for medicine and the creatures we love.

Finally, we must work with people around the world to make this globe safe for all life. You are a child of many races, religions, and cultures. You are our future. I pray that I do not let you down.

• • •

This book suggests ways in which a student can help achieve these goals. There are many different avenues, and the one chosen should depend on life-style and personality.

There is a need for an environmental ethic, and for philosophers and religious leaders who will develop and preach that ethic.

There is a need for economists who will help speed the acceptance of the idea that production of a healthy environment should have equal footing with production of cars and television.

There is a need for resource managers who rely less on conquering nature, and more on working with nature.

There is a need for scientists to educate the public that species cannot be preserved unless ecosystems and landscapes are preserved.

There is a need for lawyers and politicians who will work for environmental laws, regulations, and incentives that achieve conservation objectives.

There is a need for environmental activists and watchdog groups to ensure that laws are workable and are obeyed.

There is a need for development workers sensitive to the importance of culture for the success of conservation and development.

There is a need for teachers who will convey the ideas of conservation to the new generations.

There are a lot of things an aspiring conservationist can do.

REFERENCES

A

Acheson, J. M. 1987. The Lobster Fiefs Revisited: Economic and Ecological Effects of Territoriality in Maine Lobster Fishing. Pp. 37–65 *in* B. J. McCay and J. M. Acheson, eds. *The Question of the Commons: The Culture and Ecology of Communal Resources.* University of Arizona Press, Tucson.

Ae, N., J. Arihara, K. Okada, T. Yoshihara, and C. Johansen. 1990. Phosphorus Uptake by Pigeon Pea and Its Role in Cropping Systems of the Indian Subcontinent. *Science* 248: 477–480.

Alper, J. 1992. Everglades Rebound from Andrew. News and Comment, *Science* 257: 1852–1854.

Altieri, M. A. 1987. *Agroecology.* Westview Press, Boulder, Colo.

Alvim, P. T. 1981. A Perspective Appraisal of Perennial Crops in the Amazon. *Interciencia* 6: 139–145.

American Society of Agronomy. 1990. *Organic Farming: Current Technology and Its Role in a Sustainable Agriculture.* American Society of Agronomy, Madison, Wis.

Anderson, A. B., ed. 1990. *Alternatives to Deforestation: Steps Toward Sustainable Use of the Amazon Rain Forests.* Columbia University Press, New York.

Anderson, A. N. 1987. A Malaysian Tragedy of the Commons. Pp. 327–343 *in* B. J. McCay and J. M. Acheson, eds. *The Question of the Commons: The Culture and Ecology of Communal Resources.* University of Arizona Press, Tucson.

Anderson, G. 1985. Foreword. P. vii *in* D. Vaughan and R. E. Malcolm, eds., *Soil Organic Matter and Biological Activity.* Kluwer Academic Publishers, Dordrecht.

Arnold, D. 1988. Famine: *Social Crisis and Historical Change.* Basil Blackwell, Oxford.

Arnon, I. 1987, *Modernization of Agriculture in Developing Countries.* Wiley, Chichester.

Ascher, W., and R. Healy. 1990. *Natural Resource Policymaking in Developing Countries.* Duke University Press, Durham, N.C.

B

Baker, P. T. 1984. The Adaptive Limits of Human Populations. Man 19: 1–14.

Baker, R. 1993. *Environmental Management in the Tropics. An Historical Perspective.* CRC Press, Boca Raton, Fla.

Barney, G. O. 1982. *The Global 2000 Report to the President.* Penguin Books, New York.

Barney, G. O. 1988. *Global 2000: Entering the Twenty-first Century. The Report to the President.* Seven Locks Press, Washington, D.C.

Bayliss-Smith, T. P., and S. Wanmali, eds. 1984. *Understanding Green Revolutions.* Cambridge University Press, Cambridge.

Bean, M. J. 1993. Fortify the Act. *National Parks* 67(5–6): 22–23.

Bennett, G. 1992. *Dilemmas: Coping with Environmental Problems.* Earthscan Publications, London.

Berger, J. J. 1990. *Environmental Restoration: Science and Strategies for Restoring the Earth.* Island Press, Washington, D.C.

Best, L. B. 1990. Sustaining Wildlife in Agroecosystems. Pp. 43–49 *in Farming Systems for Iowa: Seeking Alternatives.* Leopold Center for Sustainable Agriculture, Iowa State University, Ames, Iowa.

Bierregaard, R. O., T. E. Lovejoy, V. Kapos, A. Augusto dos Santos, and R. W. Hutchings. 1992. The Biological Dynamics of Tropical Rainforest Fragments. *BioScience* 42: 859–866.

Bingham, G., E. H. Clark II, L. V. Haygood, and M. Leslie. 1990. *Issues in Wetlands Protection: Background Papers Prepared for the National Wetlands Policy Forum.* The Conservation Foundation, Washington, D.C.

Bissonette, J.A. ed. 1986. Is Good Forestry Good Wildlife Management? Maine Agricultural Experiment Station, Misc. Pub. No. 689.

Biswell, H. 1989. *Prescribed Burning in California Wildlands Vegetation Management.* University of California Press, Berkeley.

Blanford, H. R. 1958. Highlights of One Hundred Years of Forestry in Burma. *Empire Forestry Review* 37(1): 33–42.

Botkin, D. B. 1990. *Discordant Harmonies. A New Ecology for the Twenty-First Century.* Oxford University Press, New York.

Botts, L. 1981. The Great Lakes Basin: What Costs Environmental Quality? Pp. 290–294 in W. E. Jeske, ed. *Economics, Ethics, Ecology: Roots of Productive Conservation.* Soil Conservation Society of America, Ankeny, Iowa.

Boucher, D. H. 1985. Mutualism in Agriculture. Pp. 375–386 *in* D. H. Boucher, ed. *The Biology of Mutualism.* Oxford University Press, New York.

Boucher, D. S., S. James, and K. H. Keeler. 1982. The Ecology of Mutualism. *Annual Review Ecology Systematics* 13: 315–347.

Bratton, S. P. 1983. The ecotheology of James Watt. *Environmental Ethics* 5(3): 225–236.

Bratton, S. P. 1984. Christian Ecoethology and the Old Testament. *Environmental Ethics* 6: 195–210.

Broad, R., and J. Cavanagh. 1989. Marcos's ghost. *The Amicus Journal* (Fall): 18–29.

Bronowski, J. 1973. *The Ascent of Man.* Little and Brown. Boston, pp 282–285.

Browder, J. O. 1988. Public Policy and Deforestation in the Brazilian Amazon. Pp. 247–297 *in* R. Repetto and M. Gillis, eds. *Public Policies and the Misuse of Forest Resources.* Cambridge University Press, Cambridge.

Brown, B. 1982. Productivity and Herbivory in High and Low Diversity Tropical Successional Ecosystems in Costa Rica. Ph.D. diss., University of Florida, Gainesville.

Brown, L. R., C. Flavin, and H. Kane. 1992. *Vital Signs, 1992: The Trends That Are Shaping Our Future.* Worldwatch Institute. W. W. Norton, New York.

Brown, L. R., C. Flavin, and S. Postel. 1991. *Saving the Planet: How to Shape an Environmentally Sustainable Economy.* W. W. Norton, New York.

Brown, L. R., H. Kane, and E. Ayres. 1993. *Vital Signs, 1993: The Trends That Are Shaping Our Future.* Worldwatch Institute. W. W. Norton, New York.

Brown, L. R., et al. 1992. *State of the World: 1992.* A Worldwatch Institute Report. W. W. Norton, New York.

Brown, L. R., et al. 1993. *State of the World: 1993.* A Worldwatch Institute Report. W. W. Norton, New York.

Buschbacher, R. 1992. An Assessment of International Policies on Tropical Forest Conservation. *Tropinet* 3(4): 1.

C

Callicott, J. B. 1987. The Land Aesthetic. Pp. 157–171 *in* J. B. Callicott, ed. *Companion to a Sand County Almanac.* University of Wisconsin Press, Madison.

Callicott, J. B., and R. T. Ames. 1989. Introduction: The Asian Traditions as a Conceptual Resource for Environmental Philosophy. Pp. 1–21 *in* J. B. Callicott and R. T. Ames, eds. *Nature in Asian Traditions of Thought: Essays in Environmental Philosophy.* State University of New York Press, Albany.

Campen, J. T. 1986. *Benefit, Cost, and Beyond. The Political Economy of Benefit-Cost Analysis.* Ballinger (Harper and Row), Cambridge, Mass.

Capel, S. W. 1988. Design of Windbreaks for Wildlife in the Great Plains of North America. *Agriculture, Ecosystems and Environment.* 22/23: 337–347.

Capistrano, L. N., J. Durno, and I. Moeliono. 1990. *Resource Book on Sustainable Agriculture for the Uplands.* International Institute of Rural Reconstruction, The Philippines.

Carpenter, R. A. 1987. What to Do While Waiting for an Environmental Ethic. *The Environmental Professional* 9: 327–335.

Carroll, C. R. 1992. Ecological Management of Sensitive Natural Areas. Pp. 347–372 *in* P. L. Fiedler and S. K. Jain, eds. *Conservation Biology: The Theory and Practice of Nature Conservation, Preservation and Management.* Chapman and Hall, New York.

Carson, R. 1962. *Silent Spring.* Houghton Mifflin, Boston.

Cavalcanti, C. 1991. Government Policy and Ecological Concerns: Some Lessons from the Brazilian Experience. Pp. 474–485 *in* R. Costanza, ed. *Ecological Economics.* Columbia University Press, New York.

Chambers, R. 1984. Beyond the Green Revolution: A Selective Essay. Pp. 362–379 *in* T. P. Bayliss Smith and S. Wanmali, eds. *Understanding Green Revolutions.* Cambridge University Press, Cambridge.

CIMMYT 1989. *Toward the 21st Century: Strategic Issues and the Operational Strategies of CIMMYT.* International Center for the Improvement of Maize and Wheat, Mexico City.

Clark, C. W. 1991. Economic Biases Against Sustainable Development. Pp. 319–330 *in* R. Costanza, ed. *Ecological Economics: The Science and Management of Sustainability.* Columbia University Press, New York.

Clark, W. C., and R. E. Munn. 1986. *Sustainable Development of the Biosphere.* Cambridge University Press, Cambridge.

Cody, M. L. 1985. An Introduction to Habitat Selection in Birds. Pp. 4–46 *in* M. L. Cody, ed. *Habitat Selection in Birds.* Academic Press, Orlando, Fla.

Colby, M. E. 1990. *Environmental Management in Development: The Evolution of Paradigms.* World Bank Discussion Paper 80. The World Bank, Washington, D.C. Cited *in* C. Folke, and T. Kaberger. 1991. Recent Trends in Linking the Natural Environment and the Economy. Pp. 273–300 *in* C. Folke and T. Kaberger, eds. 1991. *Linking the Natural Environment and the Economy: Essays from the Eco-Eco Group.* Kluwer Academic Publishers, Dordrecht.

Commoner, B. 1971. *The Closing Circle: Nature, Man, and Technology.* Alfred A. Knopf, New York.

Commoner, B. 1975. *Making Peace with the Planet.* New Press, New York.

Connell, J. H., and R. O. Slatyer. 1977. Mechanisms of Succession in Natural Communities and Their Role in Community Stability and Organization. *American Naturalist* 111: 1119–1144.

Conway, G. R. 1987. The Properties of Agroecosystems. *Agricultural Systems* 24: 95–107.

Conway, G. R., and E. B. Barbier. 1990. *After the Green Revolution: Sustainable Agriculture for Development.* Earthscan Pubs., London.

Cortner, H. J., and D. L. Schweitzer. 1993. Below-cost Timber Sales and the Political Marketplace. *Environmental Management* 17: 7–14.

Costanza, R. 1990. Ecological Economics as a Framework for Developing Sustainable National Policies. *In* B. Aniansson and S. Svedin, eds. *Towards an Ecologically Sustainable Economy.* Report from a Policy Seminar. The Swedish Council for

Planning and Coordination of Research (FRN), Stockholm. Cited *in* C. Folke, and T. Kaberger, eds. 1991. *Linking the Natural Environment and the Economy: Essays from the Eco-Eco Group.* Kluwer Academic Publishers, Dordrecht.

Costanza, R. 1991. Assuring Sustainability of Ecological Economic Systems. Pp. 331–343 *in* R. Costanza, ed. *Ecological Economics.* Columbia University Press, New York.

Costanza, R., and H. E. Daly. 1990. Natural Capital and Sustainable Development. Workshop on Natural Capital, March 15–16, 1990. Canadian Environmental Assessment Research Council, Vancouver, Canada. Cited *in* C. Folke and T. Kaberger, eds. 1991. *Linking the Natural Environment and the Economy: Essays from the Eco-Eco Group.* Kluwer Academic Publishers, Dordrecht.

Costanza, R., H. E. Daly, and J. A. Bartholomew. 1991. Goals, Agenda, and Policy Recommendations for Ecological Economics. Pp. 1–20 *in* R. Costanza, ed. *Ecological Economics: The Science and Management of Sustainability.* Columbia University Press, New York.

Culver, D. C. 1986. Cave Faunas. Pp. 427–443 *in* M. E. Soulé, ed. *Conservation Biology: The Science of Scarcity and Diversity.* Sinauer Associates Inc., Sunderland, Mass.

D

Dahl, D. W., S. Pyne, E. V. Anderson, and T. Crow. 1978. Fire and Public Policy. Pp. 557–569 *in Fire Regimes and Ecosystem Properties.* U.S.D.A. Forest Service, General Technical Report WO-26.

Daily, G. C., and P. R. Ehrlich. 1992. Population, Sustainability, and Earth's Carrying Capacity. *BioScience* 42(10): 761–771.

Daly, H. E. 1984. Alternative Strategies for Integrating Economics and Ecology. Pp. 19–29 *in* A. M. Jannson, ed. *Integration of Economy and Ecology: An Outlook for the Eighties.* Proceedings from the Wallenberg Symposia, Asko Laboratory, University of Stockholm. Cited *in* C. Folke and T. Kaberger, eds. 1991. *Linking the Natural Environment and the Economy: Essays from the Eco-Eco Group.* Kluwer Academic Publishers, Dordrecht.

Daly, H. E. 1989. Toward a Measure of Sustainable Social Net National Product. Pp. 8–9 *in* Y. J. Ahmad, S. El Serafy, and E. Lutz, eds. *Environmental Accounting for Sustainable Development.* The World Bank, Washington, D.C.

Daly, H. E. 1991a. *Steady-State Economics.* 2nd ed. Island Press, Washington, D.C.

Daly, H. E. 1991b. Elements of Environmental Macroeconomics. Pp. 32–46 *in* R. Costanza, ed. *Ecological Economics: The Science and Management of Sustainability.* Columbia University Press, New York.

Daly, H. E. 1991c. From Empty-World Economics to Full-World Economics. Pp. 29–38 *in Environmentally Sustainable Economic Development: Building on Brundtland.* UNESCO, Paris.

Daly, H. 1992. Natural Capital and Sustainable Development. Pp. 46–47 *in* A. Jansson, C. Folke, and M. Hammer, eds. *Investing in Natural Capital: A Prerequisite for Sustainability.* Abstracts of the Second Meeting of the International Society for

Ecological Economics, Stockholm, Sweden, August 3–6, 1992.

Daly, H. E. 1993. The Steady-State Economy: Toward a Political Economy of Biophysical Equilibrium and Moral Growth. Pp. 325–363 *in* H. E. Daly and K. N. Townsend, eds. *Valuing the Earth: Economics, Ecology, Ethics.* MIT Press, Cambridge, Mass.

Daly, H. E., and J. G. Cobb Jr. 1989. *For the Common Good: Redirecting the Economy Toward Community, the Environment, and a Sustainable Future.* Beacon Press, Boston. Cited *in* R. Costanza, H. E. Daly, and J. A. Bartholomew. 1991. Goals, Agenda, and Policy Recommendations for Ecological Economics. Pp. 1–20 *in* R. Costanza, ed. *Ecological Economics: The Science and Management of Sustainability.* Columbia University Press, New York.

Darwin, C. 1845. *Journal of Researches into the Natural History and Geology of the Countries Visited during the Voyage of H.M.S. Beagle Round the World.* A new edition reprinted in 1909 by D. Appleton and Co., New York.

Dasmann, R. F. 1968. *Environmental Conservation.* John Wiley, New York.

de Groot, R. 1992. *Functions of Nature.* Wolters-Noordhoff. The Netherlands.

Diamond, J. M. 1975. The Island Dilemma: Lessons of Modern BioGeographic Studies for the Design of Natural Reserves. *Biological Conservation* 7: 129–146.

Douglas-Hamilton, I., and O. Douglas Hamilton. 1992. *Battle for the Elephants.* Viking, New York.

Douglis, C. 1990. The Last Global Commons. *Wilderness* 53(189): 12–57.

Dowdeswell, W. H. 1987. *Hedgerows and Verges.* Allen and Unwin, London.

Dubos, R. 1968. *So Human an Animal.* Scribner's, New York.

Duke, S. O., J. J. Menn, and J. R. Plimmier. 1993. Challenges of Pest Control Which Enhances Toxicological and Environmental Safety. *Proceedings of American Chemical Symposium Series* No. 524. American Chemical Society, Washington, D.C.

Durno, J., I. Moeliono, and R. Prasertcharoensuk, 1992. *Resource Book on Sustainable Agriculture for the Lowlands.* Southeast Asia Sustainable Agriculture Network. The Philippines.

Dyer, M. I., C. L. Turner, and T. R. Seastedt. 1992. Herbivory and Its Consequences. *Ecological Applications* 3: 10–16.

E

Ehrlich, P. R. 1969. *The Population Bomb.* Illust. ed. The Sierra Club.

Ehrlich, P., and A. Ehrlich. 1981. *Extinction. The Causes of the Disappearance of Species.* Random House, New York.

Elfring, C. 1989. Preserving Land Through Local Land Trusts. *BioScience* 39: 71–74.

El Serafy, S. 1991. Sustainability, Income Measurement and Growth. Pp. 59–69 *in* UNESCO, *Environmentally Sustainable Economic Development; Building on Brundtland.* UNESCO, Paris.

Engelberg, J., and L. L. Boyarsky. 1979. The Noncybernetic Nature of Ecosystems. *American Naturalist* 114: 317–324.

Erdle, T. A., and G. L. Baskerville. 1986. Optimizing Timber Yields in New Brunswick Forests. Pp. 275–300 in *Ecological Knowledge and Environmental Problem Solving.* National Academy Press, Washington, D.C.

Ewel, J., F. Benedict, C. Berish, and B. Brown. 1982. Leaf Area, Light Transmission, Roots, and Leaf Damage in Nine Tropical Plant Communities. *Agro-ecosystems* 7: 305–326.

F

Falinski, J. B. 1986. *Vegetation Dynamics in Temperate Lowland Primeval Forests: Ecological Studies in Bialowieza Forest.* Junk, Dordrecht, The Netherlands.

FAO. 1985. *Tropical Forestry Action Plan.* Food and Agriculture Organization of the United Nations, Rome.

Farnworth, E. G., T. H. Tidrick, C. F. Jordan, and W. M. Smathers. 1981. The Value of Natural Ecosystems: An Economic and Ecological Framework. *Environmental Conservation* 8: 275–282.

Farnworth, E. G., T. H. Tidrick, W. M. Smathers, and C. F. Jordan 1983. A Synthesis of Ecological and Economic Theory Toward a More Complete Valuation of Tropical Moist Forests. *The International Journal of Environmental Studies* 21: 11–28.

Fearnside, P. 1988. Jari at Age 19: Lessons for Brazil's Silvicultural Plans at Carajas. *Interciencia* 13(1): 12–24.

Fearnside, P. M. 1989. Forest Management in Amazonia: The Need for New Criteria in Evaluating Development Options. *Forest Ecology and Mangement* 27: 61–79.

Fernandes, E. C. M., and P. K. R. Nair. 1986. An Evaluation of the Structure and Function of Tropical Homegardens. *Agricultural Systems* 21: 279–310.

Ferré, F. 1993. Persons in Nature: Toward an Applicable and Unified Environmental Ethics. *Zygon: Journal of Religion and Science* 28: 417–453.

Field, J. O., ed. 1993. *The Challenge of Famine: Recent Experience, Lessons Learned.* Kumarien Press, West Hartford, Conn.

Firor, J. 1990. *The Changing Atmosphere: A Global Challenge.* Yale University Press, New Haven, Conn.

Folke, C., and T. Kaberger. 1991. Recent Trends in Linking the Natural Environment and the Economy. Pp. 273–300 *in* C. Folke, and T. Kaberger, eds. *Linking the Natural Environment and the Economy: Essays from the Eco-Eco Group.* Kluwer Academic Publishers, Dordrecht.

Folke, C., and T. Kaberger, eds. 1991. *Linking the Natural Environment and the Economy: Essays from the Eco-Eco Group.* Kluwer Academic Publishers, Dordrecht.

Folan, W. J., L. A. Fletcher, and E. R. Kintz. 1979. Fruit, Fiber, Bark, and Resin: Social Organization of a Maya Urban Center. *Science* 204: 697–701.

Fox, J. E. D. 1976. Constraints on the Natural Regeneration of Tropical Moist Forest. *Forest Ecology and Management* 1: 37–65.

Fox, M. W. 1992. *Superpigs and Wondercorn*. Lyons and Burford. New York.

Francis, C. A. 1986. *Multiple Cropping Systems*. Macmillan, New York.

Francis, C. A., and J. W. King, 1988. Cropping systems based on farm-derived, renewable resources. *Agricultural Systems* 27: 67–75.

Frankel, O. H., and M. E. Soulé. 1981. *Conservation and Evolution*. Cambridge University Press, London.

Franklin, J. F. 1993. Preserving Biodiversity: Species, Ecosystems, or Landscapes. *Ecological Applications* 3: 202–205.

Frederickson, L. H., and F. A. Reid. 1990. Impacts of Hydrologic Alteration on Management of Freshwater Wetlands. Pp. 72–90 *in* J. M. Sweeney. *Management of Dynamic Ecosystems*. North Cent. Sect., Wildlife Society, West Lafayette, Ind.

French, H. F. 1990. *Clearing the Air: A Global Agenda*. Worldwatch Paper 94. Worldwatch Institute, Washington, D.C.

Fujisaka, S. 1991. Thirteen Reasons Why Farmers Do not Adopt Innovations Intended to Improve the Sustainability of Upland Agriculture. Pp. 509–522 *in Evaluation for Sustainable Land Management in the Developing World*. Vol. 2: Technical Papers. Bangkok, Thailand: International Board for Soil Research and Management, 1991. *IBSRAM Proceedings* No. 12(2).

G

Gajaseni, J. 1992. Socio-economic Aspects of Taungya. *In* C. F. Jordan, J. Gajaseni, and H. Watanabe, eds. *Taungya: Forest Plantations with Agriculture in Southeast Asia*. CAB International, Wallingford, U.K.

Gilbert, L. E. 1980. Food Web Organization and the Conservation of Neotropical Diversity. Pp. 11–33 *in* M. E. Soulé and B. A. Wilcox, eds. *Conservation Biology: An Evolutionary-Ecological Perspective*. Sinauer Associates, Sunderland, Mass.

Glick, W. 1973. *The Writings of Henry D. Thoreau: Reform Papers*. Princeton University Press, Princeton, N.J.

Gliessman, S. R., R. Garcia, and M. Amador. 1981. The Ecological Basis for the Application of Traditional Agricultural Technology in the Management of Tropical Agro-ecosystems. *Agro-Ecosystems* 7: 173–185.

Goldsmith, E. 1972. A Blueprint for Survival. *The Ecologist*, London.

Goldsmith, E. 1990. GATT and Gunboat Diplomacy. *The Ecologist* 20(6): 204.

Golley, F. B. 1987. Deep Ecology. *Environmental Ethics* 9: 45–55.

Golley, F. B. 1993a. Environmental Attitudes in North America. Pp. 20–32 *in* R. J. Berry, ed. *Environmental Dilemmas*. Chapman and Hall, London.

Golley, F. B. 1993b. *A History of the Ecosystem Concepts in Ecology*. Yale University Press, New Haven, Conn.

Gómez-Pompa, A., J. Salvador Flores, and V. Sosa. 1987. The "Pet Kot": A Man-made Tropical Forest of the Maya. *Interciencia* 12(1): 10–15.

Gómez-Pompa, A., T. C. Whitmore, and M. Hadley. 1991. *Rain Forest Regeneration and Management*. UNESCO, Paris, and the Parthenon Publishing Group, U.K.

Gonzalez Niño, E. Undated. *Amazonas. El Medio—El Hombre*. Oficina de Informacion y Relaciones Publicas. Ministerio de Obras Publicas, Caracas, Venezuela.

Goodland, R. 1978. *Environmental Assessment of the Tucurui Hydroproject, Rio Tocantins, Amazonia, Brazil*. Eletronorte, S.A. Brasilia, Brazil.

Gordon, H. S. 1954. The Economic Theory of a Common-Property Resource: The Fishery. *Journal of Political Economy* 62: 124–142.

Gore, A. 1992. *Earth in the Balance: Ecology and the Human Spirit*. Houghton Mifflin, Boston.

Graedel, T. E., and P. J. Crutzen. 1989. The Changing Atmosphere. *Scientific American* 261(3) (September): 58–68.

Graham, E., D. M. Pendergast, and G. D. Jones. 1989. On the Fringes of Conquest: Maya-Spanish Contact in Colonial Belize. *Science* 246: 1254–1259.

Graham, F. 1971. *Man's Dominion: The Story of Conservation in America*. M. Evans Co. and J.B. Lippincott, Philadelphia and New York.

Grigg, D. B. 1984. The Agricultural Revolution in Western Europe. Pp. 1–17 *in* T. P. Bayliss Smith and S. Wanmali, eds. *Understanding Green Revolutions*. Cambridge University Press, Cambridge.

Grossman, G. M., and A. B. Kruger. 1991. *Environmental impacts of a North American Trade Agreement*. Discussion Paper #158, Woodrow Wilson School, Princeton University, Princeton, N.J.

Grove, R. H. 1992. Origins of Western Environmentalism. *Scientific American* (July): 42–47.

Guedes, R. 1993. Phosphorus Mobilization by Root Exudates of Pigeon Pea *(Cajanus cajan)*. Ph.D. diss. University of Georgia, Athens, Ga.

H

Haas, P., M. A. Levy, and E. A. Parson. 1992. Appraising the Earth Summit: How Should We Judge UNCED's Success. *Environment* 34(8): 7 and ff.

Haffer, J. 1982. General Aspects of the Refuge Theory. Pp. 6–24 *in* G. T. Prance, ed. *Biological Diversification in the Tropics*. Columbia University Press, New York.

Hammond, N. 1982. The Exploration of the Maya World. *American Scientist* 70: 482–495.

Hanke, S. H., and R. A. Walker. 1974. Benefit-cost Analysis Reconsidered: An Evaluation of the Mid-State Project. *Water Resources Research* 10(5): 898–908.

Hannon, B. 1992. Measures of Economic and Ecological Health. Pp. 207–222 *in* R. Costanza, B. G. Norton, and B. D. Haskell, eds. *Ecosystem Health: New Goals for Environmental Management*. Island Press, Washington, D.C.

Hansen, M. K. 1983. Interactions among Natural Enemies, Herbivores, and Yield in Monocultures and Polycultures of Corn, Bean, and Squash. Ph.D. diss., University of Michigan, Ann Arbor.

Hanski, I., and J. Tiainen. 1988. Populations and Communities in Changing Agro-ecosystems in Finland. *Ecological Bulletins* 39: 159–168.

Hardin, G. 1968. The Tragedy of the Commons. *Science* 162: 1243–1248.

Hardin, G. 1974. Living on a Lifeboat. *BioScience* 24: 561–568.

Hargrove, E. C. 1989. *Foundations of Environmental Ethics*. Prentice Hall, Englewood Cliffs, N.J.

Harrison, G. A., ed. 1988. *Famine*. Oxford University Press, New York.

Harrison, L. E. 1992. Who Prospers? How Cultural Values Shape Economic and Political Success. *New York Times*, Sunday, June 20, 1993, p. F11.

Hart, J. H. 1990. Nothing Is Permanent Except Change. Pp. 1–17 *in* J. M. Sweeney. 1990. *Management of Dynamic Ecosystems.* North Cent. Sect., Wildlife Society, West Lafayette, Ind.

Hart, R. D. 1980. A Natural Ecosystem Analog Approach to the Design of a Successional Crop System for Tropical Forest Environments. *Biotropica* 122 (Supplement): 73–82.

Hartshorn, G. S. 1990. Natural Forest Management by the Yanesha Forestry Cooperative in Peruvian Amazonia. Pp. 128–151 *in* A. B. Anderson, ed. *Alternatives to Deforestation: Steps Toward Sustainable Use of the Amazon Rain Forest.* Columbia University Press, New York.

Hays, S. P. 1993. Environmental Philosophies. *Science* 258: 1822–1823.

Heilbroner, R. 1993. The Worst Is Yet to Come. Review of *Preparing for the Twenty-first Century*, by P. Kennedy, Random House, New York. *The New York Times Book Review,* February 14, 1993, p. 1.

Herfindahl, O. C. 1965. What Is Conservation? Pp. 229–236 *in* I. Burton and R. W. Kates, eds. *Readings in Resource Management and Conservation.* University of Chicago Press, Chicago.

Hobbs, R., and K. Wallace. 1991. Remnant Vegetation on Farms Is a Valuable Resource. *Western Australia Journal of Agriculture* 32: 43–45.

Hoi-Sen, Y. 1978. Natural Habitats and Distribution of Malayan Mammals. Pp. 3–9 *in* BIOTROP Special Pub. No. 8. SEAMEO-BIOTROP, Indonesia.

Holmes, B. 1994. Biologists Sort the Lessons of Fisheries Collapse. *Science* 264: 1252–1253.

Homer-Dixon, T. F., J. H. Boutwell, and G. W. Rathjens. 1993. Environmental Change and Violent Conflict. *Scientific American* 268(2): 38–45.

Hughes, J. D. 1975. Ecology in Ancient Greece. *Inquiry* 18: 115–125.

Hultkrans, A. N. 1988. Greenbacks for Greenery. *Sierra* 73(6): 43–47.

Hunter, M. L. 1990. *Wildlife, Forests, and Forestry: Principles of Managing Forests for Biological Diversity*. Regents/Prentice Hall. Englewood Cliffs, N.J.

Huston, M. 1993. Biological Diversity, Soils, and Economics. *Science* 262: 1676–1680.

I

Institute of Petroleum. Data cited *in* R. Barr. 1993. Clinton's World: Globe More Crowded But in Some Ways Safer. Associated Press Dispatch (*Athens Banner Herald,* 162–12; January 19, 1993, p. 1).

Isherwood, J. 1982. In Defense of Woodlots. *American Forests* 88(12): 30–61.

J

Janzen, D. H. 1974. Tropical Blackwater Rivers, Animals, and Mast Fruiting by the *Dipterocarpaceae. Biotropica* 6: 69–103.

Janzen, D. H. 1985. The Natural History of Mutualisms. Pp. 40–99 *in* D. H. Boucher, ed. *The Biology of Mutualism.* Oxford University Press, New York.

Jeske, W. E., ed. 1981. *Economics, Ethics, Ecology: Roots of Productive Conservation.* Soil Conservation Society of America, Ankeny, Iowa.

Johannes, R. E., and B. G. Hatcher. 1986. Shallow Tropical Marine Environments. Pp. 371–382 *in* M. E. Soulé, ed. *Conservation Biology: The Science of Scarcity and Diversity.* Sinauer Associates, Sunderland, Mass.

Johnson, R. J., and M. M. Beck. 1988. Influences of Shelterbelts on Wildlife Management and Biology. *Agriculture, Ecosystems and Environment* 22/23: 301–335.

Jordan, C. F. 1982. Amazon Rain Forests. *American Scientist* 70: 394–401.

Jordan, C. F. 1985a. Jari: A Development Project for Pulp in the Brazilian Amazon. *The Environmental Professional* 7: 135–142.

Jordan, C. F. 1985b. *Nutrient Cycling in Tropical Forest Ecosystems.* Wiley, Chichester.

Jordan, C. F. 1986. Ecological Effects of Clear Cutting. Pp. 345–355 *in Ecological Knowledge and Environmental Problem-Solving.* National Academy Press, Washington, D.C.

Jordan, C. F. 1987a. Conclusion. Comparison and Evaluation of Case Studies. Pp. 100–121 *in* C. F. Jordan, ed. *Amazonian Rain Forests: Ecosystem Disturbance and Recovery.* Springer Verlag, New York.

Jordan, C. F. 1987b. Agroecology at Tome-Assú, Brazil. Pp. 70–73 *in* C. F. Jordan, ed. *Amazonian Rain Forests: Ecosystem Disturbance and Recovery.* Springer Verlag, New York.

Jordan, C. F. 1989. *An Amazon Rain Forest: The Structure and Function of a Nutrient Stressed Ecosystem and the Impact of Slash-and-Burn Agriculture.* Man and the Biosphere Series, Vol. 2. UNESCO, Paris, and the Parthenon Publishing Group, U.K.

Jordan, C. F. 1992. River of Rains. Manuscript.

Jordan, C. F., J. Gajaseni, and H. Watanabe. 1992. *Taungya: Forest Plantations with Agriculture in Southeast Asia.* CAB International, Wallingford, U.K.

Jordan, C. F., and C. E. Russell. 1989. Jari: A Pulp Plantation in the Brazilian Amazon. *GeoJournal* 19(4): 429–435.

Jordan, E. L. 1981. The Birth of Ecology: An Account of Alexander von Humboldt's Voyage to the Equatorial Regions of the New World. Pp. 4–15 *in* C. F. Jordan, ed. *Tropical Ecology*. Benchmark Papers in Ecology/10. Hutchinson Ross, Stroudsburg, Pennsylvania. (Humboldt's original travel notebooks were considered lost for many decades but have been retrieved from various sources and are now kept in the manuscript collection of the Deutche Staatsbibliothek in Berlin. The excerpts on conservation were reprinted in this chapter.)

Jordan, W. R., III, M. E. Gilpin, and J. D. Aber. 1987. *Restoration Ecology: A Synthetic Approach to Ecological Research*. Cambridge University Press, Cambridge.

K

Kasting, J. F. 1993. Earth's Early Atmosphere. *Science* 259: 920–926.

Kennedy, P. 1993. *Preparing for the Twenty-First Century*. Random House, New York.

Keever, C. 1973. Distribution of Major Forest Species in Southeastern Pennsylvania. *Ecological Monographs* 43: 303–327.

Kelly, P. 1989. Swapping Debt for Nature. *Orion Nature Quarterly* 8(3): 16–17.

Keswani, C. L., and B. J. Ndunguru, eds. 1982. *Intercropping*. Proceedings of the Second Symposium on Intercropping. Morogoro, Tanzania. International Development Research Centre, Ottawa.

Kolb, S. R. 1993. Islands of Secondary Vegetation in Degraded Pastures of Brazil: Their Role in Reestablishing Atlantic Coastal Forest. Ph.D. diss., University of Georgia, Athens, Ga.

Koshland, D. E. 1994. A Milk-free Zone. *Science* 264: 11.

Kothari, A., and R. Bhartari. 1984. *Narmada Valley Project: Development or Destruction? Economic and Political Weekly* (Bombay) June 19: 907–920. Cited *in* W. Ascher and R. Healy. 1990. *Natural Resource Policymaking in Developing Countries*. Duke University Press, Durham, N.C.

Kummer, D. M. 1991. *Deforestation in the Postwar Philippines*. University of Chicago Press, Chicago.

Kusler, J. A., W. J. Mitsch, and J. S. Larson. 1994. Wetlands. *Scientific American* 270(1): 64B–70.

L

Laarman, J. G., and R. A. Sedjo. 1992. *Global Forests. Issues for Six Billion People*. McGraw-Hill, New York.

Lack, P. C. 1988. Nesting Success of Birds in Trees and Shrubs in Farmland Hedges. *Ecological Bulletins* 39: 191–193.

Lal, R. 1991. Myths and Scientific Realities of Agroforestry as a Strategy for Sustainable Management for Soils in the Tropics. *Advances in Soil Science* 15: 91–137.

Lal, R., and P. A. Sanchez. 1992. *Myths and Science of Soils of the Tropics*. SSSA Special Pub. No. 29. Soil Science Society of America, American Society of Agronomy, Madison, Wis.

Larkin, A. 1981. The Ethical Problem of Economic Growth vs. Environmental Degradation. Pp. 208–220 *in* K. S. Shrader-Frechette, ed. *Environmental Ethics.* Boxwood Press, Pacific Grove, Calif.

Le Houérou, H., and H. Gillet. 1986. Conservation Versus Desertization in African Arid Lands. Pp. 444–461 *in* M. E. Soulé, ed. *Conservation Biology: The Science of Scarcity and Diversity.* Sinauer Associates, Sunderland, Mass.

Leopold, A. 1949. The Land Ethic. Pp. 201–226 *in A Sand County Almanac, and Sketches Here and There.* Oxford University Press, New York (reprinted in 1987).

Lewis, D. H. 1985. Symbiosis and Mutualism: Crisp Concepts and Soggy Semantics. Pp. 29–39 *in* D. H. Boucher, ed. *The Biology of Mutualism.* Oxford University Press, New York.

Lincoln, C., and D. Isley. 1947. Corn as a Trap Crop for the Cotton Bollworm. *Journal of Economic Entomology* 40: 437–438.

Losos, E. 1993. The Future of the U.S. Endangered Species Act. *Trends in Ecology and Evolution* 8(9): 332–336.

Lovejoy, T. E. 1980. A Projection of Species Extinctions. Pp. 328–331 *in* G. O. Barney (study director). *The Global 2000 Report to the President. Entering the Twenty-First Century.* Council on Environmental Quality. U.S. Government Printing Office. Washington, D.C.

Lovejoy, T. E. 1981. Prepared statement. Pp. 175–180 in *Tropical Deforestation, An Overview. The Role of International Organizations, The Role of Multinational Corporations.* Hearings before the Subcommittee on International Organizations of the Committee on Foreign Affairs. House of Representatives, 96th Congress, second session, May 7, June 19, and September 19, 1980. U.S. Government Printing Office, Washington, D.C.

Lovelock, J. E. 1979. *Gaia: A New Look at Life on Earth.* Oxford University Press, Oxford.

Lowe, R. G. 1977. Experience with the Tropical Shelterwood System of Regeneration in Natural Forest in Nigeria. *Forest Ecology and Management* 1: 193–212.

Ludwig, D., R. Hilborn, and C. Walters. 1993. Uncertainty, Resource Exploitation, and Conservation: Lessons from History. *Science* 260: 17–36.

Lugo, A. E. 1988. Estimating Reductions in the Diversity of Tropical Forest Species. Pp. 58–70 *in* E. O. Wilson, ed. *Biodiversity.* National Academy Press, Washington, D.C.

Luken, J. O. 1990. *Directing Ecological Succession.* Chapman and Hall, London.

M

MacArthur, R. H. 1972. *Geographical Ecology: Patterns in the Distribution of Species.* Harper and Row, New York.

MacArthur, R., and E. O. Wilson. 1967. *The Theory of Island Biogeography.* Princeton University Press, Princeton, N.J.

MacDicken, K. G., and N. T. Vergara. 1990. *Agroforestry: Classification and Management.* John Wiley, New York.

MacKinnon, J., K. MacKinnon, G. Child, and J. Thorsell. 1986. *Managing Protected Areas in the Tropics.* International Union for Conservation of Nature and Natural Resources. Gland, Switzerland.

Maddox, J. R. 1972. *The Doomsday Syndrome.* McGraw-Hill, New York.

Mader, H. J. 1988. Effects of Increased Spatial Heterogeneity on the Biocenosis in Rural Landscapes. *Ecological Bulletins* 39: 169–179.

Mahar, D. J. 1989. *Government Policies and Deforestation in Brazil's Amazon Region.* The World Bank, Washington, D.C.

Makarabhirom, P. 1989. *Agroforestry Species for Forest Extension.* Agroforestry Technical Paper No. 30. Royal Forestry Department, Thailand.

Mann, C. C., and M. L. Plummer. 1993. The High Cost of Biodiversity. *Science* 260: 1868–1871.

Marsh, G. P. 1864. *Man and Nature.* Scribner's, New York.

Marshall, A. 1892. *Elements of Economics.* Macmillan, London.

Matta Machado, R. 1993. A Comparison of Alley Cropping and Conventional Green Manuring on an Ultisol of the Georgia Piedmont. Ph.D. diss., University of Georgia, Athens, Ga.

Matthews, E. 1989. The Metaphysics of Environmentalism. Pp. 38–56 *in* N. Dower, ed. *Ethics and the Environmental Responsibility.* Avebury, Brookfield, Vt.

May, K. W., and R. Misangu. 1982. Some Observations on the Effects of Plant Arrangements for Intercropping. Pp. 37–42 *in* C. L. Keswani and B. J. Ndunguru, eds. *Intercropping.* Proceedings of the Second Symposium on Intercropping. Morogoro, Tanzania. International Development Research Centre, Ottawa.

May, R. M. 1972. Will a Large Complex System Be Stable? *Nature* 238: 413–414.

May, R. M. 1973a. Qualitative Stability in Model Ecosystems. *Ecology* 54: 638–641.

May, R. M. 1973b. *Stability and Complexity in Model Ecosystems.* Monographs in Population Biology 6. Princeton University Press, Princeton, N.J.

McConnell, G. 1960. The Conservation Movement—Past and Present. Pp. 189–201 *in* I. Burton and R. W. Kates, eds. *Readings in Resource Management and Conservation.* University of Chicago Press, Chicago.

McGoodwin, J. R. 1990. *Crisis in the World's Fisheries.* Stanford University Press, Stanford, Calif.

McHarg, I. L. 1969. *Design with Nature.* The American Museum of Natural History. Natural History Press, Garden City, N.Y.

McNaughton, S. J. 1977. Diversity and Stability of Ecological Communities; A Comment on the Role of Empiricisms in Ecology. *American Naturalist* 111: 515–525.

McNaughton, S. J. 1978. Stability and Diversity of Ecological Communities. *Nature* 274: 251–253.

McNaughton, S. J. 1992. Grasses and Grazers, Science and Management. *Ecological Applications* 3: 17–20.

McNaughton, S. J. 1993. Biodiversity and Function of Grazing Ecosystems. Pp. 361–383 *in* Ernst-Detlef Schulze and H. A. Mooney eds. *Biodiversity and Ecosystem Function.* Springer-Verlag, Berlin.

McNeely, J. A. 1988. *Economics and Biological Diversity: Developing and Using Economic Incentives to Conserve Biological Resources*. IUCN, Gland, Switzerland.

McNeely, J. A., K. R. Miller, W. V. Reid, R. A. Mittermeier, and T. B. Werner. 1990. *Conserving the World's Biological Diversity*. International Union for Conservation of Nature, World Resources Institute, Conservation International, World Wildlife Fund—U.S., and the World Bank.

Meadows, D. H., D. L. Meadows, and J. Randers. 1992. *Beyond the Limits. Confronting Global Collapse; Envisioning a Sustainable Future*. Chelsea Green Publishing Co., Post Mills, Vt.

Meggars, B. J. 1971. *Amazonia: Man and Culture in a Counterfeit Paradise*. Aldine Atherton, Chicago.

Mellos, K. 1988. *Perspectives on Ecology*. St. Martin's Press. New York.

Mervis, J. 1993. Science Cedes Ground to Environmental Concerns. *Science* 261: 676.

Miller, G. T. 1990. *Resource Conservation and Management*. Wadsworth, Belmont, Calif.

Miller, K. R. 1982. Parks and Protected Areas: Considerations for the Future. *Ambio* 11: 315–317.

Miller, K., and L. Tangley. 1991. *Trees of Life: Saving Tropical Forests and Their Biological Wealth*. Beacon Press, Boston.

Mills, L. S., M. E. Soulé, and D. F. Doak. 1993. The Keystone-Species Concept in Ecology and Conservation. *BioScience* 43: 219–224.

Milward, A. S. 1984. *The Reconstruction of Western Europe: 1945–51*. University of California Press, Berkeley.

Modis, T. 1992. *Predictions. Society's Telltale Signature Reveals the Past and Forecasts the Future*. Simon and Schuster, New York.

Moffat, A. S. 1992. Crop Scientists Break Down Barriers at Ames Meeting. *Science* 257: 1347–1348.

Moffat, A. S. 1993. Predators, Prey, and Natural Disasters Attract Ecologists. Meeting Briefs. *Science* 261: 1115.

Moran, E. F. 1981. *Developing the Amazon*. Indiana University Press, Bloomington.

Myers, N. 1979. *The Sinking Ark. A New Look at the Problem of Disappearing Species*. Pergamon, New York.

Myers, N. 1982. Forest Refuges and Conservation in Africa with Some Appraisal of Survival Prospects for Tropical Moist Forests Throughout the Biome. Pp. 658–672 *in* G. T. Prance, ed. *Biological Diversification in the Tropics*. Columbia University Press, New York.

Myers, N. 1983. Conservation of Rain Forests for Scientific Research, for Wildlife Conservation, and for Recreation and Tourism. Pp. 325–334 *in* F. B. Golley, ed. *Tropical Rain Forest Ecosystems, Structure and Function*. Elsevier, Amsterdam.

Myers, N. 1988. Threatened Biotas: "Hotspots" in Tropical Forests. *Environmentalist* 8(3): 1–20.

Myers, N. 1992. The Primary Source: Tropical Forests and Our Future ("updated for the 1990s"). W. W. Norton, New York.

N

Naess, A. 1973. The Shallow and the Deep, Long-Range Ecology Movement. A Summary. *Inquiry* 16: 95–100.

Naess, A. 1984. A Defense of the Deep Ecology Movement. *Environmental Ethics* 6: 265–270.

Nash, R. 1967. *Wilderness and the American Mind.* Yale University Press, New Haven, Conn.

Nash, R. F. 1990. The Civilian Conservation Corps. Pp. 140–143 *in* R. F. Nash, ed. *American Environmentalism: Readings in Conservation History*, McGraw-Hill, New York.

National Research Council. 1980. *Research Priorities in Tropical Biology*. National Academy of Science, Washington, D.C.

National Research Council. 1982. *Ecological Aspects of Development in the Humid Tropics.* National Academy Press, Washington, D.C.

National Research Council. 1989. *Alternative Agriculture*. National Academy of Science, Washington, D.C.

National Research Council (U.S.) 1992. *Conserving Biodiversity: A Research Agenda for Development Agencies: Report of a Panel of the Board of Science and Technology for International Development.* National Academy Press, Washington, D.C.

National Wildlife Federation Staff. 1993. A Report Card on the Environment. *National Wildlife* 31(2): 34–41.

Neher, D. 1992. Ecological Sustainability in Agricultural Systems: Definition and Measurement. Pp. 51–61 *in* R. K. Olson, ed. *Integrating Sustainable Agriculture, Ecology, and Environmental Policy.* Haworth Press, New York.

Nepstad, D., C. Uhl, and E. A. Serrao. 1990. Surmounting Barriers to Forest Regeneration in Abandoned, Highly Degraded Pastures: A Case Study from Paragominas, Pará, Brazil. Pp. 215–229 *in* A. B. Anderson, ed. *Alternatives to Deforestation: Steps Toward Sustainable Use of the Amazon Rain Forest*. Columbia University Press, New York.

Neumann, H. E. 1993. Mexico Adopts a Bottom-up Environmental Policy. *The Wall Street Journal,* June 25, p. A11.

Norton, B. G., ed. 1986. *The Preservation of Species*. Princeton University Press, Princeton, N.J.

Norton, B. G. 1991. *Toward Unity Among Environmentalists.* Oxford University Press, New York.

O

O'Callaghan, K. 1992. Whose Agenda for America? *Audubon* 94(5): 80–91.

O'Connor, A. 1993. Puerto Ricans Say No to Becoming 51st State. *The Atlanta Constitution,* November 15, 1993, p. 1.

Odum, E. P. 1989. *Ecology and Our Endangered Life-Support Systems.* Sinauer Assoc., Sunderland, Mass.

Odum, E. P. 1992. Great Ideas in Ecology for the 1990s. *BioScience* 42: 542–545.

Odum, H. T. 1988. Self-organization, Transformity, and Information. *Science* 242: 1132–1139.

Odum, H. T., and R. C. Pinkerton. 1955. Time's Speed Regulator: The Optimum Efficiency for Maximum Power Output in Physical and Biological Systems. *American Scientist* 43: 331–343.

Odum, W. E. 1982. Environmental Degradation and the Tyranny of Small Decisions. *BioScience* 32: 728–729.

Oldfield, M. I. 1984. T*he Value of Conserving Genetic Resources*. U.S. Department of the Interior, National Park Service, Washington, D.C.

Oldfield, S. 1989. The Tropical Chainsaw Massacre. *New Scientist* 23 (September): 55–57.

O'Neill, R. V., D. L. DeAngelis, J. B. Waide, and T. F. H. Allen. 1986. *A Hierarchical Concept of Ecosystems*. Monographs in Population Biology 23. Princeton University Press, Princeton, N.J.

Orians, G. H. 1993. Endangered at What Level? *Ecological Applications* 3: 206–208.

Ostrom, E. 1990. *Governing the Commons: The Evolution of Institutions for Collective Action.* Cambridge University Press, Cambridge.

O'Toole, R. 1988. *Reforming the Forest Service.* Island Press, Washington, D.C.

P

Palmer, J. R. 1986. Jari: Lessons for Land Managers in the Tropics. Manuscript presented at the International Workshop on Rainforest Regeneration and Management, Guri, Venezuela, November 24–28, 1986. Under the auspices of the UNESCO Man and the Biosphere Programs.

Partridge, E., ed. 1981. *Responsibilities to Future Generations: Environmental Ethics.* Prometheus Books, Buffalo, N.Y., p. 22.

Passmore, J. 1974. *Man's Responsibility for Nature: Ecological Problems and Western Traditions*. Duckworth, London.

Patten, B. C. 1991. Network Ecology: Indirect Determination of the Live-Environment Relationship in Ecosystems. Pp. 288–351 *in* M. Higashi and T. P. Burns, eds. *Theoretical Studies of Ecosystems: The Network Perspective.* Cambridge University Press, Cambridge.

Patten, B. C., and E. P. Odum. 1981. The Cybernetic Nature of Ecosystems. *American Naturalist*, 118: 886–895.

Pearce, D., and K. Turner. 1990. Economics of Natural Resources and the Environment. Harvester-Wheatsheaf, Hemel Hempstead, England. Cited on p. 288 *in* C. Folke and T. Kaberger, eds. *Linking the Natural Environment and the Economy: Essays from the Eco-Eco Group.* Kluwer Academic Publishers, Dordrecht.

Peluso, N. L. 1992. *Rich Forests, Poor People.* University of California Press, Berkeley.

Perl, M. A., M. J. Kiernan, D. McCaffrey, R. J. Buschbacher, and G. J. Batmanian. 1991. *Views from the Forest: Natural Forest Management Initiatives in Latin America.* Report on a Workshop. World Wildlife Fund, Washington, D.C.

Pianka, E. R. 1966. Latitudinal Gradients in Species Diversity: A Review of Concepts. *American Naturalist* 100: 33–46.

Pigou, A. C. 1920. *The Economics of Welfare.* Macmillan, London.

Pimentel, D., H. Acquay, M. Biltonen, P. Rice, M. Silva, J. Nelson, V. Lipner, S. Giordano, A. Horowitz, and M. D'Amore. 1992. Environmental and Economic Costs of Pesticide Use. *BioScience* 42: 750–760.

Pimentel, D., L. E. Hurd, A. C. Bellotti, M. J. Forster, I. N. Oka, O. D. Sholes, and R. J. Whitman. 1973. Food Production and the Energy Crisis. *Science* 182: 443–449.

Pimentel, D., U. Stachow, D. A. Takacs, H. W. Brubaker, A. R. Dumas, J. J. Meany, J.A.S. O'Neal, D. E. Onsi, and D. B. Corzilius. 1992. Conserving Biological Diversity in Agricultural/Forestry Systems. *BioScience* 42: 354–362.

Pimm, S. L. 1982. *Food Webs.* Chapman and Hall, London.

Pimm, S. L. 1991. *The Balance of Nature.* University of Chicago Press, Chicago.

Pinkett, H. T. 1970. *Gifford Pinchot: Private and Public Forester.* University of Illinois Press, Urbana.

Plucknett, D. L. 1993. International Agricultural Research for the Next Century. *BioScience* 43: 432–440.

Ponting, C. 1990. Historical Perspectives on Sustainable Development. *Environment* 32(9): 4–33.

Popkin, R. 1986. Two Killer Smogs the Headlines Missed. *Environmental Protection Agency Journal* 12: 27–29.

Posey, D. A. 1982. The Keepers of the Forest. *Garden* 6: 18–24.

Prance, G. T., ed. 1982. *Biological Diversification in the Tropics.* Columbia University Press, New York.

R

Rancourt, L. M. 1992. Saving the Endangered Species Act. *National Parks* 66(3–4): 28–33.

Rees, J. 1988. Pollution Control Objectives and the Regulatory Framework. Pp. 170–189 *in* R. K. Turner, ed. *Sustainable Environmental Management.* Belhaven Press, Westview Press, Boulder, Colo.

Reganold, J. P., A. S. Palmer, J. C. Lockhart, and A. N. Macgregor. 1993. Soil Quality and Financial Performance of Biodynamic and Conventional Farms in New Zealand. *Science* 260: 344–349.

Reiger, H. A., and G. L. Baskerville. 1986. Sustainable Redevelopment of Regional Ecosystems Degraded by Exploitative Development. Pp. 75–103 *in* W. C. Clark and R. E.

Munn, eds. *Sustainable Development of the Biosphere*. Cambridge University Press, Cambridge.

Reilly, A. E. 1991. The Effects of Hurricane Hugo on three tropical forests in the U.S. Virgin Islands. *Biotropica* 24: 414–419.

Reisner, M. 1991. *Game Wars: The Undercover Pursuit of Wildlife Poachers*. Viking Press, New York.

Repetto, R. 1987. Creating Incentives for Sustainable Forest Development. *Ambio* 16(2–3): 94–99.

Repetto, R. 1992. Accounting for Environmental Assets. *Scientific American,* June 1992: 94–100.

Repetto, R., R. C. Dower, R. Jenkins, and J. Geoghegan. 1992. *Green Fees: How a Tax Shift Can Work for the Environment and the Economy.* World Resources Institute, Washington, D.C.

Repetto, R., W. Magrath, M. Wells, C. Beer, and F. Rossini. 1989. *Wasting Assets: Natural Resources in the National Income Accounts*. World Resources Institute, Washington, D.C.

Rhoades, R. E. 1992. Ms. presented at a Workshop on Sustainable Agriculture and Natural Resource Management, University of Georgia, Athens, Ga.

Richards, E. M. 1991. The Forest *ejidos* of South-east Mexico: A Case Study of Community Based Sustained Yield Management. *The Commonwealth Forestry Review* 704(4): 290–311.

Ricklefs, R. E. 1993. *The Economy of Nature.* W. H. Freeman, New York.

Risch, S. J. 1981. Insect Herbivore Abundance in Tropical Monocultures and Polycultures: An Experimental Test of Two Hypotheses. *Ecology* 62: 1325–1340.

Rivière, J. W. 1989. Threats to the World's Water. *Scientific American* 261(3): 80–94.

Robinson, G. O. 1956. *And What of Tomorrow*. Comet Press, New York.

Robinson, J. 1962. *Economic Philosophy.* Aldine Press, Chicago.

Robinson, J. G., and K. H. Redford, eds. 1991. *Neotropical Wildlife Use and Conservation.* University of Chicago Press, Chicago.

Rolston, H. III. 1988. *Environmental Ethics: Duties to and Values in the Natural World.* Temple University Press, Philadelphia.

Russell, C. E. 1983. Nutrient Cycling and Productivity of Native and Plantation Forests at Jari Florestal, Para, Brazil. Ph.D. diss., Institute of Ecology, University of Georgia, Athens, Ga.

S

Saetersdal, M., J. M. Lime, and H.J.B. Birks. 1993. How to Maximize Biological Diversity in Nature Reserve Selection: Vascular Plants and Breeding Birds in Deciduous Woodlands, Western Norway. *Biological Conservation* 66: 131–138.

Sagoff, M. 1988. *The Economy of the Earth*. Cambridge University Press, New York.

Sanchez, P. A., D. E. Bandy, J. H. Villachica, and J. J. Nicholaides. 1982. Amazon Basin Soils: Management for Continuous Crop Production. *Science* 216: 821–827.

Saffirio, J., and R. Hames. 1983. The Forest and the Highway. Pp. 1–52 *in* Report No. 11. *The Impact of Contact: Two Yanomamo Case Studies. Cultural Survival.* Cambridge, Mass.

Savory, A. 1988. *Holistic Resource Management.* Island Press, Covelo, Calif.

Schmidheiny, S. 1992. *Changing Course: A Global Business Perspective on Development and the Environment.* MIT Press, Cambridge, Mass.

Schmink, M. 1992. Building Institutions for Sustainable Development in Acre, Brazil. Pp. 276–304 *in* K. H. Redford and C. Padoch, eds. *Conservation of Neotropical Forests: Working from Traditional Resource Use.* Columbia University Press, New York.

Schulze, E., and H. Mooney, eds. 1993. *Biodiversity and Ecosystem Function.* Springer Verlag, Berlin.

Selle, R. 1983. A "Divine Land Ethic" on Subduing the Earth. *American Forests* 89(12): 6–7.

Sen, A. 1981. *Poverty and Famines: An Essay on Entitlement and Deprivation.* Clarendon Press, Oxford.

Shafer, C. L. 1990. *Nature Reserves: Island Theory and Conservation Practice.* Smithsonian Institution Press, Washington, D.C.

Shands, W. E., and R. G. Healy. 1977. *The Lands Nobody Wanted.* The Conservation Foundation, Washington, D.C.

Shrader-Frechette, K. S. 1981. Technology, the Environment, and Intergenerational Equity. Pp. 67–81 *in* K. S. Shrader-Frechette, ed. *Environmental Ethics.* Boxwood Press, Pacific Grove, Calif.

Simberloff, D. 1983. Are We on the Verge of Mass Extinction in Tropical Rain Forests? Unpublished Manuscript. July 1983.

Simberloff, D. 1986. Are We on the Verge of a Mass Extinction in Tropical Rain Forests? Pp. 165–180 *in* D. K. Elliot, ed. *Dynamics of Extinction.* John Wiley, New York.

Simberloff, D., and J. Cox. 1987. Consequences and Costs of Conservation Corridors. *Conservation Biology* 1: 63–71.

Simon, J. L. 1980. Resources, Population, Environment: An Oversupply of Bad News. *Science* 208: 1431–1437.

Simon, J. L. 1990. There Is No Environmental, Population, or Resource Crisis. Pp. 24–25 *in* G. T. Miller, 1990. *Resource Conservation and Management.* Wadsworth, Belmont, Calif.

Simon, J. L., and H. Kahn, eds. 1984. *The Resourceful Earth: A Response to Global 2000.* Oxford University Press, New York.

Sioli, H. 1973. Recent Human Activities in the Brazilian Amazon Region, and Their Ecological Effects. Pp. 321–324 *in* B. J. Meggars, E. S. Ayensu, and W. D. Duckworth, eds. *Tropical Forest Ecosystems in Africa and South America: A Comparative Review.* Smithsonian Institution Press, Washington, D.C.

Sioli, H. 1986. Tropical Continental Aquatic Habitats. Pp. 383–393 *in* M. E. Soulé, ed. *Conservation Biology: The Science of Scarcity and Diversity.* Sinauer Assoc., Sunderland, Mass.

Smith, A. 1776. *An Inquiry into the Nature and Causes of the Wealth of Nations.* (1976 ed.) University of Chicago Press, Chicago.

Smith, F. E. 1971. *Conservation in the United States, A Documentary History: Land and Water, 1492–1900.* Van Nostrand Reinhold, New York.

Soulé, M. E., ed. 1986. *Conservation Biology: The Science of Scarcity and Diversity.* Sinauer Assoc., Sunderland, Mass.

Soulé, M. E. 1991. Conservation: Tactics for a Constant Crisis. *Science* 253: 744–750.

Soule, J. D., and J. K. Piper. 1992. *Farming in Nature's Image: An Ecological Approach to Agriculture.* Island Press, Washington, D.C.

Spero, J. E. 1990. *The Politics of International Relations.* St. Martin's Press, New York.

Staples, P. 1988. Patterns of Purification: The New England Puritans. Pp. 65–87 *in* W.E.A. van Beek, ed. *The Quest for Purity.* Mouton de Gruyter, Berlin.

Steinhart, C. E., and J. S. Steinhart. 1974. *Energy: Sources, Use, and Role in Human Affairs.* Duxbury Press, North Scituate, Mass.

Stocks, B. J. 1987. Fire Potential in the Spruce Budworm-Damaged Forests of Ontario. *The Forestry Chronicle* (February): 8–14.

Stone, C. D. 1992. A Proposal in Observation of Earth Day: Repairing the Biosphere Through a Global Commons Trust Fund. *Environmental Conservation* 19(1) : 3–5.

Strickland, D. A. 1968. *Scientists in Politics: The Atomic Scientists Movement.* Purdue University Studies. Lafayette, Ind.

Strong, D. H., and E. S. Rosenfield. 1981. Ethics or Expediency: An Environmental Question. Pp. 5–15 *in* K. S. Shrader-Frechette, ed. *Environmental Ethics.* Boxwood Press, Pacific Grove, Calif.

Subler, S., and C. Uhl. 1990. Japanese Agroforestry in Amazonia: A Case Study in Tomé-Açu, Brazil. Pp. 152–166 *in* A. B. Anderson, ed. *Alternatives to Deforestation: Steps Toward Sustainable Use of the Amazon Rain Forest.* Columbia University Press, New York.

T

Takeda, S. 1992. Origins of Taungya. Pp. 9–17 *in* C. F. Jordan, J. Gajaseni, and H. Watanabe, eds. *Taungya: Forest Plantations with Agriculture in Southeast Asia.* CAB International, Wallingford, U. K.

Taylor, P. W. 1986. *Respect for Nature: A Theory of Environmental Ethics.* Princeton University Press, Princeton, N. J..

Terborgh, J. 1992. Why American Songbirds Are Vanishing. *Scientific American,* May 1992: 98–104.

Tierney, J. 1990. Betting the Planet. *The New York Times Magazine,* December 2, 1990: 52 & ff.

Tilman, D., and J. A. Downey. 1994. Biodiversity and Stability in Grasslands. *Nature* 367: 363–365.

Tonkinson, R. 1974. *The Jigalong Mob: Aboriginal Victors of the Desert Crusade.* Cummings Publishing Co., Menlo Park, Calif.

Townsend, R., and J. A. Wilson. 1987. An Economic View of the Tragedy of the Commons. Pp. 311–326 *in* B. J. McCay and J. M. Acheson, eds. *The Question of the Commons: The Culture and Ecology of Communal Resources.* University of Arizona Press, Tucson.

Turner, B. L. 1974. Prehistoric Intensive Agriculture in the Mayan Lowlands. *Science* 185: 118–124.

Turner, B. L., and P. D. Harrison. 1981. Prehistoric Raised-Field Agriculture in the Maya Lowlands. *Science* 213: 399–405.

U

Udall, S. L. 1963. *The Quiet Crisis.* Holt, Rinehart and Winston, New York.

Urban, D. I., and T. M. Smith. 1989. Micro-Habitat Pattern and the Structure of Forest Bird Communities. *American Naturalist* 133: 811–829.

V

Vandermeer, J. H. 1989. *The Ecology of Intercropping.* Cambridge University Press, Cambridge.

Van Lavieren, L. P. 1983. *Wildlife Management in the Tropics with Special Emphasis on South-East Asia: A Guidebook for the Warden.* Handbook prepared for Ciawi School of Environmental Conservation Management. Bogor, Indonesia. 3 vols.

Vickers, W. T. 1991. Hunting Yields and Game Composition over Ten Years in an Amazon Indian Territory. Pp. 53–81 *in* J. G. Robinson and K. H. Redford, eds. *Neotropical Wildlife Use and Conservation.* University of Chicago Press, Chicago.

Vincent, J. R. 1992. The Tropical Timber Trade and Sustainable Development. *Science* 256: 1651–1655.

Vitousek, P. M., P. R. Ehrlich, A. H. Ehrlich, and P. A. Matson. 1986. Human Appropriation of the Products of Photosynthesis. *BioScience* 36: 368–373.

Volkman, J. M. 1992. Making Room in the Ark. *Environment* 34 (4): 18–20, 37–43.

W

Wagner, W. C. 1981. Future Morality. Pp. 62–66 *in* K. S. Shrader-Frechette, ed. *Environmental Ethics.* Boxwood Press, Pacific Grove, Calif.

Wang, Z., and R. D. Nyland. 1993. Tree Species Richness Increased by Clearcutting of Northern Hardwoods in Central New York. *Forest Ecology and Management* 57: 71–84.

Weber, P. 1983. *Abandoned Seas: Reversing the Decline of the Oceans.* Worldwatch Paper 116. Worldwatch Institute, Washington, D.C.

Weiss, H., M. A. Courty, W. Wetterstrom, F. Guichard, L. Senior, R. Meadow, and A. Curnow. 1993. The Genesis and Collapse of Third Millennium North Mesopotanian Civilization. *Science* 261: 995–1004.

White, L., Jr. 1967. The Historical Roots of Our Ecologic Crisis. *Science* 155: 1203–1207.

Whittaker, R. H. 1975. *Communities and Ecosystems.* Macmillan, New York.

Wicksell, K. 1935. *Lectures on Political Economy,* Vol. 1. *General Theory.* Macmillan, New York.

Willis, K. G., and J. F. Benson. 1988. Valuation of Wildlife: A Case Study on the Upper Teesdale Site of Special Scientific Interest and Comparison of Methods in Environmental Economics. Pp. 243–264 *in* R. K. Turner, ed. *Sustainable Environmental Management.* Westview Press, Boulder, Colo.

Wilson, E. O. 1992. *The Diversity of Life.* Belknap Press of Harvard University Press, Cambridge, Mass.

Wilson, R. J., and S. G. Diver. 1991. The Role of Birds in Agroforestry Systems. Pp. 256–273 *in* H. E. Garrett, ed. Proceedings of the Second Conference on Agroforestry in North America. School of Natural Resources, Springfield, Mo.

Wirth, T. E., and J. Heinz. 1991. *Project 88—Round II. Incentives for Action: Designing Market-Based Environmental Strategies.* A Public Policy Study sponsored by Senator Timothy E. Wirth, Colorado, and Senator John Heinz, Pennsylvania.

Wolf, R. 1981. God, James Watt, and the Public Land. *Audubon* 83(3): 58–65.

Wolfe, L. M. 1945. *Son of the Wilderness. The Life of John Muir.* Alfred A. Knopf, New York.

Woodwell, G. M. 1970. Effects of Pollution on the Structure and Physiology of Ecosystems. *Science* 168: 429–433.

World Commission on Environment and Development. (Gro Harlem Bruntland, Chairman). 1987. *Our Common Future.* Oxford University Press, Oxford.

World Resources Institute. 1992. *World Resources 1992–93.* A Report by the World Resources Institute, in collaboration with the United Nations Environment Programme and the United Nations Development Program. Oxford University Press, New York.

Y

Young, A. 1989. *Agroforestry for Soil Conservation.* CAB International, Wallingford, U. K.

Young, J. 1990. *Sustaining the Earth.* Harvard University Press, Cambridge, Mass.

Youngs, M. 1984. The Basis of Romanticism. Pp. 76–81 *in* D. Pepper, ed. *The Roots of Modern Environmentalism.* Croom Helm, London.

Subject Index

T

Author Index

A

B

T

U

V

W

Y

Z